Ancient Astronomy and Celestial Divination

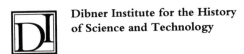

**Dibner Institute for the History
of Science and Technology**

Dibner Institute Studies in the History of Science and Technology
Jed Buchwald, general editor

Anthony Grafton and Nancy Siraisi, editors, *Natural Particulars: Nature and the Disciplines in Renaissance Europe*

Frederic L. Holmes and Trevor H. Levere, editors, *Instruments and Experimentation in the History of Chemistry*

N. M. Swerdlow, editor, *Ancient Astronomy and Celestial Divination*

ANCIENT ASTRONOMY AND CELESTIAL DIVINATION

———

Edited by N. M. Swerdlow

The MIT Press
Cambridge, Massachusetts
London, England

This book was set in Bembo by Asco Typesetters, Hong Kong, and printed and bound in the United States of America

Library of Congress Cataloging-in-Publication Data

Ancient astronomy and celestial divination / edited by N. M. Swerdlow.
   p.   cm. — (Dibner Institute studies in the history of science and technology)
   Includes bibliographical references and index.
   ISBN 0-262-19422-8 (alk. paper)
   1. Astronomy, Assyro-Babylonian. 2. Astrology, Assyro-Babylonian.
3. Astronomy-Greek. I. Swerdlow, N. M. (Noel M.) II. Series.
QB19.A53   1999
520′.935—dc21
                         99-15241
                         CIP

CONTENTS

PREFACE

In 1937 O. Neugebauer, then at the University of Copenhagen, set out a plan for the publication of the principal classes of texts of Babylonian astronomy: mathematical, observational, and astrological. It was a plan that has taken many years to realize, and it has not yet been completed, although the last forty years have seen great progress. Neugebauer's own *Astronomical Cuneiform Texts* (ACT), containing the mathematical texts, appeared in 1955, and in the same year Abraham Sachs published copies of mathematical and observational texts in the British Museum, the largest single collection, in *Late Babylonian Astronomical Texts* (LBAT). The editing and translating of the nonmathematical texts, as all but ACT are called, by Sachs and Hermann Hunger began to appear in 1989 in *Astronomical Diaries and Related Texts From Babylon*, of which three volumes have thus far been published. Following the initial publication of copies early in the century, the last twenty-five years have seen extensive publication of the astrological texts by a number of scholars including Erica Reiner, David Pingree, Francesca Rochberg, Simo Parpola, and Hermann Hunger. And in 1975 both Babylonian and Greek astronomy received their most extensive study ever in Neugebauer's *A History of Ancient Mathematical Astronomy* (HAMA).

Thus, when in the summer of 1993 Jed Buchwald and Evelyn Simha asked me to arrange a conference at the Dibner Institute for the History of Science and Technology at MIT, the subject of ancient astronomy and celestial divination, the sources of which were now so readily accessible in recent editions, came immediately to mind. The conference was held May 6–8, 1994, at the Dibner Institute.

I am grateful to the MIT Press, and particularly to Michael Sims and Chryseis Fox, for their efficient and attentive handling of material of exceptional technical difficulty. I would also like to thank Jed Buchwald, Evelyn Simha, and the Dibner Institute for initiating and sponsoring the conference and arranging for publication of its proceedings. Finally, I am very grateful to the contributors for the high technical quality of their papers and for their patience. It is my hope that the final publication will be considered worthy of the effort that has been required of everyone and that has been so generously forthcoming.

Bibliographical and Other Standard Abbreviations

The following list contains short explanations of many of the standard abbreviations in this volume; additional abbreviations and greater bibliographical detail are given in individual papers.

ACT   O. Neugebauer, *Astronomical Cuneiform Texts*

ADT   A. J. Sachs and H. Hunger, *Astronomical Diaries and Related Texts from Babylonia*

AfO   *Archiv für Orientforschung*

AOAT   *Alter Orient und Altes Testament*

BiOr   *Bibliotheca Orientalis*

BM   British Museum

BPO   E. Reiner and D. Pingree, *Babylonian Planetary Omens*

CAD   *The Assyrian Dictionary of the Oriental Institute of the University of Chicago*

CCAG   *Catalogus Codicum Astrologorum Graecorum*

CT   *Cuneiform Texts from Babylonian Tablets in the British Museum*

EAE   *Enūma Anu Enlil*

HAMA   O. Neugebauer, *A History of Ancient Mathematical Astronomy*

JCS   *Journal of Cuneiform Studies*

JNES   *Journal of Near Eastern Studies*

LAS   S. Parpola, *Letters from Assyrian Scholars to the Kings Esarhaddon and Assurbanipal*

LBAT   A. J. Sachs, *Late Babylonian Astronomical and Related Texts*

MCT   O. Neugebauer and A. J. Sachs, *Mathematical Cuneiform Texts*

MKT   O. Neugebauer, *Mathematische Keilschrift-Texte*

RA   *Revue d'assyriologie et d'archéologie orientale*

SAA   *State Archives of Assyria*

SSB   F. X. Kugler, *Sternkunde und Sterndienst in Babel*

WAI   H. C. Rawlinson, *Cuneiform Inscriptions of Western Asia*

ZA   *Zeitschrift für Assyriologie*

It was believed in antiquity that astronomy, including celestial divination, was the most ancient science, and this with good reason. According to Josephus (*Ant. Jud.* 1.69–72), the knowledge of the heavens was discovered by the descendants of Seth, who inscribed all they found on two pillars, one of brick, the other of stone, so that if the former were destroyed in the Deluge, as in fact came to pass, the latter would survive to teach men what was written on it. And God so loved these Patriarchs of wisdom that for the sake of the usefulness of their knowledge of astronomy and geometry he granted them very long lives, for had they not lived for 600 years, the period of a "Great Year," they could have foretold nothing with certainty (1.106). Abraham too (1.156–157) studied the sun and moon and all the phenomena of the heavens, from which he was led to declare that God is one, for from their irregularities he concluded that they were unable to regulate themselves, so that those benefits they grant us arise not from their own volition but in service to a commanding authority, to whom alone it is proper to give homage and gratitude. For this reason the Chaldaeans rose against him so that, with the aid of God, he left Mesopotamia and emigrated to the land of Canaan. Later, when he traveled to Egypt, he taught arithmetic and astronomy to the Egyptians, who previously knew nothing of them, and in this way these sciences passed from the Chaldaeans to the Egyptians, from whom they in turn passed to the Greeks.

So even Abraham learned astronomy from the Chaldaeans. And who might they be? According to a detailed account by Diodorus (2.29–31), who seems to be drawing upon a good source even if he does not get everything quite right, the Chaldaeans, like the priests of Egypt, are dedicated to service of the gods, spend their lives in pursuit of knowledge, for which they are relieved of all other duties, and hand down their learning within their families. They are most celebrated for their knowledge of astrology, but they also practice every sort of divination, including augury, prognostication through dreams and portents, and inspection of entrails, and they are able to avert the fulfillment of evil omens by their skill in rituals of purification, sacrifices, and incantations. They hold that the phenomena in the heavens occur, not by chance, but by the immutable decree of the gods. Of all the stars, the greatest influence belongs to the five wanderers, which they call "interpreters," for by their risings and settings they make known the future, interpreting the will of the gods, signifying winds or rain or heat, comets, eclipses of the sun and moon, earthquakes, and all things occurring in the air, both beneficial and harmful, for all nations, rulers, and people. The star the Greeks call Kronos they call the star of Helios—meaning that Saturn is identified with the sun—and the stars of Ares, Aphrodite, Hermes, and Zeus are named as among Greek astrologers. They recognize thirty stars, called "gods of council," which rise ten days apart, and to each of the twelve greatest of these gods they attribute a month and a zodiacal sign. Beyond the zodiac they recognize twelve northern and twelve southern stars; those that are visible they consider stars of the living, and those that are invisible stars of the dead. They believe the moon to suffer eclipse from the shadow of the earth, but only imperfectly do they explain or predict solar eclipses and their durations. It is nevertheless certain that the Chaldaeans have of all men the greatest skill in astrology, and this is because they have applied themselves to it with the greatest dedication for the longest time, for from their first observations until Alexander crossed into Asia they count 473,000 years. Lest the reader doubt this great figure, either on the grounds that it precedes the age of the Patriarchs or for any other reason, other sources show it to be modest, for Pliny (*NH* 7.193) reports, on the authority of Epigenes, that the Babylonians inscribed on baked tablets observations of the stars for 720,000 (3,20,0,0) years, a figure doubled by Simplicius (*in De caelo* 475B) to the extraordinary number of 1,440,000 (6,40,0,0) years.

We have told these stories of the early history of astronomy as it was understood in antiquity, and which were repeated again and again when little more was known, because no one has paid a whit of attention to them for over a hundred years. But that is not such a long time, for it was only in the second half of the last century that the discovery of original sources enlarged upon the reports of Diodorus and a few other ancient writers. Alas, of the doubtless excellent astronomy of the Patriarchs, there is little to add to Josephus's account, unless one wishes to consider the *Book of Enoch* an authentic relic of the antediluvian age. But of the astronomy and astrology of the Chaldaeans, or, as we shall now call them, the Babylonians, although many vestiges in fact survive in Greek and also in Sanskrit, these could not be identified as Babylonian until the original cuneiform sources were themselves discovered and deciphered. This began in 1870 with the publication by Sir Henry Rawlinson and George Smith of the third volume of *Cuneiform Inscriptions of Western Asia* (WAI)—some texts had already been published in WAI II, XLVII–IL—of which Plates LI–LXIV of Kuyunjik tablets from the library of Assurbanipal concern celestial divination, including fragments of a series of "upwards of 70 tablets treating of the Moon, the Sun, the Clouds, and lastly, the Stars and Planets," that is, the four principal divisions of the omen series Enūma Anu Enlil. Among the texts published were, in Plate LXIII (by coincidence?) Tablet 63 of the series, the Venus tablet of Ammiṣaduqa, several other substantial fragments from all parts of the series, and about thirty reports and letters to the Assyrian kings Esarhaddon and Assurbanipal, principally concerning lunar visibility. The following year Jules Oppert translated in *Journal Asiatique* 18 (1871) from his own copy a letter (WAI III.51.9, now LAS 290) from Mar-Issar to Esarhaddon concerning a watch for a solar eclipse that did not occur, correcting Rawlinson's identification of Jupiter in the text to Mercury. Rawlinson was right after all—it was Jupiter—but Oppert did find the correct name of Saturn. In 1874 a considerable number of Rawlinson's texts were translated in *The Astronomy and Astrology of the Babylonians* by Archibald Henry Sayce, an extraordinary work of scholarship and ingenuity when one considers the difficulty of attempting to make sense of these obscure texts for the first time. Sayce identified the series Enūma Anu Enlil under the name Namar-Bili or Enu-Bili, the "Illumination of Bel" or the "Eye of Bel," from which he published omens concerning the moon, eclipses, stars, planets, including the first translation of Tablet 63, and meteorological phenomena, along with many of the reports and letters to Esarhaddon and Assurbanipal. At the end of the century James A. Craig published copies of sections of Enūma Anu Enlil in *Astrological-Astronomical Texts copied from the original tablets in the British Museum* (1899), Reginald Campbell Thompson published more than three hundred reports, most with translation, in *The Reports of the Magicians and Astrologers of Nineveh and Babylon* (1900), Robert Francis Harper published more than three hundred letters in copies in *Assyrian and Babylonian Letters* (1892–1914), and Charles Virolleaud published extensive copies and transcriptions of Enūma Anu Enlil in *L'astrologie chaldéene, le livre intitulé "enuma (Anu) ^{ilu}Bêl"* (1905–12). Thus, by early in the present century, a great deal of material on celestial divination had become available.

The long career of the next important scholar of this subject, Ernst Weidner, takes us nearly up to the present day. Weidner began with the study of early, premathematical astronomy—*very* early, he was a Pan-babylonist, for which he was soundly criticized by Kugler—but then turned to astrology, investigating, among much else, the ordering of Enūma Anu Enlil, establishing the contents of the first fifty tablets based principally upon Virolleaud's copies. Tablet 63, edited from all sources then known by Stephen Langdon in *The Venus Tablets of Ammizaduga* (1928), has been re-edited from yet more by Erica Reiner and David Pingree, followed by Tablets 50–51, the first of those on stars and planets, and the remaining omens for Venus in *Babylonian Planetary Omens* 1, 2 and 3 (1975, 1981, 1998). The lunar eclipse omens of Tablets 15–22 have been published by Francesca Rochberg in *Aspects of Babylonian Celestial Divination: The Lunar Eclipse Tablets of Enūma Anu Enlil* (1988), and Wilfried H. van Soldt has published the solar omens in *Solar Omens of Enuma Anu Enlil Tablets 23(24)–29(30)* (1995). The Assyrian reports and letters, the most intelligible of the celestial divination texts, have received what must be their definitive editions, the letters by Simo Parpola in *Letters from Assyrian Scholar to the Kings Esarhaddon and Assurbanipal* (1970, 1983) and more recently *Letters from Assyrian and Babylonian Scholars* (1993), and the reports by Hermann Hunger in *Astrological Reports to Assyrian Kings* (1992). For many aspects of the study of antiquity—scientific, political, ritual—these are among the most interesting sources ever discovered. Such is the current state of the recovery of Babylonian celestial divination, our knowledge of which may someday seem to be only a beginning. The difficulty, the obscurity, of these texts can scarcely be exaggerated; they were the first to be published in copies by Rawlinson and the last to be adequately deciphered following Sayce's heroic first effort.

Sayce also initiated the modern study of Babylonian astronomy in three articles published with R. H. M. Bosanquet in the *Monthly Notices of the Royal Astronomical Society* 39–40 (1879–80), of which the first concerned the calendar, the second two "planispheres," one well known as planisphere or astrolabe A, and the third an analysis of the periods of visibility and invisibility in the Venus tablet. The authors described a method of dating the observations considering the days of the month on which the visibility phases occurred, and although they did not believe that a fixed dating was possible owing to the absence of historical evidence fixing the date to within, say, 100 years, they assumed that the date was anterior to 1700 B.C. There have been several attempts at more rigorous dating. The first was by Schiaparelli (1906), who assumed that the observations could not long precede the Nineveh provenance of the tablet and found possible dates in the ninth or seventh centuries. Weidner (1911), then a Panbabylonist, argued for a date in the late fifth millennium. The crucial historical evidence was found by Kugler (1912) in correctly reading the reference to the "year of the golden throne," the eighth year of Ammiṣaduqa, which placed the dating in the First Babylonian Dynasty. Since then the text has been used several times in attempts to establish the chronology of the First Babylonian Dynasty, notably by Kugler; Langdon, Fotheringham, and Schoch (1928); and most recently by Peter Huber et al. in *Astronomical Dating of Babylon I and Ur III* (1982). The difficulties in using the observations, which contain many errors, remain very great, perhaps insurmountable, and one may suppose that the chronology will eventually be fixed by completely nonastronomical means.

The Venus text is unique in the attention it has received for other than astronomical purposes. The analysis of astronomical cuneiform texts, both mathematical and observational, which offer the advantage over astrological texts of internal, mathematical consistency and external, astronomical control, came a decade after Rawlinson's publication and a year after Sayce's and Bosanquet's first efforts. The initial decipherment was the work of the Jesuit Fathers Johann Strassmaier, who transcribed texts in the British Museum, and Joseph Epping, then in Quito, Peru, who worked out their astronomical and mathematical content. Their first paper, "Zur Entzifferung der astronomischen Tafeln der Chaldäer," appeared in the Catholic journal *Stimmen aus Maria-Laach* 21 (also called *Stimmen der Zeit*) in 1881. Strassmaier pointed out the relation between the Assyrian reports of lunar visibility, of which the known examples now numbered over one hundred, and fragments of a hemerology giving the god and rituals for the days of Ulūlu II and Araḫsamnu published by Rawlinson (WAI IV.32–33). From the additional evidence of an eponym list and the date in the inscription of Tiglath-Pileser I—the horrendous chronicle that, when translated independently for the Royal Asiatic Society by Rawlinson, Fox Talbot, Hincks, and Oppert in 1857, showed that Akkadian had indeed been deciphered and gave the Assyrians a rather bad name—Strassmaier concluded that the Babylonian calendar could be extended back beyond the tenth century, and further, that the astronomical knowledge of the Assyrians was imported from Babylon. He described fragments of a series of astronomical observations from the Seleucid period that appear to belong to a great collection from a very early period, tablets containing lists of numbers that appear to apply to the computation of the heliacal risings and setting of the planets and to new and full moon, and some with names of stars with numbers that appear to contain observations and calculations of the motions of the planets or of lunar eclipses. As presently identified, these would appear to be Diaries, mathematical or ACT texts, and normal-star almanacs. Unfortunately, he wrote, these precious relics of antiquity are so broken and fragmentary that they may remain forever undeciphered.

Nevertheless, an impressive beginning was made in the next section by Epping, on the relation of thirteen lines of what he saw, correctly, to be calculations of the new moon—in fact columns K and L of the System B lunar ephemeris ACT 122—in which he showed that column K plus 29 days gives the differences for the dates in column L, and that column K is formed from two columns to the left, evidently G plus J. With the computation of new moon, he remarked, goes hand in hand the computation of full moon and of solar and lunar eclipses, so the fragment not only confirms the old tradition of the science of the Chaldaeans—indeed, he wrote, this small tablet gives us more information concerning the science of the Babylonians than all surviving reports from antiquity—but also shows that with reason we may seek in them the source from which all peoples of antiquity drew their astronomical knowledge. In *Astronomisches aus Babylon* (1889) and "Die babylonische Berechnung des Neumondes," *Stimmen aus Maria-Laach* 31 (1890), Epping and Strassmaier published thirteen lines of columns F through P_3 of ACT 122 and of K through Q of ACT 120. Epping showed the method of computation and significance of columns F through L with partial analysis of the remaining columns, found the implied lengths of the anomalistic month from F, of the synodic month from G, of the solar year from H, and drew the first ever "linear zigzag" to illustrate G. *Astronomisches aus Babylon* and following papers in *Zeitschrift für Assyriologie* 5–10 (1890–95) contained publication and analysis of at least parts of Diaries LBAT 243–244, 284, 498, normal-star almanacs 1055, 1059, goal-year texts 1237, 1295, 1297, the "Saros Canon" 1428 along with part of ACT 60, and the "Cambyses Inscription" LBAT 1477. Epping provided a detailed astronomical analysis, and his achievement was little short of miraculous, in particular analyzing the lunar rising and setting times in the almanacs and showing that they agree with modern computation within a very few minutes, identifying 28 "standard" (in German, "normal") stars, most of them correctly, and explaining the application of planetary periods to both heliacal phenomena and conjunctions with normal stars in the goal-year texts. In *ZA* 6 (1891) Oppert pointed out that the lunar eclipse of −522 in the Cambyses

Inscription was used by Ptolemy in *Almagest* 5.14. This remains the only eclipse mentioned by Ptolemy for which a Babylonian report is known.

Epping died in 1894, and his place was soon taken by Father Franz Xaver Kugler, then in his early thirties, whose work was to revolutionize not only the study of Babylonian astronomy but, when properly understood—something that I fear is still known to few—the entire history of science in antiquity. Working from transcriptions by the indefatigable Strassmaier, in 1900 Kugler published *Die Babylonische Mondrechnung*, in which he analyzed most of what he called Systems I and II, now known as Systems B and A, of the lunar ephemerides. The analysis of System B was based upon a complete reconstruction of ACT 122 columns A through L, now correctly dated to Seleucid Era 208–210 (−103 to −101)—the date column T is missing—and 120, 121, 123, 126. The detail and insight of the analysis are extraordinary: the method of computation, the underlying parameters, and the astronomical interpretation of every column are determined. The only doubtful point is that Kugler's column E, ACT column Ψ, is now interpreted as eclipse magnitude rather than lunar latitude. Among his many observations, Kugler noted for the first time the relation of the parameters of System B to the lunar theory of Hipparchus and Ptolemy.

System A was far more difficult, both because Kugler was starting from scratch and because there was no master text, as ACT 122 provided for System B. Nevertheless, from parts, sometimes very small parts, of ACT 4, 5, 7, 9, 10, 11, 13, 15, 16, 60, 70, of which 5 and 9 provide parts of all columns, and the procedure text ACT 200, he managed to reconstruct the rules, parameters and significance of, in ACT nomenclature, columns Φ through M, showing, among much else, the relation of Φ, F, and G, and the computation of G from Φ, the latter being interpreted, with some reservation, as a function for the apparent diameter of the moon in units of 0;15°. Without taking anything from Epping, Kugler's first book is, in my opinion, the finest, the most original, and the most difficult study carried out in the history of science up to the date of its publication. An entire lost world of extremely sophisticated science had been discovered, something that has happened only once. Although the far more complete restorations of ACT may now be consulted, we must never underestimate the magnitude of Kugler's achievement.

He was soon to equal if not surpass it. *Sternkunde und Sterndienst in Babel* began to appear in 1907, and its two large volumes and three supplements, the last by Johann Schaumberger, continued to appear until

1935, following Kugler's death in 1929. The first volume is a treatise on planetary theory in much the way that he had earlier treated lunar theory; the second volume and supplements are more a collection of papers on a variety of subjects. The work is on far more than astronomy, and it remains to this day the fundamental treatment of Babylonian and Assyrian astronomical chronology, astral religion, and mythology, though less so of celestial divination, which Kugler seems to have avoided. Here our concern is exclusively with astronomy. Kugler published larger and smaller fragments of twelve almanacs, four normal-star almanacs, one goal-year text—three more were published from Kugler's papers by Schaumberger in *Orientalia* NS 2 (1933)—and for the first time planetary ephemerides, which he had referred to at the end of *Die Babylonische Mondrechnung*, where he briefly described three systems for Jupiter. Now in the first volume he analyzed what we call Systems A, A′, and B of Jupiter from ACT 602, 603, 607, 610, 611, 612, 621, 622, 623, and procedure texts ACT 810 and 813; System B of Saturn from ACT 704 and 704a; parts of Systems A_1 and A_2 of Venus from ACT 410, 421, and 421a; and System A_1 of Mercury from ACT 301 and 302, from which he analyzed the morning phases. In an appendix in the second volume, using two Louvre texts from Uruk published by Thureau-Dangin, from the procedure text ACT 801, he extended his analysis to Mercury's evening phases and System A of Saturn, and from ACT 501 System A of Mars. He analyzed texts concerned with fixed stars in detail, providing many more identifications than had previously been known, and he investigated the risings and settings of stars. Kugler also had some fun in these volumes. He was a sharp critic, writing scathing refutations of beliefs in the Babylonian discovery of precession, the great age of Babylonian astronomy proposed by Panbabylonists, and Panbabylonism in general, a subject also considered in the book *Im Bannkreis Babels* (1910). Finally, in the third supplement Schaumberger published completely all seventeen columns of ACT 122, for the first time analyzing the visibility columns N through P_3. He thanked O. Neugebauer for providing a collation of the first ten columns and a figure (Plate XVII) illustrating the connection of the various fragments of the tablet. Here now was a scholar of the younger generation.

At the time, Neugebauer, who was at the Mathematical Institute of the University of Copenhagen, was only beginning to work on Babylonian astronomy, and it was hardly at the center of his interests. He was still seeing *Mathematische Keilschrift-Texte* (MKT) through the press, and the preceding year had published *Vorgriechische Mathematik*, on Egyptian and Babylonian mathematics, the first volume of *Vorlesungen über Geschichte der antiken mathematischen Wissenschaften*, which was to continue with one volume on Greek mathematics and one on Babylonian and Greek astronomy. For the last volume, it was his intention to summarize Kugler's work and extend it to the few more recently published texts, about fifty texts in all, something he thought could be done expeditiously. But then he began to analyze the texts on his own, developing the technique of using linear diophantine equations to restore missing sections and check for possible connections between fragments. He found that fragments could indeed be joined and dated and that some of the lunar columns ran for very long periods, and he realized that it was not only possible but necessary to reconsider the entire corpus of ephemerides and related texts using a methodologically consistent analysis leading to a more systematic mathematical control of the material. He published papers on lunar and eclipse theory in *Quellen und Studien zur Geschichte der Mathematik, Astronomie und Physik* (1937–38), setting out his methods, but the entire task was to take far longer than he had anticipated. In 1939 he came to Brown University, where he was joined in 1941 by Abe Sachs, who had been working at the Oriental Institute of the University of Chicago. Together they published *Mathematical Cuneiform Texts* in 1945, a considerable supplement to MKT based principally upon Sachs's investigations of the collections of Yale and the University of Pennsylvania. The number of astronomical texts also increased significantly. In 1945 a film of more than a hundred fragments from Uruk was sent by F. R. Kraus in Istanbul, and in 1949 Strassmaier's notebooks became available for study, in which Sachs found another hundred fragments. Each of these required nearly a complete rewriting of Neugebauer's manuscript. In 1952 Sachs worked at the British Museum, where he was given access to copies made by Theophilus G. Pinches at the end of the last century and personally examined many more tablets. This provided about sixty new fragments of mathematical astronomy and also increased greatly the number of known astronomical texts of all kinds.

In 1955, after twenty years of work and many delays, Neugebauer published *Astronomical Cuneiform Texts* (ACT), and in the same year Sachs published *Late Babylonian Astronomical and Related Texts* (LBAT), based upon the copies made by Strassmaier and Pinches, containing about 1500 texts in addition to those in ACT. Since that time the study of Babylonian astronomy has been founded upon these two publications. ACT contains all texts concerned with mathematical astronomy, that is, ephemerides and procedure texts, now called ACT texts, known at the time of publication, about 300 tablets and fragments, four or five times as many as had previously been published, of which nearly two-thirds concern the moon and eclipses and the remainder the planets. The first volume is devoted to lunar theory and eclipses according to Systems A and B with pertinent procedure texts, the second to planetary theory, again with procedure texts, and the third to transcriptions of all ephemerides, with a vast amount of restoration and enlargement by computation, and copies or photographs of the procedure texts and some of the ephemerides. LBAT is quite different in that after the cataloguing and dating of the texts at the beginning it consists entirely of copies. In addition to about 160 ACT texts, of those texts that contain enough information to be dated, the approximate numbers of fragments are: 360 of Diaries (plus about 500 undated), 70 of normal-star almanacs (plus 50 undated), 80 of almanacs (plus 10 undated), 90 of goal-year texts (plus 60 undated), and 90 of collections of observations of the moon, eclipses, and planets; in addition there are 40 of various kinds of early, pre-ACT astronomy, 20 of horoscopes, 60 of Enūma Anu Enlil, and 50 of later astrology.

Research in Babylonian astronomy has been more or less continuous since the publication of ACT and LBAT. Both Neugebauer and Sachs published additional texts with analyses, A. Aaboe has both published texts and written extensively on lunar and planetary theory, and B. L. van der Waerden has written on a variety of subjects, including a general treatment of astronomy and astral religion and mythology in *Erwachende Wissenschaft II. Die Anfänge der Astronomie* (1966, Eng. ed. 1974). The most comprehensive treatment of mathematical astronomy is again the work of Neugebauer in Book II, some two hundred pages, of *A History of Ancient Mathematical Astronomy* (HAMA), which appeared in 1975. In a way this is the long-delayed third volume of the *Vorlesungen*, now grown far larger than its originally anticipated size, and along with ACT it is the indispensable guide to the subject. Among other scholars whose contributions to mathematical and observational astronomy and celestial divination have appeared in the last forty years are Lis Brack-Bernsen, John Britton, Norman Hamilton, Janice Henderson, Peter Huber, Hermann Hunger, Alexander Jones, Johannes Koch, Yasukatsu Maeyama, Kristian Moesgaard, A. Leo Oppenheim, David Pingree, Erica Reiner, Francesca Rochberg, F. Richard Stephenson, Wilfried van Soldt, and Christopher Walker. Hence, research is very much alive, as was shown by the papers in *A Scientific Humanist. Studies in Memory of Abraham Sachs* (1988) and *Die Rolle der Astronomie in den Kulturen Mesopotamiens* (1993), the proceedings of a conference held in Graz in 1991. And some part of this scholarship would certainly meet with Kugler's and Neugebauer's approval. However, by far the most important publication since ACT, LBAT, and HAMA is Sachs's and Hunger's *Astronomical Diaries and Related Texts from Babylonia*, of which three volumes, containing the dated Diaries from −651 to −60, appeared in 1988–96. These are the continuous observational reports upon which, it is believed, Babylonian mathematical astronomy is built, and their publication was one reason for holding the conference on which this publication is based.

The modern recovery of Babylonian astronomy and celestial divination is one of the great achievements of the study of antiquity, comparable in difficulty with the decipherment of cuneiform itself. Greek astronomy and astrology have had a different history. The most important work of all, that of Ptolemy, has never been lost; it was preserved and studied, in Arabic and Latin translations as well as in Greek, although only among the Arabs could it be said to have been understood. However, most earlier Greek astronomy and astrology disappeared by late antiquity, due principally to its being displaced by Ptolemy's, and, aside from the discovery of some few papyrus fragments, the loss has never been made up. Thus, we shall never know as much of pre-Ptolemaic Greek astronomy as we do of Babylonian, for which we must be grateful to the durability of clay tablets compared to papyrus and even parchment. The historical study of Greek astronomy, like Greek mathematics, began as part of its recovery as living science in the early modern period. The *Epitome of the Almagest* by George Peurbach and Johannes Regiomontanus, written circa 1460–63 and first printed in 1496, although not intended as a historical work, marks the beginning of the modern technical understanding of Ptolemy, and along with the Latin versions of the *Almagest*—from the Arabic by Gerard of Cremona, made in 1175 and printed in 1515, and from the Greek by George of Trebizond, made in 1451 and first printed in 1528—it formed the basis of the recovery and development of Ptolemy's methods seen in the work of, to name only the most prominent, Copernicus, Tycho, and Kepler. There was yet more to Regiomontanus's contribution to the recovery of Ptolemy, for in the sixteenth century his manuscripts were the basis of the first editions of the Greek texts of the *Tetrabiblos* by Joachim Camerarius (1535) and of the *Almagest* by Simon Grynäus with the commentaries of Theon and Pappus by Camerarius (1538). One early, and excellent, publication that does appear historical in intent is the first edition of the Greek text of the *Planetary Hypotheses* with Latin translation by John Bainbridge (1620), a publication that meets, and in some ways exceeds, modern standards in the editing of classical texts.

Among the publications of the Renaissance and early modern period that can truly be called historical, Joseph Scaliger's great works on chronology *De emendatione temporum* (1583) and *Thesaurus temporum* (1606) and the commentaries in his editions of Manilius (1579, 1600) are also the most extended discussions and the largest collections of materials of their age for the history of ancient astronomy and astrology, near-eastern as well as Greek, although it must be added that stray remarks on these subjects in Kepler's works are more likely to be correct. But all earlier historical scholarship was surpassed by two monuments of scientific and historical erudition of the seventeenth century, the *Opus de doctrina temporum* (1627) of Denys Petau (Petavius) and the *Almagestum novum astronomiam veterem novamque complectens* (1651) of Giambattista Riccioli, the latter not surpassed, or even equaled, until Jean Baptiste Joseph Delambre's *Histoire de l'astronomie ancienne* (1817). Contemporary with Delambre are the Abbé N. Halma's publications of the *Almagest*, the *Handy Tables*, and other works of Ptolemy with French translation, which, while not reaching the highest standards, are the first editions produced for historical purposes with translation into a modern language. Hence, by the middle of the seventeenth century, and certainly with Delambre's and Halma's work at the beginning of the nineteenth century, although many curious stories were still told, nearly as much was known of Greek astronomy as is known today. But since little more was known of Babylonian astronomy than the account by Diodorus, which means essentially nothing of substance, there was no way of grasping the magnitude of the debt of Greek astronomy to the Babylonians. This became clear only in our own century through the research of, above all, Kugler and Neugebauer. Since Delambre wrote his great history, the principal works of comparable importance have been N. Herz's *Geschichte der Bahnbestimmung von Planeten und Cometen* (1887–94), J. L. Heiberg's edition of Ptolemy's works (1898–1907), K. Manitius's translation of the *Almagest* (1912–13), as well as his earlier editions and translations of Geminus (1898), Hipparchus's *Commentary on Aratus* (1894), and Proclus's *Hypotyposis* (1909), and more recently A. Rome's edition of the commentaries of Pappus and Theon on the *Almagest* (1931, 1936, 1943), O. Pedersen's *A Survey of the Almagest* (1974), Neugebauer's *A History of Ancient Mathematical Astronomy* (1975), G. J. Toomer's translation of the *Almagest* (1984), and J. Mogenet and A. Tihon's editions of Theon's small and large commentaries on the *Handy Tables* (1979, 1985, 1991).

So much for astronomy. Astrology fell into such disrepute in the course of the seventeenth century that even its historical study was for the most part neglected until the late nineteenth century, aside from the occasional edition of Manilius. The revival of serious interest was due to a handful of scholars who grasped the crucial role of astrology, as well as other forms of divination, in Hellenistic civilization. The first, and still the only, worthwhile survey of Hellenistic astrology is A. Bouché-Leclercq's *L'Astrologie Grecque* (1899), a continuation of his *Histoire de la Divination dans l'Antiquité* (4 vols., 1879–82), among its many virtues, such as taking advantage of the recent publications of Strassmaier, Epping, and others on Babylonian astronomy and astrology, the most amusing book ever written on astrology. Bouché-Leclercq's work, as excellent as it is, and it is hard to imagine its being superseded, has little on the mathematical aspects of astrology, but no mind, for an explanation of the mathematics was soon provided in the extensive notes to C. A. Nallino's edition of al-Battānī's *az-Zīj aṣ-Ṣābi* (parts 1 and 2, 1903–07), the foundation of the modern study of Arabic astronomy and an invaluable source for the history of astrology and its technical methods that has unfortunately been little used for the latter purpose. Among the notable scholars who published texts and studies of Hellenistic astrology before and after the turn of the century are F. Boll, W. Kroll, E. Maass, W. Gundel, A. E. Housman, and F. Cumont, of whom the two last require special comment. Housman's contribution is his annotated edition of Manilius, at least as well known for its two irreverent prefaces—mostly on the failings of German classical philology, as amusing as Bouché-Leclercq but in a different way—as for its scholarship, which is stronger in philology than mathematics, although it is also true that Manilius does not require a highly technical elucidation. Franz Cumont is in a class by himself. He was an archaeologist and epigrapher, the founder of the modern study of eastern religion in the Roman Empire, in which he has been followed by A.-J. Festugière and M. J. Vermaseren, and wrote extensively on astrology, his most well-known work being the elegant *L'Égypte des Astrologues* (1937). His greatest contribution, however, was founding and for many years editing the *Catalogus Codicum Astrologorum Graecorum*, CCAG (1900–40, 1951–53), descriptions of astrological manuscripts and editions of texts from many of the great European libraries, most catalogued and edited for the first time. The period since CCAG has seen Festugière's *La Révélation d'Hermès Trismégeste I. L'Astrologie et les Sciences Occultes* (2d ed. 1950), Neugebauer and H. B. Van Hoesen's *Greek Horoscopes* (1959), editions with translation and analysis of all horoscopes then known from papyri and literary sources, E. Boer's editions of Paulus Alexandrinus (1958) and Heliodorus (1962)—she had earlier completed Boll's edition of the *Tetrabiblos*

(1940)—G. P. Goold's Manilius (1977), and David Pingree's editions of Hephaestio (1973–74), Dorotheus (1976), and Vettius Valens (1986) and various papers on later Greek and Byzantine astrology. The renewed popularity of astrology in recent years has produced a fair quantity of writing on ancient astrology and astrological religion by adepts and enthusiasts, along with a few curious efforts at translation, and although some of it actually seems to be taken seriously by classicists and historians of religion, the kindest observation on this literature is silence.

This may suffice as a brief sketch of the historiography of our subjects, bringing us more or less up to the time of the papers in this volume. The editor requested that the participants in the conference write on any subject of their choosing with the object of assembling a collection of papers on the most interesting aspects of ancient astronomy and celestial divination as seen by scholars currently working on the subjects. The term "celestial divination" was used rather than "astrology" because it is a wider term, taking in Babylonian practices of receiving omens from anything that happens in the heavens, whether astronomical or meteorological, and differing from Hellenistic and later astrology, which is mostly horoscopic. It should be noted, however, that celestial divination based upon omens continued to flourish in India. That the conference turned out to be concerned more with Babylonian than Greek subjects was due simply to the papers presented by the participants, since many of them have worked in both areas.

Whatever the sophistication of ancient mathematical astronomy, both Babylonian and Greek, and even though at its highest level it was a science practiced for its own sake and guided by the curiosity and ingenuity of its practitioners, there is no reason to doubt that its origins lie in celestial divination. There is no contradiction between an original astrological motivation and a later purely scientific development. Hence we begin with the most important early evidence for the development of celestial divination, the omen series Enūma Anu Enlil treated by Erica Reiner in "Babylonian Celestial Divination." Peter Huber has presented evidence that the earliest layers of this vast collection go back to lunar eclipse omens from the Dynasty of Akkad and Ur III late in the third millennium, and Reiner quotes a poem referring to a solar eclipse during the reign of Sargon of Akkad. Reiner also considers relations between Enūma Anu Enlil and the hemerological series Iqqur īpuš of the second millennium, which appears to preserve omens of an early date, omens in Hittite found at Boghazköy, and the contents of the tablets of the series concerning planets and stars that she and David Pingree have been editing.

From the earliest celestial divination texts, Francesca Rochberg, in "Babylonian Horoscopes: The Texts and Their Relations," takes us to the latest, personal horoscopes that begin to appear about 400 B.C. Examples have been published by Strassmaier and Epping (ZA 3–4), Kugler (SSB 2, 554–562), and most notably Sachs (JCS 6, LBAT 1458–1476); Rochberg has published the entire known corpus through the American Philosophical Society (1998). Horoscopes differ from celestial omens most obviously in that they are concerned with individual nativities rather than the king and the state. But the astronomical differences go deeper. Celestial omens are taken from lunar and planetary phenomena and conjunctions with stars, the dates and locations of which are given by the heavens. The object of ACT mathematical astronomy and other computational or observational texts is to compute or record those dates and locations. In horoscopes, the dates are given by the nativities and the object is to find significant phenomena or the locations of bodies in the heavens on or close to such dates, which seems to have been done using a combination of observational records and computation, with the locations of planets given in degrees of zodiacal signs, the computation of which, although not obvious, was possibly based upon methods of interpolation between phases, whether observed or computed, of the kind described in ACT procedure texts, although probably not the most sophisticated methods known for Jupiter and Mercury. Even if these texts represent an early stage, one may presume that the origins of Hellenistic horoscopic astrology are Babylonian.

As is evident from the earliest eclipse omens and the Venus Tablet, the foundation of divination through celestial omens was observation. In "Babylonian Observations of Saturn during the Reign of Kandalanu," Christopher Walker here publishes a very early observational text, a series of Saturn's heliacal risings and settings, dated to month and day with locations by stars and parts of zodiacal constellations, for −646 to −633, beginning only five years after the earliest surviving Diary. While we do not know the function of such a collection at so early a date, it is precisely such records of phenomena for a single planet that would have been the most directly useful for the later development of the ACT procedures for computing planetary phenomena.

The entire corpus of what are called "nonmathematical" astronomical texts, to distinguish them from ACT texts, following the classification by Sachs (JCS 2), who devoted many years to editing them, is being completed by Hermann Hunger. In addition to Diaries, the largest group, this consists of goal-year texts, almanacs, normal-star almanacs, and col-

lections of observations of planets. The relation of observation and computation in these texts is neither simple nor certain, something made all
the more difficult by differences in duplicated information in different
texts. Hunger has found that while Diaries and goal-year texts, which are
both principally observational, generally agree, a comparison of goal-year
texts with almanacs and normal-star almanacs shows notable differences.
The paper also includes translations, the first into English, of a goal-year
text for Seleucid Era 140 ($-171/-170$) and an almanac for Seleucid Era
236 ($-75/-74$), both particularly well preserved.

The most frequent observations of the moon and planets in the
Diaries are of distances "above" or "below" and "in front of" or "behind"
normal stars by so many "cubits" and/or "fingers." The principal difficulties in interpreting such observations are in the meaning of the
directions, whether there is an implied coordinate system, and in the
equivalent of the units, if there is a consistent equivalent, in degrees and
minutes. In "Normal Star Observations in Late Babylonian Astronomical
Diaries," Gerd Graßhoff has undertaken the first extensive analysis of these
observations to provide definitive answers to these questions. His conclusions are that "in front of" and "behind" are not separations but differences in longitude, "above" and "below" are differences in latitude—
hence implying ecliptic coordinates—and that the cubit used in these
observations, which may be divided into both 24 and 30 fingers, has a
mean value of about $2.4°$, implying that it was defined as $2\frac{1}{2}°$. The paper
also is a demonstration of a method of analyzing quantities of uncertain
meaning, such as the reports of distances in the Diaries, with a minimum
of assumptions by separately testing every alternative hypothesis for each
quantity and for the meaning of each term.

After conjunctions of the moon and planets with normal stars, the
most frequent observations in the Diaries are six quantities measured in
UŠ, "degrees of time" equal to $1/360$ of a day, of the rising and setting
times of the moon, used to determine the dates of first and last visibility
and full moon. Lis Brack-Bernsen has investigated these in previous
papers, one written with Olaf Schmidt, and shown that from the sum
of the four quantities ŠÚ, NA, ME, and GE$_6$ measured before and after
full moon may be derived the period of column Φ of System A lunar
ephemerides, which in turn shows how the period of the anomalistic
month itself was determined. In "Goal-Year Tablets: Lunar Data and
Predictions," she considers the sums of the setting arcs ŠÚ + NA and the
rising arcs ME + GE$_6$—the first of each pair before, the second after, full

moon—showing that each individual quantity can be predicted at intervals of one Saros, 223 synodic months ≈ 18 years, from its value and the
sums one Saros earlier, the interval for these quantities recorded in goal-
year texts. Further, the setting time at first visibility NA and the rising
time at last visibility KUR may be found from its value and the sums at
intervals of one Saros plus 6 months, 229 synodic months, the period for
which the sums are also recorded in goal-year texts. These discoveries
bring us close to the empirical and analytical foundation of Babylonian
lunar theory.

The writers of the astronomical texts, the scribes of Enūma Anu
Enlil, may also have been responsible for instruction in mathematics, as
shown by two procedure texts for Jupiter, ACT 813 and 815, that also
contain a problem concerning a trapezoid. In "A New Mathematical Text
from the Astronomical Archive in Babylon: BM 36849," Asger Aaboe
publishes a multiplication table from a collection, containing many astronomical tablets, that reached the British Museum on 17 June 1880. It may
be paired with a related tablet from Uruk, published by Aaboe in JCS 12.
Both seem to be copies of the same table containing squares of 1 to 59
and multiples of at least 15 numbers, only one in the standard list, and 4 of
them irregular, that is, having prime factors other than 2, 3, 5, and thus
nonterminating reciprocals.

John Britton's "Lunar Anomaly in Babylonian Astronomy: Portrait
of an Original Theory" is a theoretical investigation of the origin of
column Φ, the excess of 223 synodic months over 6585 days as a function
of and with the period of the lunar anomaly, and related functions F for
lunar velocity and G, W, and Λ for the length of one, six, and twelve
months in System A lunar theory. He shows that both the Saros and the
19-year cycle can be taken as the basis of a hypothetical function S, in
units of the "sarosly change in the length of a saros," which in turn
underlies Φ. System A is considerably more sophisticated and accurate
than System B, and it is shown here that its curiously high mean value for
F is actually an excellent approximation to lunar velocity at syzygy due to
the contribution of the variation, the third largest lunar inequality,
amounting to about $+0;16^{o/d}$ at syzygy. The development of these
functions is surprisingly early, Φ and F appearing in a text known as S for
38 solar eclipse possibilities in the Saros from −474 to −456, in which the
functions for longitude of conjunction and eclipse magnitude are far more
primitive than both Systems A and B. This suggests that the origin of Φ
and F, their structure and relation a defining event in the development of

mathematical astronomy, may be placed only slightly later, in the second half of the fifth century, while the complete longitude theory does not appear until about the middle of the fourth century.

The Astronomical Diaries give the time of heliacal phenomena of the planets to the day of the calendar month and the location by zodiacal sign, sometimes specifying beginning or end of zodiacal sign. In "The Derivation of the Parameters of Babylonian Planetary Theory with Time as the Principal Independent Variable," N. M. Swerdlow, the editor shows how this information, in particular the maximum and minimum synodic times given by the dates and, for Mercury, one or two additional synodic times, can be used to derive the parameters of the planetary theory of the ACT ephemerides.

Aside from clay tablets and a very few inscriptions, the only original astronomical and astrological documents surviving from antiquity are papyrus fragments, found mostly in Egypt at the end of the last century. Neugebauer published a good number of these—many papyrologists eventually sent him anything with numbers on it on the chance that it was astronomical—and now Alexander Jones is preparing an edition of the Oxyrhynchus papyri in the Ashmolean, a collection containing more astronomical and astrological fragments than all hitherto published. In "A Classification of Astronomical Tables on Papyrus," he reports on the types and numbers of tables found, preliminary to his publication of the entire corpus in the *Memoirs of the American Philosophical Society*. Among the more remarkable discoveries are tables for each planet using ACT arithmetic methods and parameters, of which examples of first visibility in System A_1 of Mercury and System B of Saturn are published here. Along with a fragment of column G of a System B lunar ephemeris published by Neugebauer in *A Scientific Humanist* (1988), these show a more extensive transmission of Babylonian methods than had previously been imagined, and also a later use since these tables are roughly contemporary with papyrus fragments of Ptolemy's *Handy Tables*.

Ptolemy's observations have been the subject of intense scrutiny, mainly with respect to modern recomputations of them. In "The Role of Observations in Ptolemy's Lunar Theories," Bernard R. Goldstein and Alan C. Bowen address a different set of issues. They seek to assess the reliability of the lunar observations in the *Almagest* as perceived in Ptolemy's time, and to indicate the reasons that Ptolemy had for citing specific dated observations. They conclude that Ptolemy gained confidence in his first lunar theory on the basis of his own observations, and that his confidence was enhanced (not established) by agreement of ancient observations with the theory that was derived from his own observations. The case of Ptolemy's second lunar theory is more complicated, but again, in their view, Ptolemy depended in a fundamental way on his own observations. In particular, they argue that Ptolemy's observations of elongations from the moon to a planet confirmed his second lunar theory and his parallax theory, taken as a whole. This essay challenges a number of standard claims about ancient astronomy and is likely to be controversial.

It is well known that in Robert Simson's edition of Euclid (1756) "the errors by which Theon, or others, have long ago vitiated these books are corrected," and wherever Simson comes upon anything in Euclid that does not please him, poor Theon is taken to task. This is of course the same Theon who wrote a commentary on the *Almagest* and two commentaries on the *Handy Tables*, a "Great Commentary" in five books and a "Small Commentary" in one book, and there has been some uncertainty about whether he also "vitiated" the *Handy Tables*. The discovery of the pre-Theon Peyrard manuscript of Euclid, Vat. Gr. 190, showed that Theon had been unfairly accused, that his alterations were mostly additions for the sake of clarity, that Euclid's faults were, alas, his own. Anne Tihon has edited the Small Commentary on the *Handy Tables* and is in the course of completing the edition of the Great Commentary begun by J. Mogenet. In "Theon of Alexandria and Ptolemy's *Handy Tables*," she describes just what it is that Theon did. It turns out that he did not alter the *Handy Tables* at all, and in his commentaries he did his best to explain how the tables were computed and how they differ from the *Almagest*, with nothing to guide him besides the texts themselves, that is, his situation was no different from our own. And while there were certainly parts of the computation of the *Handy Tables* that he did not understand, the same goes for us, so Theon may once again be considered vindicated as an honest and hard-working guide to the works of Ptolemy in all their difficulty, much like us.

1

BABYLONIAN CELESTIAL DIVINATION
Erica Reiner

"On the thirteenth of the month of Ulūlu, the moon became eclipsed and set while eclipsed. It was a sign that the moon god requests a high priestess." So speaks Nabonidus, the last king of the Babylonian empire before its conquest by Cyrus, in his account of the selection and installation of his daughter as high priestess of the moon god, Sin.[1] The date of the eclipse can be established as 26 September 554 B.C., and its interpretation was given to the king by the scholars in his entourage who had at their disposal a compendium of celestial omina in which they were expected to look up the significance and the prediction associated with the event observed.

Eclipses were not the only phenomena that the Babylonians observed in the belief that they signalled the will of the gods or foretold future events. A fortuitous occurrence and a subsequent good fortune or misfortune were linked in the Mesopotamian mind, as they were in many early cultures and still are in primitive societies, not so much as cause and effect, but as a forewarning and a subsequent event. Such linked pairs, consisting of a protasis (if-clause) and an apodosis (forecast), a pair called by the technical term "omen," were collected in lists, and these lists eventually developed into large compendia that we call omen series. Portents from lunar and solar eclipses, as well as from other lunar, solar, stellar, and meteorological phenomena, were collected and more or less standardized in the omen collection called from its incipit Enūma Anu Enlil, "When Anu (and) Enlil" (henceforth EAE), comprising seventy books (called, in Akkadian, "tablets"[2]) along with various excerpts from these and commentaries on them.[3]

The scholars who made astronomical observations and recorded the forecasts derived from celestial phenomena were called *ṭupšar Enūma Anu Enlil*, a term meaning "scribe of (the celestial omen series entitled) 'When Anu, Enlil …'" An abundant correspondence from the Neo-Assyrian empire attests the importance at the royal court of the diviners and the astronomers who apprised the king of the portents, and of the exorcists who were expert in averting ill-boding forecasts by their rituals.[4] The astronomers regularly conveyed to the king such routine reports as the monthly sighting of the new moon and the date of the opposition of sun and moon as well as special reports on various celestial and meteorological phenomena, such as conjunctions, occultations, and rain and thunder.[5] More than a thousand years earlier, in Mari, the correspondents reported on such extraordinary events as torrential rains and thunder.[6]

The seventy books of EAE deal with a variety of celestial and meteorological phenomena. The predictions affect king and country, not individuals, and are for the most part very general, involving the fall, revolt, or conquest of one of the four countries into which their world was divided—that is, the home country, normally called Akkad; the land to the east or Elam; the land to the west or Amurru; and to the north, called Guti or Subartu. They also predict the death of the king of one of these lands, the success or failure of crops, and the like. The series consists of four main sections: the first deals with the moon, the second with the sun, the third with the weather, and the fourth with stars and planets. The first twenty-two books derive predictions for the fate of king and country from lunar phenomena: the moon's general appearance, such as shape of the horns, halos, conjunctions, and especially eclipses. Books 23–37 deal with solar phenomena: halos, colors, and solar eclipses. Books 37–49 treat meteorological phenomena—thunder, lightning, rainbows, winds—and earthquakes. Books 50–70, the final section, derive omina from planets and fixed stars.

It is this last section that David Pingree and I have started to publish in a series of fascicles, as *Babylonian Planetary Omens*. Fascicle 2 of this series summarized the contents of the series, as far as it was then known to us. While it was possible to establish that Book 70[7] was indeed the last book of the series, the catch line—the line that serves as the title or incipit of the tablet that follows—was not identified at that time, although it was evident that it did not belong to EAE or indeed to any omen series. This line, *ēnu Anu rabû abi ilī*, "when great Anu, father of the gods," is now attested as the incipit of a text published by E. von Weiher.[8] It is this text, then, that followed Book 70 of EAE.[9] The practice of adding one series as the continuation of another is by no means unusual; we know of other series that were also concatenated, for whatever reasons. Perhaps the sequence in which they are listed represents the curriculum of the scribe or some other overarching organization of the corpus of cuneiform literature.

Most of the sources for celestial omina come from the library of Assurbanipal at Nineveh from about the seventh century B.C. and, somewhat later, from Babylon in the Neo-Babylonian period. The earliest material comes from the Old Babylonian period, that is, the eighteenth or seventeenth centuries. The two groups of sources are bridged only by a few texts from Assur, from around 1000 B.C. However, texts from outside Mesopotamia proper, from Syria and from Anatolia (in the Hittite language as well as in Akkadian, and even a sole text in the Ugaritic language found at Ras Shamra in Syria), can give us some idea of the growth of the celestial omina in the second millennium, in late Old Babylonian and Middle Babylonian times. All these peripheral texts deal with eclipses and with various configurations affecting the moon.

In all cultures, eclipses were terrifying events, inexplicable as long as the conditions for their occurrences were not recognized. Since astronomical knowledge in the second millennium was not adequate to predict eclipses, even lunar eclipses, and even in the early seventh century B.C. lunar eclipses could be predicted only shortly before the time of their occurrence, these celestial events were considered on a par with conquests, defeats, and other events affecting the land and were similarly predicted through the observation of sheep entrails or through lecanomancy, that is, the shape taken by a drop of oil poured into a water basin. These predicted eclipses are sometimes specified as lunar or solar but often remain unspecific.

Eventually, divination was extended from lamb entrails and various terrestrial events to celestial phenomena. The first of the celestial phenomena from which predictions were derived were eclipses. In fact, the theory has been advanced (by P. Huber[10]) that omen literature had its origins when a connection was perceived between certain significant events, such as changes in the reigns of dynasties, and lunar eclipses.

Rare are celestial phenomena other than eclipses predicted in noncelestial omen texts. One is the rising, presumably the heliacal rising, of Mars. Mars, the hostile planet, presides over destruction and is equated with Nergal, god of pestilence. His rising is a bad omen, portending death of the herds: "Mars will rise and destroy the herd." This prognosis appears as apodosis in several lunar eclipse omina,[11] but it also appears as apodosis in a solar eclipse text and a liver omen.[12] The only other celestial phenomenon predicted in an omen text is a meteor; this occurs in a teratological omen (*Izbu* XVII 21) and says: MUL GAL ŠUB-*ut*, "a large star will fall."

Eclipses were, and remained even until premodern times, mysterious events, and allusions to them appear in Mesopotamian poetic texts and historical texts as well. In addition to Nabonidus—whose account qualifies as a historical text but, as I have maintained in another context, a highly poetic one—other kings' reigns are associated with the miraculous occurrences of eclipses.

The so-called King of Battle (*šar tamhāri*), an Old Babylonian poem dealing with the exploits of King Sargon of Akkad (ca. 2400 b.c.),[13] relates how "the sun became obscured, and the stars came forth for the enemy."[14] This description no doubt refers to a solar eclipse during which the stars became visible, an event that was evidently to be interpreted as the stars portending victory for Sargon against his enemy.

Allusions to celestial portents given to Sargon also appear in liver omina. For example, a particular configuration of the sheep's liver is said to refer to the same Sargon of Akkad as one who "made an incursion during darkness and light came out for him."[15] A similar omen states that Sargon "marched during darkness and saw light." These descriptions in Old Babylonian liver omina must record an eclipse occurring under Sargon, in spite of their peculiar phraseology. This interpretation seems to be confirmed by the account of Sargon's later namesake, Sargon II of Assyria, no doubt intentionally referring to his illustrious predecessor's example, that a lunar eclipse portended victory for him on his campaign against Urartu in 714 b.c.[16]

Sargon reports that while he was en route to Urartu the moon became eclipsed and the darkness lasted from the first night watch into the second; the haruspex, who as usual accompanied the king on his campaign, was called upon to interpret the meaning of the eclipse.[17] Sargon continued his route only upon being assured that the portent presaged victory. Similarly, the interpretation of the lunar eclipse occurring under King Nabonidus also had to be confirmed by the haruspex.

Between the early, Old Babylonian lunar eclipse omina and the full-fledged, serialized compendium of celestial omina of the first millennium we have only a few texts, and these do not come from Mesopotamia proper and are not part of the EAE tradition. F. Rochberg has summarized in her book the lunar eclipse omina attested before the more or less standardized first-millennium versions, in texts from Susa in the East and from Syria in the West. (Texts from Anatolia, both in Akkadian and in Hittite, are somewhat later.)

Apart from the two lunar eclipse texts in Akkadian found at Boghaz-köy, there is a third Akkadian text discussed by Rochberg, KUB 37, 162, which deals with the horns of the moon. Moreover, a bilingual Hittite and Akkadian text from Boghazköy, edited by H. G. Güterbock,[18] also deals with the horns of the moon. If we would appear to be too hasty in judging from these two sources alone that the shape of the moon's horns was the phenomenon that attracted the attention of diviners after the observation of the time and extent of an eclipse, we can now adduce a group of texts from a recently excavated site in Syria, the ancient city of Emar, that also treat the same subject.

The second paragraph of the Boghazköy text (KUB 37, 162), from the protasis of which only the predicate *lawi*, "is sourrounded," is preserved, must deal with the lunar halo that is normally designated with the logogram GIŠ.HUR; this phenomenon appears, apart from the celestial omen collection, also in the second half of the compendium *Iqqur īpuš*, "he tore down, he built."[19] This compendium, also called a Babylonian calendar, lists in each of its over one hundred paragraphs an activity along with the prognosis if the activity is performed in a particular month. Each paragraph consists normally of twelve entries, one for each month of the year. The fragmentary paragraph of the Boghazköy text can, with a high degree of probability, be identified with paragraph 77 of *Iqqur īpuš*, since the apodoses of months I, III, and IV seem to correspond to this paragraph, and only the apodosis for month II differs.[20]

In the main recension of *Iqqur īpuš* the celestial phenomena begin with paragraph 67; they are essentially the same as those treated in the EAE series, but include only a selection. The list of ominous celestial events varies with the different sources, ranging from the always present predictions from eclipses of the moon and of the sun, from halos of the moon, from first appearance of Venus, and from phenomena such as thunder, rain, and earthquakes, to risings of other planets[21] not attested in every source. Some of the *Iqqur īpuš* texts arranged by months include solar phenomena that had not found their way into the main recension, such as obscuration (a phenomenon that is not identified but is different from eclipse since it is denoted with a term different from that used for eclipse), redness at sunrise (expressed with the phrase "if the sunrise is spattered with blood"[22]), and standing still in the middle of the afternoon;[23] this latter entry is also listed in STT 306 r. 10 (month VI). Among the Šamaš omina this phrase occurs as *Šamaš ina* IZI.AN.NE *izziz*,[24] "the sun stands (still) in the heat of the day²," a phenomenon listed for each month; the logogram IZI.AN.NE may correspond to Akkadian *muṣlalu* or *anqullu* (an atmospheric phenomenon). The same spelling, *Šamaš* IZI.AN.NE [. . .] (to be emended to ⟨*ina*⟩ IZI.AN.NE), occurs in the monthly series for month VIII.[25] The sun standing still of course reminds one of Joshua 10 : 12, which some have tried to explain as a reference to a total solar eclipse (in 1131 B.C.).

The phenomena listed in *Iqqur īpuš* are among the earliest celestial phenomena recorded, as was pointed out in an unpublished study by K. Riemschneider. His study was written before the discoveries in Emar, so that he could speak only of the texts found at Boghazköy and Nuzi. Today, we can add to Riemschneider's list the recensions of *Iqqur īpuš* found in Emar.[26] Their subject matter includes the usual topics of celestial divination: lunar eclipse (in some months adding lunar halo), solar eclipse, the sun standing still at noon, and SUR (instead of the expected *šamû*) *īrup*, "the sky darkens" (no. 610); these are preceded by thunder in months VIII and IX, and expanded in month XI by mud (*im.gú*), rain, and a star falling upon the man. In number 611 the enumeration is, for month I: lunar eclipse, lunar halo, solar eclipse, thunder, rain, [solar] eclipse on the 23d, south wind, earthquake, mud, eclipse of Venus, the sun standing still at noon, and a star falling upon the man.

Celestial omina in the Hittite language include, apart from the omina taken from the moon's horns mentioned above, some omina from stars, planets, comets, and meteors, even though the meanings of the Hittite words are not always exactly known. It is worth noting that the omen that so far was unique in these texts, namely "if in month x a star falls from heaven" (KUB VIII, 25), can now be paralleled by the Emar texts that speak of a star falling upon a man. Two Hittite texts (KUB VIII, 14 r. 8ff. and KUB XXXIV, 16 ii 7–9) contain omina on planets passing through the constellation Wagon (Ursa Maior), phenomena that even I, with my modicum of astronomical knowledge, must doubt. If the omina found at Boghazköy reflect Mesopotamian omina from the Kassite period in the second half of the second millennium, this fact would indicate that not every omen that was attested in second millennium Mesopotamia was taken over into the standard compilation of celestial omina, or better, into the first millennium compilations, since no canonical text of EAE seems ever to have been composed.

Predictions of an eclipse also appear as apodoses of lunar omina from Emar, in a text that deals with the horns and halos of the moon. In addition to the apodoses that are expected that affect the fate of the country and the king, in number 651 : 32 and 42 an eclipse (of the moon or sun?) is predicted for that very month, and in number 651 : 33 a solar eclipse for day 30.

In the first millennium series, that is, in EAE, the books on lunar and solar omina are followed, as I have mentioned, by meteorological omina and omina from stars and planets. For the history of astronomy, obviously it is the planetary phenomena that are of the greatest interest. Apart from Venus, to which Books 59–63 are devoted, only Jupiter is the subject of extensive treatment, in Books 64–65. It is remarkable that the treatment of these two planets is quite dissimilar. Many omina concern Venus's color and similar horizon phenomena, and the planet's relation to the moon and to other planets and constellations, but they also refer to such unique features as wearing a crown and of course, as in the famous Venus Tablets of Ammiṣaduqa, the length of the periods of visibility and invisibility. For Jupiter, on the other hand, the most commonly occurring features are its position with respect to a constellation. Even these descriptions seem to have been standardized without particular thought given to logic or verisimilitude.

For example, several texts describe Jupiter's position in regard to the constellation called Scorpion, more or less corresponding to Scorpius:

19 [¶ MUL SAG.ME.GAR *ana* S]AG MUL GÍR.TAB *ik-ta-šad ina* KUR URI.KI KI.LAM GÁL-*ú ana* 2 HA.LA

20 [¶ MUL SAG.ME.GAR *ana* M]URUB$_4$ MUL GÍR.TAB *ik-ta-šad ina* KUR URI.KI KI.LAM 1 GUR *ana* 1 (BÁN) ŠE GUR

21 [¶ MUL SAG.ME.GAR *ana* K]UN MUL GÍR.TAB *ik-ta-šad ina* KUR URI.KI KI.LAM $\frac{1}{2}$ SÌLA *ana* 1 GÍN GUR

22 [¶ MUL SAG.ME.GAR ana] ú-ru-ud: ú-ru-uh MUL GÍR.TAB *ik-ta-šad*

23 [*ina* KUR NIM.MA].KI KI.LAM GÁL-*ú ana* 2 HA.[LA]

24 [¶ MUL SAG.ME.GAR *ana* L]I.DUR MUL GÍR.TAB *i[k-ta-ša]d ina* KUR NIM.MA.KI KI.LAM 1 GU[R *ana* 1 (BÁN) ŠE GUR]

25 [¶ MUL SAG.ME.GAR *ana zi-q*]*it* MUL.G[ÍR.TAB … (traces)

If Jupiter reaches the head of the Scorpion, in the land of Akkad the market will be halved.

If Jupiter reaches the middle of the Scorpion, in the land of Akkad the market [price] of one kor of barley will be reduced to one bushel [i.e., one fifth]

If Jupiter reaches the tail of the Scorpion, in the land of Akkad the market (price) of half a liter of barley will become one shekel [of silver]

If Jupiter reaches the throat of the Scorpion, in the land of Elam the market will be halved.

If Jupiter reaches the navel of the Scorpion, in the land of Elam the market [price] of one kor of barley will be reduced to one bushel [i.e., one fifth]

If Jupiter reaches the sting of the Scorpion, in the land of Elam the market will be halved.

Once you have established the parts of the constellation and their proper sequence, you can apply these terms to other constellations as well, dissimilar though their shapes may be. Thus Jupiter can pass to the right and to the left of the constellation Goat (our Lyra), but also by the navel, the sting, and the throat of the Goat, or stand not only at the head or the tail of the Fish, but also at its navel, its throat, and its sting.

As to the other planets, we have little material before the first millennium. Not even the place of tablets dealing with planets in the series is certain, with a few exceptions. A tentative ordering may be as follows:

EAE 56 may deal with planets in general, since the writing, ^{d}UDU.BAD (= Akk. *bibbu*) denotes any planet; possibly the tablet treats one of the two planets whose names are composed with ^{d}UDU.BAD, that is, Mercury (^{d}UDU.BAD.GUD.UD) and Saturn (^{d}UDU.BAD.SAG.UŠ). Only toward the end of the tablet is there a section explicitly referring to Saturn and Mercury, giving a period of visibility and invisibility for each.

This tablet is flanked by two tablets, Books 55 and 57, that deal with various constellations, singling out their brightness or some other light phenomenon and including such features as position, for example, facing upward or downward, features that were predicated of the moon's horns too. Often the relationship to the moon is given in terms similar to those in the other tablets of EAE, that is, inside (*ina libbi*), in front of, behind, in the right or left horn, and in the flank (*ina naglabi*). However, what stands in the various parts of the moon is not a single star but the constellation as a whole, for example, the Scorpion, the Goat, or the Wagon.

It is when these tablets deal with the relationships of constellations with one another, such as Books 52 and 53, that we enter the astronomically impossible area. Just as predictions are based on one planet approaching another, or nearing, reaching, or passing another, so the same movements are predicated of one constellation in regard to another, as if fixed stars had the ability of moving with respect to and among other fixed stars. Even the shape of the constellation is said to be subject to change; for example, the omen is different if the stars of ŠU.GI (Perseus?) are close together or wide apart.

Some ancient scholia, aware of the astronomical impossibility of the phenomena described, try to avoid the difficulty by equating the constellation with a planet or substituting planets for one or both of the fixed stars; for example, the protasis "if the Kidney star comes close to a planet (UDU.BAD)" is given the explanation "Mercury in Aquarius comes close to Saturn" (K.2064:6f.). When constellations approach or reach other constellations, the comments are similar. For example, Tablet 53 of EAE, K.3558, begins: If the Pleiades reach the Yoke—Saturn reaches Mars. If the Pleiades reach the Star of Marduk—Mercury or Jupiter reach the Pleiades. If the Pleiades reach the Kidney—the Pleiades and Mars. If Orion stands in the head of the Pleiades—Saturn stands in front of the Pleiades. If the Pleiades and the Wagon [i.e., Ursa Maior] stand together—Venus rises with the Pleiades; and so on.[27]

Unfortunately, according to certain texts it is one set of constellations that ought to designate a planet, and according to others, another set. For example, Book 52 lists the various constellations and planets (Venus, Mars, and "planet"—either planet in general, or short for Saturn or Mercury—are mentioned) that enter the constellation Field, which corresponds to the Square of Pegasus. According to one commentary to Book 52, the Square of Pegasus is replaced by a planet; Jupiter and Venus are preserved. The commentary therefore understood the protasis to refer to a conjunction of two planets. According to another commentary to the same text, both celestial bodies (the Square of Pegasus and the other constellation or planet) are replaced by a planet and a constellation (not always identical to the one which is commented on) or by two planets.

The statements about the movements of fixed stars may have originally referred to constellations; Pingree prefers to surmise that it was rather the various verbs of motion that denoted some phenomenon other than actual movement. We do not have any early material that could confirm or invalidate this supposition. Whatever was originally meant by the movements of fixed stars and constellations when the texts were composed, it is clear that by the time the commentaries were written, this was considered nonsense and various explanations were devised to "save" not the "phenomena" but the text.

Many omina are derived from light phenomena such as colors or brightness. Some of the terms we cannot identify (*mešhu*, *šabihu*, *ṣirhu*, etc.). Again, some commentaries relate these phenomena to the movements of planets, even though there is nothing astronomically impossible in them. For example, if one star in ŠU.GI is very red, the commentary explains it as "Mars [is seen] in ŠU.GI with Venus"; if it is very dark, the commentary explains it as "Mercury is seen in ŠU.GI."

The early inclusion in omina of the shape of the moon's horns and of the stars that appear within and around them has no doubt directed attention to the movement of planets, presumably first in relation to the moon and then in relation to the fixed stars and other planets. Whether it is from this interest that the observation of the planets' paths developed is hard to say; the fact that a large portion of all the omen texts dealing with planets is devoted to their position with respect to the moon suggests such a connection.

In addition to the relation of planets to the moon, quite a number of fragments (which all may belong to EAE, Book 53) deal with the Pleiades, including the relationship between the Pleiades and the moon. No doubt it is from such omina that the Pleiades intercalation rule, which establishes whether an intercalary month is to be added to the year depending on the date of the conjunction of the moon and the Pleiades, was abstracted.[28]

Sargon of Akkad, we have seen, heeded portents taken from eclipses, as did his later namesake, Sargon II of Assyria, in the first millennium. At the time of the Sargonids, however, the movements of planets also began to be considered ominous. Two such events are recorded under the Sargonid kings. The same letter of Sargon to the god Assur that reports on the lunar eclipse in 714 B.C., during Sargon's eighth campaign, contains another allusion to a favorable portent given the king by a certain, not specified, phenomenon of Jupiter, here called "the star of Marduk."[29]

King Esarhaddon, Sargon's grandson, reports on how he secured the throne for himself in the midst of the struggle for power among the sons of Sennacherib following that king's murder. His rightful succession was foretold in the stars: among other favorable signs the "secret place" reached by the planet Venus is mentioned.[30] When Jupiter shone exceptionally brightly and reached its "secret place" in the beginning of his reign,[31] a place that seems to correspond to what in Greek astrology was called the planet's "exaltation" (hypsoma), the sign of the zodiac in which it has the greatest influence, this sign was interpreted as a favorable portent for the rebuilding of Babylon.

When a celestial phenomenon predicted an unfavorable future, there were still means to avoid the evil consequences of the omen. These are the namburbi rituals, to which Stefan Maul has devoted a monumental book.[32] Namburbi rituals exist to every imaginable omen but are hardly ever mentioned for averting the future foretold by celestial events. That such did exist nevertheless, we gather from those texts that list in catalog form[33] the events against whose evil consequences such rituals could be invoked. The rarity of such apotropaia is perhaps to be explained by the fact that forecasts based on celestial omina affect the king and were written down for his use exclusively, as we also see from references in the royal correspondence from Assyria.

For example, one of the letters from the scholars to the king mentions the evil (portended) by Mars that missed its appointed time and entered Aries.[34] Another speaks of the late rising of Jupiter,[35] and a third of its dark aspect at rising.[36] This last example provides an additional proof, if such were still necessary, that the scholars who reported to Esarhaddon were consulting the omen series, since the partly broken omen that the writer cites can be restored from a number of celestial omen tablets, some of them as yet unpublished.[37] The ritual for averting the evil portended by an earthquake is also mentioned in the letters of exorcists to the Assyrian king.[38] The conjunction of Mars and the moon was also an occasion for an apotropaic ritual, but only the incipit is extant in the subscript of a namburbi: "if Mars and the moon go side by side."[39]

The catalogs of ill-portending celestial phenomena follow the same sequence as the compendium of celestial omina EAE: moon, sun, weather phenomena (including earthquakes), and stars and planets. One catalog[40] enumerates eclipses of, first (in a broken section) the moon, then of the sun and of Venus, as well as flaring up of stars, earthquakes, and various cloud configurations. A parallel enumeration is contained in a royal ritual in which the king lists "either an eclipse of the moon [or of the sun?] or of Jupiter or an eclipse of . . . , [or . . .] the roar of Adad that came down from the sky, [or . . .] or a flaring? star or a scintillating? star or [. . .] which came close to the stars of the (three) paths"[41] and continues with "any evil that is in my land and my palace."

An important and as yet unsolved problem is the relationship of celestial omina to the mathematical astronomy developed in Babylonia from the fifth century B.C. onward. I am not competent to deal with this question, since I approach the celestial omina as a philologist. Not only am I ignorant in matters astronomical, but I also believe that the two disciplines, omen astronomy and mathematical astronomy, were parts of different traditions of scholarship. How else can one explain the fact that Babylonian scholars still assembled personal libraries by copying EAE with all its nonsensical statements, albeit inventing such explanations as I have cited for these, while there was in place, from at least the seventh century B.C. onward, a system for accurately describing astronomical phenomena in the so-called Astronomical Diaries, based on regular daily observations? Still, while I do not think that celestial divination can be regarded as some kind of forerunner to the mathematical astronomy developed in Babylonia, there are a number of as yet insufficiently explored cuneiform texts that may eventually point to a connection on the one hand to Babylonian astronomical texts and on the other hand to Greek astrology, with its predictions for individuals based on planetary positions in the signs of the zodiac and on their mutual configurations.

A. SAMPLE FROM TABLET 57, K.2330 AND DUPLICATES

1 [If the Raven star] reaches [the Path of the Sun ...]

2 [If the Raven star] passes [the Path of the Sun ...]

3 [If the Raven star] comes close to Jupiter [...]

4 If the Raven star is upside down²: in the midst [...]

5 If the Raven star disappears² in the South [...]

6 If the Raven star ditto in the North: the market will be stable, the sesame harvest [will (not) prosper]

7 If the Raven star—its star is very red: the sesame harvest [...]

8 If the Raven star—its star is not red: [there will be] an epidemic [...]

[continuing with six more omens about the Raven star, then omens about the Eagle and the Fox].

B. SAMPLE FROM A COMMENTED TEXT, K.35

1 If Venus has a ṣirhu, not favorable—in her progress she ascends quickly

2 If Venus does not have a ṣirhu, favorable—she completely reaches her position slowly and stands there

3 If Venus flashes and goes around the Yoke star, an observer observes her, someone sees her: the land will be scattered, the reign will change, women will fall through weapons—Venus goes around Jupiter

4 If Venus flashes and goes toward the Wagon, and someone sees her: a storm² (SÙH) will rise in three days and cover the land—Venus stands in front of Old Man

5 If Venus turns around within the [Bull² of] Heaven: [...] will die

6 [If Venus [...] ... sets [...] will die, variant: the mind of the land will change

7 [If Venus in] the morning is steady: enemy kings will become reconciled

8 [If Venus ...] rises [...] three months: there will be hostilities, the crop of the land will succeed—she stands [in² ...] the Bull of Heaven or the True Shepherd of Anu

9 [If Venus rises in the morning watch: there will be sugagāti [variant:] massacres in the land: morning watch = late watch—she exceeds her appointed time in the sky

10 If Venus in the month of the Harvest Furrow the Stars stand at her left: there will be famine, variant: confusion in the land—Mars stands at her left

11 If Venus sets at the neomenia of month Tamhiri and rises in month XII: there will be famine in the land—she exceeds her appointed time in the sky

12 If Venus as soon as she rises goes progressively higher: rains in the sky, floods in the springs will cease—like Mars, she goes very high

13 If Venus at her appearance is high up [...] variant: great stars [...]: the angry gods will return to the land—great stars = Jupiter and [...]

14 If Venus rises and changes her position: his servants will rebel against the king and another will stand in his position

15 If Venus rises? and her position is complete: the gods will have mercy toward the land [...] she completes her appointed days and stands there

16 [If Venus's] position is red: downfall of horses, [...] will be in the land—Mars stands with her

17 [If Venus's] position is green: pregnant women will die with their fetuses—Saturn stands with her.

ABBREVIATIONS

ABL	R. F. Harper, *Assyrian and Babylonian Letters*. 14 vols., Chicago: University of Chicago Press, 1892–1914
ACh	C. Virolleaud, *L'Astrologie chaldéenne*. Paris: Geuthner, 1905–12
AfO	*Archiv für Orientforschung*. Berlin, 1926–
AOAT	*Alter Orient und Altes Testament*
ARMT	*Archives Royales de Mari* (texts in transliteration and translation). Paris: Imprimerie Nationale, 1950–
Borger Esarh.	R. Borger, *Die Inschriften Asarhaddons, Königs von Assyrien* (AfO Beiheft 9). Osnabrück: Biblio, 1967
BMS	L. W. King, *Babylonian Magic and Sorcery*, London: Luzac & Co., 1896
BPO	E. Reiner and D. Pingree, *Babylonian Planetary Omens* (Bibliotheca Mesopotamica, vol. 2). Malibu: Undena, 1975–
CAD	*The Assyrian Dictionary of the Oriental Institute of the University of Chicago*. Chicago and Glückstadt: J. J. Augustin, 1956–
CT	*Cuneiform Texts from Babylonian Tablets in the British Museum*. London, 1896–
JNES	*Journal of Near Eastern Studies*. Chicago, 1942–
KAR	E. Ebeling, *Keilschrifttexte aus Assur religiösen Inhalts (WVDOG XXVIII/1–4, XXXIV/1–5)*, Leipzig: Hinrichs, 1915–23

LAS	S. Parpola, *Letters from Assyrian Scholars to the Kings Esarhaddon and Assurbanipal* (AOAT, 5/1 and 5/2) Kevelaer: Butzon & Bercker, 1970 and 1983
LKA	E. Ebeling, *Literarische Keilschrifttexte aus Assur*. Berlin: Akademie-Verlag, 1953
Or. NS	*Orientalia*, Nova Series, Rome, 1932–
RA	*Revue d'assyriologie et d'archéologie orientale*. Paris, 1884–
SAA	*State Archives of Assyria*. Helsinki: Helsinki University Press, 1987–
SpTU	*Spätbabylonische Texte aus Uruk* (I–III: *Ausgrabungen der Deutschen Forschungsgemeinschaft in Uruk-Warka*, vols. 9, 10, 12, Berlin: Gebr. Mann, 1976–; IV: *Ausgrabungen in Uruk-Warka, Endberichte*, vol. 12, Mainz am Rhein: von Zabern, 1993).
TCL	Musée du Louvre, Département des Antiquités Orientales. *Textes cunéiformes*. Paris: Geuthner, 1910–
ZA	*Zeitschrift für Assyriologie und verwandte Gebiete*, Leipzig, 1886–

NOTES

1. For a detailed analysis of the inscription of Nabonidus and the interpretation of the eclipse see chapter 1 of my *Your Thwarts in Pieces, Your Mooring Rope Cut. Poetry from Babylonia and Assyria*. Michigan Studies in the Humanities, vol. 5 (Ann Arbor, 1985).

2. The term *tablet* is used in Assyriology to designate a chapter, book, or other unit of a series of clay tablets, so named because it is written on one clay tablet, and normally bears at its end, as colophon, a subscript naming it the nth tablet of the series.

3. The collection is in the process of being edited by E. Reiner and D. Pingree, F. Rochberg, and W. H. van Soldt; so far published are B(abylonian) P(lanetary) O(mens) fascicles 1, 2, and 3 (by Reiner and Pingree) covering E(nūma) A(nu) E(nlil) Tablets 50–51 and 59–63. F. Rochberg-Halton's *Aspects of Babylonian Celestial Divination: The Lunar Eclipse Tablets of Enūma Anu Enlil* (Archiv für Orientforschung, Beiheft 22), Horn, Austria: Ferdinand Berger, 1989, covers Tablets 15–22. An edition of Tablet 14 was published by F. Al-Rawi and A. R. George (AfO 39 [1991–92], 52–73). EAE Tablets 23–28, that is, solar omens with the exception of solar eclipse omens, have been recently published by van Soldt *(Solar Omens of Enuma Anu Enlil: Tablets 23 [24]–29 [30]*, Nederlands Historisch-Archaeologisch Instituut te Istanbul, 1995).

4. This correspondence was recently reedited and commented by S. Parpola in his *Letters from Assyrian Scholars*, vols. 1 and 2 (= AOAT 5; 1970 and 1983), and revised in his *Letters from Assyrian and Babylonian Scholars*, SAA 10 (Helsinki University Press, 1993). See also A. L. Oppenheim, "Divination and Celestial Observation in the last Assyrian Empire," *Centaurus* 14 (1969): 97–135.

5. These are edited by H. Hunger in *Astrological Reports to Assyrian Kings*, SAA 8 (Helsinki University Press, 1992).

6. ARMT 23, pp. 90 and 102, also ARMT 13, p. 133, cited Joannès, ARMT 23, p. 100 sub a.

7. According to another system of numbering in the ancient sources, this last tablet bears the number 68.

8. E. von Weiher, SpTU, vol. 3, no. 60.

9. The obverse of SpTU, vol. 3, no. 60 enumerates the major gods with their epithets; the reverse is an excerpt from the Sumerian and Akkadian bilingual mythological epic *Lugale*, and thus the text may be an exercise tablet and cannot be used with confidence to establish what composition followed EAE.

10. P. Huber, "Dating by Lunar Eclipse Omens with Speculations on the Birth of Omen Astrology," in Acta Historica Scientiarum Naturalium et Medicinalium, vol. 39: *From Ancient Omens to Statistical Mechanics. Essays on the Exact Sciences Presented to Asger Aaboe*, ed. J. L. Berggren and B. R. Goldstein (Copenhagen: University Library, 1987), pp. 3–13. For attempts at using omens that connect lunar eclipses with historical events (the so-called "historical omens") for the purposes of dating these events, see J. Schaumberger, "Die Mondfinsternisse der Dritten Dynastie von Ur," ZA 49 (1949):50–58, and "Astronomische Untersuchung der 'historischen' Mondfinsternisse in Enūma Anu Enlil," AfO 17 (1954–56):89–92.

11. See F. Rochberg, "Benefic and Malefic Planets in Babylonian Astrology," in *A Scientific Humanist. Studies in Memory of Abraham Sachs*, ed. E. Leichty, M. deJ. Ellis, and P. Gerardi (Philadelphia: The University Museum, 1988), pp. 323–28.

12. The liver omen, RA 65 (1971) 73:62, is cited CAD, vol. 11/1, p. 266, as also noted by M. Stol, *Bibliotheca Orientalis* 47 (1990):375; in the celestial omen on solar eclipses, ACh Šamaš 8:62, the name of Mars is replaced by MUL.UDU.BAD, "planet." The apodosis of both omen texts uses the verb *napāhu* for "to rise," which occurs (written as KUR) beside IGI in celestial omens but does not necessarily refer to heliacal rising. The verb and its derivatives are used to refer to the Sun; for example, sunrise is expressed as *nipih šamši*, referring, however, to the appearance of the sundisk at its rising—its color, size, and so forth. In the Old Babylonian lunar eclipse omens adduced by F. Rochberg-Halton, the verb used is *tebû*, which in English is given the same translation, "to rise," but which refers both to the physical aspect, "to get up," and to rising as insurrection. What is meant by this verb is unclear.

13. Joan G. Westenholz, *Mesopotamian Civilizations, vol. 7: Legends of the Kings of Akkade* (Winona Lake, Ind.: Eisenbrauns, 1997).

14. *Id'im šamšum kakkabū ú-ṣú-ú ana nakrim*, RA 45 (1951):174:63–64; for a different interpretation see J.-J. Glassner, RA 79 (1985):123.

15. For omens referring to Sargon of Akkad, see H. Hirsch, AfO 20 (1963):7ff.

16. A. L. Oppenheim, "The City of Assur in 714 B.C.," JNES 19 (1960):133–47.

17. Ibid. 137f.

18. H. G. Güterbock, "Bilingual Moon Omens from Boğazköy," in *A Scientific Humanist. Studies in Memory of Abraham Sachs*, ed. Erle Leichty, M. deJ. Ellis, and P. Gerardi (Philadelphia: The University Museum, 1988), pp. 161–73.

19. R. Labat, *Un Calendrier babylonien des travaux, des saisons, et des mois* (Paris: Champion, 1965).

20. The number 77 is the paragraph number given by Labat to the main recension in his edition of *Iqqur-īpuš*, the recension that is arranged according to the activities or celestial events considered ominous; the other recension, in Labat's term "série mensuelle," is recorded on separate tablets—or on one tablet but in separate sections—for each month, and lists the pertinent activities and phenomena along with the prognosis for him who would perform them.

21. Jupiter and Mercury are attested—see Labat (note 19, above), 170f., note 6—and so is Mars in the list for month IV in BM 26185, a text communicated to me by Douglas Kennedy.

22. Akkadian: *dama salih*.

23. *ina* MURUB₄ AN.NE (or AN.BAR$_X$) *izziz*, BM 26185:76.

24. ACh Šamaš 14:73ff.

25. Labat (note 19, above), p. 222:28.

26. D. Arnaud, *Recherches au pays d'Aštata: Textes sumériens et accadiens* (Paris: Editions Recherche sur les Civilisations, 1985), nos. 610, 611, 615, and several fragmentary texts.

27. For the spatial relations expressed: in front of, behind, right, or left, see Pingree, BPO 2, p. 21.

28. See J. Schaumberger, in F. X. Kugler, *Sternkunde und Sterndienst in Babel. 3. Ergänzungsheft zum ersten und zweiten Buch* (Münster, 1935) , pp. 340ff.; see also H. Hunger and E. Reiner, "A Scheme for Intercalary Months from Babylonia," *Wiener Zeitschrift für die Kunde des Morgenlandes* 67 (1975):21–28.

29. "The star of Marduk [i.e., Jupiter], who went on to take up his position among the stars which made me resort to arms." See Oppenheim, *Centaurus* 14 (1970):121 and note 46.

30. Borger Esarh., p. 2 i 39–ii 5.

31. Borger Esarh., p. 17, Bab. Ep. 13; see Schaumberger (note 28 above), p. 311f.

32. S. M. Maul, *Zukunftsbewältigung* (*Baghdader Forschungen*, 18). Mainz on Rhine: Philipp von Zabern, 1994.

33. For such catalogs see Caplice, Or. NS 34 (1965):108ff. and 42 (1973):514f. (the latter was subsequently published in Hunger, SpTU, vol. 1 as no. 6), and Ebeling, RA 48 (1954):10ff. See also Maul (note 32 above), 469f.

34. HUL Ṣalbatānu ša adanšu ušēti[quni] u ina libbi MUL.LÚ.HUN.GÁ i[nnamiruni] K.818 (CT 53 8, = SAA 10, no. 381), and see Parpola, LAS 2, pp. 350f.

35. ACh Supp. 2 62 = SAA 10, no. 362.

36. ABL 647 = SAA 10, no. 67.

37. K.2184, K.2286, BM 36627.

38. ABL 34 = SAA 10, no. 10), and ABL 357 and 1118+ = SAA 10, nos. 202 and 203; see also Parpola, LAS 2, pp. 123ff.; the ritual itself was described in the text KAR 7, but only the prayer to Šamas mentioning HUL ri-i-b[i . . .], "the evil of the earthquake," is preserved in it. A ritual performed by the temple singer kalû to avert such evil is attested in Thureau-Dangin, Rituels accadiens (Paris, Leroux, 1921), 34ff., lines 16 through reverse 1.

39. [¶ᵈ] ⸢Ṣal-bat⸣-a-nu uᵈSin it-te-e[n-tu-ú . . .] Or. NS 40 (1971):169 r. 12 (preceding lines fragmentary).

40. LKA 108, edited by E. Ebeling, RA 50 (1956):26.

41. ₁₀′ . . . lu-u AN.MI Sin ₁₁′[. . .] lu-u AN.MI ᵈŠul-pa-è-a ₁₂′[. . .] lu-u AN.MI ši-i-qí ₁₃′[. . .] KA ᵈIM ša TA AN-e ur-da ₁₄′[. . . lu]-u mi-ših MUL lu-u ṣa-ra-ár MUL ₁₅′[. . .] ša ana MUL.MEŠ KASKAL.MEŠ is-sa-ni-qu, in K.8091 + 10628, known to me from Geers's copies, parallel BMS 62, see Ebeling, RA 48 (1954):8. A similar enumeration is found in the text published by W. G. Lambert, "A Part of the Ritual for the Substitute King," AfO 18 (1957–58) 109ff., col. A, lines 11ff.: AN.MI ᵈSin AN.MI ᵈŠamaš AN.MI ᵈŠul-pa-è-a [. . . AN].MI ᵈDil-bat AN.MI ᵈUDU.BAD.MEŠ, "eclipses of the Moon, the Sun, Jupiter, Venus, (or) of the (other) planets"; see Parpola, LAS 2, p. xxii. No eclipse of Jupiter or Mars is mentioned in the celestial omens, but eclipses of Venus are. Apotropaic rituals to avert the evil portended by a lunar eclipse are published by Ebeling, RA 48 (1954):82 as no. 2 and by Caplice, Or. NS 40 (1971):166f., no. 65; the cuneiform text was subsequently published in autograph copy as CT 51, 190.

2

Babylonian Horoscopy: The Texts and Their Relations
F. Rochberg

By the end of the second millennium B.C., the reading and interpretation of celestial signs in the form of omens had become a major feature of the learned culture of Mesopotamia, and these practices soon extended beyond the Babylonian scribal centers to those of the bordering states of Hatti and Elam. The history of Babylonian astrology began with the earliest attestation of the reading of celestial omens in Mari letters[1] and late Old Babylonian omen texts (ca. 1800 B.C.); continued through the Middle Babylonian and Middle Assyrian periods (ca. 1200 B.C.) with forerunners to the canonical celestial omen series *Enūma Anu Enlil*[2]; reached something of a peak in the seventh century as evidenced by the activities of the Sargonid court astrologers[3]; and ended with a wide variety of celestial as well as nativity omens and horoscopes in the Achaemenid, Seleucid, and Arsacid periods (ca. 500 to 50 B.C.).[4]

Historically the most recent form of astrology to develop in Babylonia, horoscopy was the form that would be decisive for the further development of Western genethlialogy through Greek, Islamic, Jewish, and Christian channels. The appearance of horoscopes after 500 B.C. is evidence that the situation of the heavens at the time of a birth had come to be regarded as significant for the future of an individual. Before this time, little evidence supports the idea that the individual had a place in the scope of traditional celestial divination, although there had been divination that derived predictions for individuals based on date of birth and on physiognomy.[5]

The relationship between personal piety and personal happiness within the divine scheme of the universe is a subject of concern in the Babylonian "wisdom" literature, and although the relationship is viewed with a certain skepticism in some ancient sources, the idea that an individual's life, as all other things, is affected by the gods seems to be a basic assumption.[6] If, as is evidenced by the celestial omen texts such as in *Enūma Anu Enlil*, celestial phenomena had been taken to indicate the future for the king and the state of affairs in the country at large, it seems a

priori possible that such a belief could be carried over and applied to the
life of an individual. Unfortunately, the laconic nature of the horoscope
texts themselves frustrates attempts to penetrate the philosophical or reli-
gious commitment behind these texts. In particular, what the appearance
of horoscopes may tell of a change in the relation conceived between the
individual and the cosmos, or between the individual and the gods, after
the mid–first millennium, is an aspect which remains strictly inferential and
speculative.

The appearance of horoscopes also coincided with a marked growth
of astronomy in the direction of abstract mathematical description and
refined computation of planetary and lunar appearances. Identifying the
cultural impetus for the development of the mathematical astronomy of
the fifth century B.C. and its relation to the forms of celestial inquiry that
existed before it, that is, celestial observation and divination, has been of
interest to Assyriologists for many years. At the 14th Rencontre Assyriologique Internationale in Strassbourg (July 1965),[7] Oppenheim raised
the issue of the role of celestial divination in the history of Babylonian
astronomy. He said, "Any serious investigation of the history of Mesopotamian civilization has to face the problem of the sudden emergence of
mathematical astronomy about 400 B.C. To put it somewhat bluntly, the
question is whether there exists a direct relationship between this development and the evolution within Mesopotamian divination, to be exact,
within astrology, or whether the genesis of Mesopotamian science, that is,
of mathematical astronomy, was released by other still unknown factors."[8]

The phenomena upon which the divination series *Enūma Anu Enlil*
is based bear relation to those of the mathematical astronomy, in that they
may be defined predominantly by the horizon phenomena of the moon
and planets. As reasons for observing the heavens, celestial divination—
the importance of which was not only intellectual but political, considering the dangers the omens portended for the state—certainly established
considerable motivation for the development of a predictive astronomy.
But the content of the mathematical astronomy that emerged around 500
B.C. cannot be justified solely on the basis of the needs of the omens
attested in the canonical tradition. Even if the impetus for the development of the particular mathematical branch of Babylonian astronomy were
"astrological," the level of sophistication of the mathematical astronomy,
in terms of its predictive range and underlying conceptual grasp of phenomena, far exceeds anything reflected in the omen literature. A disparity
between the astronomy of the horoscopes and that of the contempora-
neous mathematical astronomy must also be acknowledged, although the

gap appears to have narrowed when compared with the relatively primitive astronomy of celestial omens.

Despite the difficulties in relating methods and parameters of mathematical astronomy to those of the nonmathematical classes of astronomical texts, mathematical astronomy can no longer be singled out when asking historical questions such as Oppenheim's "whence the origin of astronomy," but has to be recognized as necessarily part of a coherent piece, inclusive of all the aspects of Babylonian astronomical, or celestial, science. Any study of the cultural history of science in late Babylonia needs not only to take into account all the forms of astronomy but also the relation of the various branches of astronomy to astrology, whether omen or horoscopic.

A descriptive analysis of the astronomical content of the small corpus of Babylonian horoscopes serves to show how interconnected all the parts of late Babylonian astronomical science were. Following some general introductory remarks I will confine my discussion to elements of the Babylonian horoscopes' astronomical content and to the derivations of these elements from diverse astronomical sources. In discussing the connections between horoscopes and other classes of astronomical texts I will utilize the now standard classification and nomenclature established by A. Sachs, of which the major categories are nontabular or nonmathematical texts, including diaries, goal-year texts and almanacs,[9] and tabular or mathematical ephemerides, also termed simply ACT after the publication of Neugebauer.[10] In the following brief description of the astronomical data recorded in the horoscopes I hope to show that the horoscopes draw upon a variety of astronomical sources that include most of the classes of astronomical texts, both nonmathematical and mathematical.[11] The implications are that the astronomical methods underlying Babylonian horoscopy do not stem from only one tradition, but include observation as well as computation, and nonmathematical as well as mathematical methods.

The discovery of the first cuneiform horoscope came as part of the general decipherment of astronomical cuneiform texts in the late nineteenth century by Fathers J. Epping, J. N. Strassmaier, and F. X. Kugler. Only twenty-eight Babylonian horoscope tablets are now extant, but they make up a well-defined class of texts belonging to the Achaemenid, Seleucid, and Arsacid periods, or roughly between the fifth and first centuries B.C. The chronological range is from the oldest at 410 B.C.[12] to the youngest at 69 B.C.[13] With five documents from the first century B.C., these are among the youngest cuneiform texts known. The youngest horoscopes, dating between 89 and 69 B.C., and all from the city of

Babylon, come from the period in which Babylon's major temple, the Esagila, or Marduk temple, begins to appear moribund.[14] A connection between astronomers, designated in the texts as "astrologers," that is, scribes of *Enūma Anu Enlil,* and the Esagila temple is supported in administrative temple texts,[15] and we may be sure that at least some of the astronomers were temple scribes.[16]

The Babylonian horoscopes were all dated to the birth of an individual. Since three texts contain more than one horoscope, it cannot be the case that a horoscope was written on the date of the birth. In no case has the writing of a horoscope tablet been dated by means of a colophon. The dates are found at the beginning of the text and refer exclusively to the birth date. Given the existence of birth notes, recording dates and times of births apparently for the purpose of later casting a horoscope, it is clear that horoscopes could have been prepared well after such dates. In the single birth note preserved with more than one birth record, two of the dates are spaced 36 years. The evidence that data were excerpted from other astronomical texts further precludes the possibility that a horoscope represents some observation, or even computation, of heavenly phenomena at the time of birth.

The purpose of the Babylonian horoscope document was to record positions of the seven planets in the zodiac on the date of a birth. Following a loosely standardized formulation of the date and time of birth, the astronomical data were given. Text A (below) is a good representative text, although it contains a reference to the position of the moon with respect to a normal star (obv.3), a datum found in several horoscopes but by no means regularly included. The horoscope texts are only in the weakest possible sense "standardized"—certain data come to be expected but, aside from one text with a known duplicate,[17] each horoscope is unique and presents different problems of dating and interpretation. Two texts (A and B) are provided here.[18]

Text A (BM 36620 = 80-6-17,350 [L*1464])
Date: 92 S.E. VII. 12(?) = −219 Oct. 21

Transcription

obv.

upper edge

  ina a-mat ᵈEN u GA[ŠAN-ía liš-lim]

1 MU.1,32.ᵣKᵀ[AM ᶦAn LUGAL]

2 ITI.DU$_6$ 30 GE$_6$ 1[2(?) ina SAG GE$_6$ sin]

3 SIG MÚL ár šá SAG HUN

4 sin $\frac{1}{2}$ KÙŠ ana NIM DIB U$_4$(?). [...]

5 LÚ a-lid ina si-ma-ni-[šú sin ina HUN(?)]

6 šamáš ina GÍR.TAB MÚL.BABBAR [ina HUN]

7 dele-bat u GENNA i[na(?) PA(?)]

8 GU$_4$.UD u AN [šá ŠÚ-ú NU IGI.MEŠ]

9 KI šamáš šú-nu [ITI.BI(?)]

rev.

1 14 NA 2[7 KUR]

2 ITI.GAN 20 [šamáš GUB]

3 ITI.ŠE GE$_6$ 1[4 AN.KU$_{10}$ sin]

4 ina RÍN TIL-tim GAR-a[n]

5 U$_4$.28 AN.[KU$_{10}$ šamáš]

6 ina HUN BAR DIB [ ]

ca. 2 lines to bottom of rev., uninscribed.

Translation

upper edge

By the command of Bēl and B[ēltīja may it go well].

obv.

1 Year 92 [(S.E.), Antiochus (III) was king.]

2 *Tašrītu* 30, night of the 1[2th(?), first part of night, the moon was]

3 below "the rear star of the head of the Hired Man (= α Arietis).

4 The moon passed $\frac{1}{2}$ cubit to the east (of α Arietis) ... [...]

5 the child was born, in [his] hour, [the moon was in Aries(?),]

6 the sun was in Scorpius, Jupiter [was in Aries],

7 Venus and Saturn (were) i[n Sagittarius],

8 Mercury and Mars [which had set were not visible.]

9 They were with the sun. [That month(?),]

rev.

1 moonset after sunrise was on the 14th [last lunar visibility before sunrise on the] 2[7th.]

2 [Winter solstice (was)] on the 20th of *Kislīmu*.

3 *Addaru*, night of the 1[4th a lunar eclipse,]

4 Totality occurr[ed] in Libra.

5 On the 28th day an ecl[ipse of the sun]

6 in Aries, one-half month having passed (since the previous eclipse).

TEXT B (BM 78089) Date: SE 186 V.24 = −125 Aug. 16

Transcription

obv.

1 ⌜MU.1.ME.22.KAM šá ši-i⌝

2 ⌜MU.1.ME.1,26.KAM⌝ ᴵAr-šá-ka-a LUGAL

3 ITI.NE 30 15 NA

4 GE₆ 24 ina ZALÁG LÚ.TUR a-lid

5 ina si-man-ni-šú sin ina MAŠ.MAŠ

6 šamáš ina A MÚL.BABBAR u GENNA

7 ⌜ina ZIB⌝.ME dele-bat ina A

8 GU₄.UD u AN šá ŠÚ-ú

9 NU IGI.MEŠ erasure

lower edge uninscribed

rev.

1 ITI.BI ⌜20+x⌝ KUR

2 MU.BI ŠU.⌜3⌝ šamáš GUB

3 ITI.KIN 14 AN.KU₁₀ sin ina ZIB.ME

4 BAR DIB 28 AN.KU₁₀ šamáš

5 ina TIL ABSIN 5 SI GAR-an

ca. 3 blank lines to bottom of rev.

Translation

1 Year 122 (A.E.), which is

2 Year 186 (S.E.) Arsaces was king.

3 *Abu*, 30. Moonset after sunrise on the 15th.

4 Night of the 24th in the last part of the night, the child was born.

5 At that time, the moon was in Gemini,

6 sun in Leo, Jupiter, and Saturn

7 in Pisces, Venus in Leo,

8 Mercury and Mars which had set

9 were not visible.

rev.

1 That month, last lunar visibility before sunrise was on the 20+[...]th.

2 That year, (summer) solstice was on *Du'ūzu* the 3d.

3 *Ulūlu* the 14th a lunar eclipse in Pisces.

4 One-half (month) passed by. (Then,) on the 28th, a solar eclipse

5 at the end of Virgo; it made 5 fingers.

As is clear from the examples provided, the longitudes of the planets (Jupiter, Venus, Mercury, Saturn, Mars, as well as the sun and moon) are the principal data collected in horoscopes. Since the date of birth is of primary concern, the planets are for the most part between synodic appearances. When, however, a planet happens to be in the same sign as the sun on the date of birth and is in or near a synodic phase, sometimes the date of the synodic phenomenon will be mentioned in the text. On the whole, the longitudes are given with respect to the names of the zodiacal signs. Degrees of longitude are not common, but do occur in eight horoscopes, five of which are from Uruk.

A comparison of the Babylonian planetary longitudes against those computed by modern methods on the various dates of these eight horoscopes gives striking evidence for the excellence of the methods which underly the Babylonian data. Tables 1 and 2 summarize these data.

Table 1 lists by text number the date and time of birth given in the text, the longitudes (λ) found in the text, then the modern computed longitudes with the time corresponding approximately to that of the time of birth stated in UT (equivalent to GCT). Column (7) shows the differences between the Babylonian and modern longitudes and in most cases, the difference is between 0° and ±3°. The $\Delta\lambda$s reflect rounded values of the modern computed longitudes. The modern longitudes tabulated in column (5) reflect an adjustment[20] by means of which the systematic deviation in longitudes that results from the different methods of counting longitude—modern tropical versus Babylonian sidereal—can be corrected. The adjustments to the modern longitudes, taking into account the effect of precession on the sidereally normed Babylonian zodiac, enable a more direct comparison between the data found in the texts and that produced by modern computation.

Table 2 presents the values from the last column of table 1 in columns according to the planet. The slashes in the table indicate an absence of data for the planet. Errors of plus or minus 1° or 2° may be considered irrelevant in this context, particularly as we do not know precisely what ancient methods were used to obtain them.

Table 1

Text	Date	Time	λ Text	λ Comp.	Time (UT)	Δλ*
5	–262 Apr. 4	(last part of night)[19]	☉ ♈ 13.5°	16.3	4	–2.8°
9	–248 Dec. 29	evening	☉ ♑ 9.5°	281.8°	16	–2.3°
			☾ ♒ 12°	315.4°		–3.4°
10	–234 Jun. 2/3	dawn	☉ ♊ 12.5°	73.5°	1	–1°
			♃ ♐ 18°	260.1°		–2.1°
			♀ ♉ 4°	27.9°		+6.2°
			♄ ♋ 6°	90.5°		+5.5°
			♂ ♋ 24°	115.6°		–1.6°
16b	–199 Jun. 5	dawn	☾ ♋ 15°	118.6°	1.75	–13.6°
			♃ ♏ 26°	237.9°		–1.9°
			♀ ♊ 5°	62.4°		+2.6°
			☿ ♊ 27°	84.4°		+2.6°
			♄ ♍ 10°	157.2°		+2.8°
			♂ ♉ 10°	38.3°		+1.7°
16a	–198 Oct. 31	dawn	♃ ♑ 10°	275.3°	3	+4.7°
			♀ ♑ 4°	267.4°		+6.6°
			☿ ♏ 8°	227.7°		–9.7°
			♄ ♎ 3°	183.1°		0°
			♂ ♐ 10°	248°		+2°
21	–124 Oct. 1	dawn	☾ ♋ 24°	113.5°	2	+0.5°
	–124 Oct. 2		☾ ♌ 9°	127.7°		+1.3°
23	–87 Jan. 5	midnight	☾ ♉ 5°	32.9°	21	+2.1°
			♃ ♈ 27°	25.9°		+1.1°
			♀ ♓ 1°	330.8°		0°
			☿ ♐ 26°	266.4°		0°
			♄ ♊ [20°]	79.4°		+0.6°
			♂ ♌ 20°	141.1°		–1.1°
27	–68 Apr. 16	9th hr.	☾ ♑ 18°	297.8°	11.5	–9.8°
			☉ ♈ 30°	27.5°		+2.5°
			♃ ♐ 24°	261.9°		+2.1°
			♀ ♊ 13°	72.7°		0°
			♄ ♒ 15°	314.2°		+0.8°
			♂ ♎ 14°	189.8°		+4.2°

* λ Babylonian – λ modern

Table 2

Date	Text	Moon	Sun	Jupiter	Venus	Mercury	Saturn	Mars
−262	5	/	−2.8	/	/	/	/	/
−248	9	−3.4	+2.3	/	/	/	/	/
−234	10	/	−1	−2	+6.2	/	+5.5	−1.6
−199	16b	−13.6	/	−1.9	+2.6	+2.6	+2.8	+1.7
−198	16a	/	/	+4.7	+6.6	−9.7	0	+2
−124	21	+0.5	/	/	/	/	/	/
−124	21	+1.3	/	/	/	/	/	/
−87	23	+2.1	/	+1.1	0	0	+0.6	−1
−68	27	−9.8	+2.5	+2.1	0	/	+0.8	+4.2

Within our material, no other source but the mathematical ephemerides provides longitudes in degrees within a zodiacal sign. But the relationship between the horoscopes' longitudes and those available in ACT tables is complicated by the fact that the ephemerides usually generate longitudes of consecutive synodic phenomena not positions on arbitrary dates (for example, ACT 600 is for first stationary points, 601 for second stationary points, 604 for oppositions, and 606 for last visibilities, and so on, as can easily be seen by looking at the catalogue in ACT). Rules for subdividing the synodic arc, passing, for example, from first visibility to first station or first station to last visibility, are given in a number of procedure texts of System A (such as ACT 801 for Mercury and Saturn; 812 for Jupiter, 811a Section 10 for Mars) and can be uncovered in some table texts as well (for example, ACT 611 for Jupiter [System A′]). Some procedure texts, such as ACT 810 and the similar 813 for Jupiter (in System A′), state the daily progress of the planet in degrees per day. Finally, there are a very few ephemerides, in the true sense of giving daily positions of the planets, such as ACT 310 for Mercury. These employ refined nonlinear interpolation schemes to obtain positions between the synodic phases.

It is clear that producing daily longitudes was not the primary focus of the Babylonian mathematical astronomy, but analyses by Neugebauer, Aaboe, and Huber, of the true ephemerides within the ACT corpus giving positions from day to day within actual lunar months,[21] show that although of a secondary nature in the context of the total production of tables, they were mathematically quite refined. But finally, any meaningful connection between the ephemerides and the horoscope's planetary longitudes falls between gaps in evidence.

For one thing, no direct comparison is possible since there are simply too few horoscopes that cite longitudes with degrees and none that correspond to the years for which our extant true ephemerides apply. More serious, though, are the questions attending the possible practical aspects of these ephemerides, given the nature of the variants permissible (e.g., in the size of the retrograde arc of Jupiter, which Aaboe has discussed[22]). Therefore, even were there an example of a value in a horoscope directly comparable with an ephemeris in which the corresponding date and phenomenon were preserved and could be checked, a discrepancy would not necessarily rule out the possibility of the astrologers' use of mathematical astronomical methods.

Regardless, I do not find the excellence of the horoscopes' longitudes to be a compelling argument in and of itself that daily motion schemes of the type represented by ACT 654–655 were employed by the astrologers—they may indeed have been, but this would have to be argued on some other basis. I have as much hesitation to see those schemes as having been created to serve astrological purposes. We may well be seeing in the horoscopes rounded values obtained from ACT schemes. We know the astrologers were not a different group from the astronomers, so we are not troubled by a question of privileged knowledge. But again, we lack the evidence needed to conclude in any positive way that ACT tables or methods were used by the scribes who prepared horoscopes. More important, the evidence we do have points toward the negative conclusion that the methods and results of the daily motion schemes, which according to Aaboe are among the most sophisticated application of mathematics to the resolution of astronomical problems in Babylonian astronomy,[23] are not reflected in the data of the horoscopes.

With the exception of one of the two fifth century horoscopes, synodic appearances are mentioned in horoscopes only on the occasion of a planet's occupying the same zodiacal sign as the sun on the birth date. Only rarely is the date of the synodic appearance given, when it comes within a day or two of the birth date. When the birth occurs during the planet's invisibility, the text expresses this with the remark "planet(s) such-and-such is (are) with the sun," as in Text A line 9 above, or with the phrase found in the diary texts, "planet such-and-such, which had set, was not visible," as in Text B lines 8–9 above.

As far as synodic phenomena are concerned, the horoscopes are interested only in first and last visibilities. A single exception to this may be seen in the fifth century horoscope just mentioned, which records the dates of the stationary points as well as opposition, actually the rising at sunset, for the planet Saturn. Otherwise, attestations of synodic phenomena are limited to first and last visibility as evening star and first and last visibility as morning star for Mercury, and only last visibility for Venus, Mars, and Saturn.

In addition to the positions of the planets on the date of birth, other astronomical events of the month or even the year in which the birth occurred are regularly included. Horoscopes from Babylon record three additional lunar phenomena termed collectively the "lunar three" by Sachs.[24] These are: Whether the previous month was full (30^d) or hollow (29^d), the date of the time interval around full moon termed *na*, which measured the interval between sunrise and moonset, and the date of another time interval termed KUR, which was the interval between moonrise and sunrise on the day of last lunar visibility.[25] The length of the month, *na,* and KUR are found in each monthly paragraph of an astronomical almanac, as well as being obtainable in diaries and other types of nonmathematical texts of the Seleucid period. The lunar three data appear to be essential for all the horoscopes from Babylon. The tradition from Uruk appears to be different, as none of the Uruk horoscopes includes the lunar three.

Statements about lunar latitude are included in three horoscopes, all from Uruk.[26] These statements use the terminology for latitude that is known otherwise only in the ACT vocabulary, namely, the technical terms NIM "positive latitude," SIG "negative latitude," and MURUB₄ "node." The lunar procedure text for System A, ACT 200,[27] contains a section for lunar latitude, referring to column E of the lunar ephemeris, the first line of which reads: *epēšu ša* nim *u* sig *ša sin* ab *ana* ab 12 dagal *malak* ^{d}*sin* 2,24 *qabalti qaqqar kiṣari* "procedure for latitude of the Moon month by month. 12 (degrees is) the width of the road of the moon. 2,24 (from) the middle is the 'nodal zone.' " Here, as Aaboe has pointed out,[28] the technical term for latitude is actually the phrase "nim *u* sig," which literally means "positive and negative (latitude)." The Uruk horoscopes recall the language of the astronomical procedure text with the statements (1) "The moon keeps going with (increasing) positive latitude." (Text 10:4 *sin* TA MURUB₄ *a-na* NIM *pa-ni-šú* GAR.MEŠ), (2) "The moon keeps going from negative latitude toward the node." (Text 16a:9, *sin* TA SIG KI(?) *pa-nu-šú ana* MURUB₄ GAR.MEŠ), and (3) "The moon keeps going from positive latitude toward the node." (Text 16b r.10 *sin* TA LAL *ana* MURUB₄ *pa-nu-šú* GAR.MEŠ). Why the lunar latitudes are found only in Uruk texts and what their meaning is in this context remains quite puzzling.

Table 3

Text	Birthdate (Julian)	Lunar eclipse date (Julian)
3:5′	−297 Feb 2–5 (?)	undatable due to broken context
4 r 3–4	−287 Sep 1	−287 Nov 22
13 r.5	−223 Jul 29	undatable due to broken context
14 r.3–4	−219 Oct 21	−218 Mar 20
20 r.3–4	−125 Aug 16	passed by
21 r.1–3	−124 Oct 1	−124 Aug 24
22a r.8′	−116 Jul 15	−116 Sep 24
22b r.14′	−114 Jun 30	−113 Jan 29
23:10–12	−87 Jan 5	−87 Mar 11
24:9–10	−82 Dec 20	
25:6–7	−80 Apr 22/23	−80 Apr 21
26 r.3–5	−75 Sep 4	−75 Jul 24
27 r.5–8	−68 Apr 16	−68 Sep 3

Horoscopes regularly record lunar and solar eclipses, even when their occurrences did not coincide with the birth dates.[29] As illustrated by the horoscopes quoted above, the majority of horoscopes in which eclipses are preserved mention both lunar and solar eclipses, in particular, those which occur one-half month apart, the lunar in mid-month followed by the solar at month's end (Texts A rev.3–6 and B rev.3–5). In the vocabulary of the Babylonian astronomical texts, a distinction between predicted and observed lunar eclipses is conveyed in the writing of "lunar eclipse," as AN.KU$_{10}$ *sin* when predicted and *sin* AN.KU$_{10}$ when observed.[30] The eclipses recorded in the horoscopes are exclusively expressed with AN.KU$_{10}$ *sin,* and indeed these passages all represent predictions. Furthermore, that the zodiacal sign in which the eclipse occurred, or, specifically, in which totality occurred, is cited, points definitively toward a prediction. Table 3 lists the lunar eclipses predicted in the horoscopes. I have listed the dates of birth and the eclipses given with their Julian dates. In each case (but one), the eclipse's occurrence is confirmed by the modern computation, either by Oppolzer's *Canon,*[31] or P. Huber's computer program LUNEC.

When one considers the content of the *Enūma Anu Enlil* lunar eclipse omens, a great many aspects of eclipses appear to have become astrologically significant at a relatively early date, since the series as a whole was formed by the Neo-Assyrian period. "Astrologically significant" means that some event of social, political, or economic significance to the state (of Assyria or Babylonia) and its population was associated with some aspect of an eclipse such that the occurrence of that eclipse phenomenon would be regarded as portending some specific mundane event. In the protases of these eclipse omens are included the date of occurrence (month, day), the time (watch of night), the magnitude (in fingers), direction of eclipse shadow, and color of the eclipse.[32] One can only speculate that in a manner similar to the celestial omens, the eclipses cited in horoscopes were incorporated for what this data might have contributed to the interpretation of the heavens on or around the birth. The details of eclipses, however, are not so prominent in the horoscope texts.

Only three of the thirteen preserved eclipses include data for magnitude, given in fingers in two instances (Texts 21 rev. 1–3, and 26 rev. 3–5) or in the fraction of the disk covered in another example (Text 27). Only one horoscope (Text 27 rev. 5–8) states the time of the eclipse, noting that the moon was already eclipsed when it rose. In these features—that is, date, zodiacal sign in which the moon was positioned when eclipsed, magnitude and time, the manner in which the eclipses are presented in the horoscope texts—is not paralleled by those found in the observational genres such as diaries, goal-year texts, or the observational eclipse report compendia.[33] In the eclipse reports that reflect observations, the zodiacal sign is never mentioned; instead, we find the *ziqpu*-star that was culminating at the beginning of the eclipse and other data relevant to the time and duration of occurrence that never find their way into horoscopes, whose eclipse passages bear resemblance to those of the predictive texts such as the almanacs.

Most horoscopes will also include the date of the solstice or equinox closest to the birth date. In fact, no solstice or equinox date is more than 2 months before or after a given birth date, as can be seen at a glance in Table 4.

This makes the solstice/equinox data useful as a limiting factor for the dating of texts in which the birth date is not well preserved. Because of the chronology of the extant horoscopes, according to which all belong to the period after the introduction of the 19-year cycle, the method of obtaining the relevant equinox or solstice date was in each case that of the so-called Uruk Scheme,[34] which computed the cardinal points of the year on the basis of the rule that the year was 12 lunar months plus 11;3,10 tithis, or 12;22,6,20 months, the value of the year underlying the 19-year cycle.

Table 4

Text	Birthdate	Date of solstice/equinox	Months apart
1	Dar (II) X.24	WS X.9	0
4	S.E. 24 V.19	AE VI.16	1
6	S.E. 54 IX.8	WS IX.20	0
	(conception date: S.E. 53 [XII$_2$] with VE [XII$_2$].12)		
8	S.E. 61 IX.8	WS X.8	1
13	S.E. 88 V.4	SS III.30	2
14	S.E. 92 VII.12	WS IX.20	2
15	S.E. 109 XI.9	WS IX.28	2
18	S.E. 169 XII.6	VE I.4	1
19	S.E. 172 [VI].13	EA VII.2	1
20	S.E. 186 V.24	SS IV.3	1
21	S.E. 187 VI.22	AE VI.17	0
22	S.E. 195 IV.2	SS III.13	1
22	S.E. 197 IV.7	SS IV.5	0
23	S.E. 223 X.9	WS IX.28	1
25	S.E. 231 I.14/15	SS III.21	2
26	S.E. 236 V.25	SS III.16	2

There is no question that horoscopes cannot contain observations made at the time of the writing of the text. In drawing on various reference sources, however, it is possible that an observation, such as from a diary or other observational text, could later be incorporated into the body of the horoscope text. This is evident in a number of references to the position of the moon in the ecliptic, which is given not with respect to the zodiac but by the ecliptical normal stars, whose use is best known from the astronomical diaries.

The importance of the normal stars in the horoscopes is exclusively in its application for citing the position of the moon.[35] In these few horoscopes, the position of the moon seems to be given with respect to a normal star when it is above the horizon at (or near) the time of the birth. The normal star reference does not replace the zodiacal one, rather it supplements it. (See Text A:2–5, above.) I cannot account for the occasional inclusion in horoscopes of a normal star position for the moon in addition to the lunar longitude given with the enumeration of planetary zodiacal positions. Since the normal procedure was to obtain positions in the ecliptic regardless of which heavenly bodies were above the horizon and visible at the moment of birth, this suggests that visibility was not a

Table 5

Horoscope	Almanac
Date of birth, year S.E.	Year S.E.
Month 30/1	Month 30/1
Day, time	
Longitudes of planets in zodiac	Longitudes of planets in zodiac at beginning of months
	Dates of entries of planets into zodiacal signs
Dates of synodic phenomena	Dates of synodic phenomena
Date of moonset after sunrise (*na*)	Date of moonset after sunrise (*na*)
Date of last lunar visibility (KUR)	Date of last lunar visibility (KUR)
Date of equinox or solstice nearest birthdate	Date of equinoxes or solstices in that month
Dates of eclipses	Dates of eclipses in that month

consideration. On the basis of the phraseology and terminology used, and after comparing the lines in the horoscopes referring to the normal star positions of the moon with corresponding statements in diaries, it is a fair assumption that the horoscopes quote these normal star positions from diaries. Such quotations, however, cannot be directly substantiated as the desired diaries from corresponding dates to these particular horoscopes are no longer extant. The surviving material is more than sufficient, though, to support the connection.

With regard to the possibilities for deriving the planetary data found in horoscopes, much of the data provided by the diaries is not ideally suited to horoscopes. Horoscopes never record when a planet passes above or below a normal star. This is limited to the occasional lunar positions just mentioned. The date of a planetary phase as often given in diaries, together with the zodiacal sign in which the planet was located at the time, would be useful for a horoscope only if the date of the phenomenon should coincide with a birth date.[36] On the other hand, at the conclusion of the day-by-day entries of a diary's month section, the summary of the zodiacal signs in which the planets were found throughout the month would be of great use to an astrologer.

The class of texts that recorded the monthly progress of the planets through the zodiac is the almanacs. The few examples of horoscopes and almanacs where dates can be matched indicate that almanacs provided a very good source of astronomical data for Babylonian horoscopes.[37] As illustrated in Table 5, a simple inventory of the general content of the almanacs as compared with that of the horoscopes shows that horoscopes contain most or all of the available data in an almanac, derived from the appropriate month of the birth or any other month paragraph containing astronomical data of importance to horoscopes, such as lunar and solar eclipses or solstice and equinox dates.

The only data that appear occasionally in horoscopes but not in almanacs are the normal star positions of the moon. And the only datum that appears as something of an organizing principle in horoscopes but not in almanacs is the specific day or date in the month, the almanacs being arranged by months, not days, and giving dates in reference to phenomena of interest. The astronomical texts organized by days of the month were, of course, the diaries. Even with the general compatibility shown above, data from almanacs seem to have been used selectively by astrologers. One example may be cited in which the planetary and lunar three data in several lines of a horoscope (Text 26 obv.4–8) are duplicated in an almanac (LBAT 1174 : 10), but where the eclipse data are entered (Text 26 rev.3–5), the reports differ. The horoscope gives the month, day, zodiacal sign, and magnitude (one finger), whereas the almanac gives month, day, the exact time (stating 8° before sunrise and that the moon set eclipsed), and the magnitude is given as over four fingers. The zodiacal sign for the eclipse does not appear in the almanac.

The content of the Babylonian horoscopes requires that the astrologer have access, either directly or by inference, to the location of planets in the zodiac on an arbitrary date. As described above, these data could have been drawn from a variety of elements collected in a diary, viz., the occasional date and position of a planetary synodic appearance or the zodiacal signs in which the planets were located during the month tallied at the end in the planetary summary section. The requisite data could equally well have come from almanacs. In view of the various ways in which elements of nonmathematical astronomical texts appear in horoscopes, we may conclude that the astrologers used a variety of these texts as reference works and used the data from them selectively.

The fact that one can demonstrate dependence of horoscopes upon texts in which astronomical phenomena are predicted—that is, almanacs, diaries, and possibly ephemerides, as we have seen in the horoscopes that give degrees within signs—raises the question of the astrological motivation for such predictions as well as for the creation of these particular classes of astronomical texts. To what extent was the demand for astronomical data by astrologers a factor in the growth of astronomical methods or the scholastic decisions to create the various classes of astronomical texts as we have them? I think in one sense, the same evidence that raises the question also answers it, by showing the great degree to which astrology was an integral part of astronomical interests in the period after 500 B.C.

There remains, however, Oppenheim's question of whether indeed there was some catalytic effect on the growth of theoretical astronomy that stemmed from the practice of astrology. As I understand it, horoscopic astrology does not give rise to astronomy. When we take account of the textual evidence for the interdependence between the two over the long parallel histories of both disciplines, the question of which gave rise to which becomes unintelligible. The computational systems of the Babylonian mathematical astronomy, which emerged at about the same time as did horoscopic astrology, cannot be accounted for by reason of their serving astrological purposes. While it is true that the goals of the mathematical astronomy seem to converge with those of horoscopy to produce zodiacal longitudes, the schemes known to us from Babylonian ephemerides and procedure texts[38] are of a complexity and produce results not evidenced in any direct way in the horoscope texts.

Such a discrepancy between the schemes available in mathematical ephemerides on the one hand and the evidence of the astrological texts on the other is paralleled in Greco-Roman astrology. In the *Anthology* of Vettius Valens and the *Tetrabiblos* of Ptolemy, Neugebauer and van Hoesen pointed out that simple arithmetical schemes are used "which belong to a period of astronomical theory which had been long surpassed at that time."[39] They go on to say, "the cliché which is so popular in histories of astronomy about the stimulating influence of astrology on exact astronomy is nowhere born out where we are able to control the details."[40] The cuneiform evidence appears consistent with the picture derived from the Hellenistic Greek sources and further supports the view that the necessary connections between astronomy and astrology in no way need be seen in terms of linear development, either from astrology to astronomy or the other way around.

In view of this, it seems that the more productive cultural question becomes not, did or did not astrology—omens as well as horoscopy—spark the development of astronomy at any point in its history, but in what precise ways did the interests and goals of astrology and astronomy converge and diverge and how did this relationship change over the course of the immensely long lifespan of Babylonian science from 2000 B.C. to the beginning of our own era.

NOTES

1. See G. Dossin, *Syria* 22 (1939), p. 101 and idem, *Seconde Ren contre Assyriologique* (Paris, 1951), pp. 46–48.

2. For the omens of *Enūma Anu Enlil*, see E. Reiner and D. Pingree, *The Venus Tablet of Ammisaduqa* (Babylonian Planetary Omens 1, Malibu, 1975), E. Reiner and D. Pingree, *Enūma Anu Enlil Tablets 50–51* (Babylonian Planetary Omens 2, Malibu, 1981), and F. Rochberg-Halton, *Aspects of Babylonian Celestial Divination: The Lunar Eclipse Tablets of Enūma Anu Enlil* (AfO Beiheft 22, 1988).

3. These texts are in the form of letters, for which see S. Parpola, *LAS I and II,* and so-called reports, for which see H. Hunger, *Astrological Reports to Assyrian Kings* (State Archives of Assyria, vol. VIII, Helsinki, 1992).

4. See Pinches-Sachs, *LBAT* 1458–1476, 1521–1577, and 1588–1593.

5. A late Babylonian commentary text from Kutha (R. D. Biggs, "An Esoteric Babylonian Commentary," *RA* 62 (1968), pp. 51–58) relates a number of omen series to astrological elements, referring to the medical diagnostic omen series SA.GIG "symptoms," the physiognomic omen series *Alandimmû* "form," as well as the malformed birth omens *Izbu,* as a group termed "secret of heaven and earth." And, as was the case for the celestial omens, authorship of the physiognomic (and medical) omens was attributed to the god Ea. Perhaps this evidence reflects the change in Babylonian divination sciences following the development of genethlialogy.

For omens from the date of birth of a child, see from the series *iqqur īpuš,* R. Labat, *Un Calendrier Babylonien des Travaux des Signes et des Mois* (Paris, 1965), pp. 132–134, 64 (K.11082). A Hittite fragment, translated from an Old Babylonian text and which derives predictions from the date of a child's birth, is cited by Oppenheim in "Man and Nature in Ancient Mesopotamia," DSB 15, p. 644; see B. Meissner, "Über Genethlialogie bei den Babyloniern," *Klio* 19 (1925), pp. 432–434; also K. Riemschneider, *Studien zu den Boghazkoy-Texten* 9 (Wiesbaden, 1970), p. 44, n. 39a; and for an Egyptian parallel, see Bakir, *The Cairo Calendar,* No. 86637 (Cairo, 1966), esp. 13–50. For the physiognomic texts, see for example, *YOS* 10, 54, and F. R. Kraus, *Texte zur babylonischen Physiognomatik* (AfO Beiheft 3).

6. See, for example, the Babylonian "Poem of the Righteous Sufferer" and the *Theodicy* in Lambert, *Babylonian Wisdom Literature* (Oxford, 1960), pp. 21–91.

7. The proceedings were published as *La Divination en Mésopotamie Ancienne et dans les Régions Voisines* (Paris, 1966).

8. A. L. Oppenheim, "Perspectives on Mesopotamian Divination," in *La Divination* (CRRA 14, 1966), p. 40.

9. A. Sachs, "A Classification of the Babylonian Astronomical Tablets of the Seleucid Period," JCS 2 (1948), pp. 271–290.

10. O. Neugebauer, *Astronomical Cuneiform Texts,* (ACT) 3 vols. (London, 1955).

11. Material in the present paper stems from the work I have done in preparing an edition of the horoscope texts. This project was proposed to me by David Pingree, to whom I owe a tremendous debt of gratitude not only for his encouragement, but also for supplying me with Sachs's notes on the horoscope corpus. See *Babylonian Horoscopes* (Philadelphia: Transactions of the American Philosophical Society vol. 88, pt. 1, 1998).

12. See AB 251, published Sachs JCS 6, pp. 54–57 (transliteration, translation, and commentary), and AO 17649, published Arnaud *TBER* 6, 52 (copy).

13. See BM 38104 (LBAT *1475), unpublished.

14. See the discussion in J. Oelsner, *Materialien zur Babylonischen Gesellschaft und Kultur in Hellenistischer Zeit* (Budapest, 1986), p. 118 and note 451. Among the youngest texts from the Esagila, according to J. Oates, *Babylon* (London, 1986), p. 142 and note 36, is a document from 93 B.C. concerning the cult at Esagila, and another text (Rm 844) dated to 88 B.C., published in Epping and Strassmaier, ZA 6 (1891), pp. 226 and 230 (see also ZA 61 [1971], p. 165).

15. CT 49 144, CT 49 186, and BOR 4 (1890), pp. 132ff. See the discussion in G. McEwan, *Priest and Temple in Hellenistic Babylonia* (Wiesbaden, *Freiburger Altorientalische Studien* 4, 1981), pp. 18–20, review of R. J. van der Spek, "The Babylonian Temple during the Macedonian and Parthian Domination," *Bibliotheca Orientalis* 42 (1985), pp. 547–562, and my "The Cultural Locus of Astronomy in Late Babylonia," in *Die Rolle der Astronomie in den Kulturen Mesopotamiens*, ed. H. Galter (Graz, *Beitrage zum 3. Grazer Morgenlandischen Symposion*, 1993), pp. 31–47.

16. See my "Scribes and scholars: The *ṭupšar Enūma Anu Enlil*" in the *Oelsner Festschrift* (in press).

17. The two texts are from Uruk, MLC 2190, published in Sachs, JCS 6, pp. 60–61 and the unpublished W 20030/143, included as texts 10 and 11 of my forthcoming edition.

18. These are texts 14 and 20, respectively, in my edition, *Babylonian Horoscopes* (Philadelphia, 1998). I thank the Trustees of the British Museum for permission to publish these texts.

19. ACT p. 279.

20. This most useful method of comparing ancient and modern data was suggested to me by J. P. Britton. Babylonian (sidereal) longitudes may accordingly be compared against modern computed (tropical) longitudes by means of a correction factor that takes into account the constant of precession and the date of the data to be compared. The correction factor is that determined by P. Huber as a mean difference between ancient and modern longitudes, that $\overline{\Delta\lambda}$ being $4°28'$ for the year -100 (see P. Huber, "Über den Nullpunkt der babylonischen Ekliptik," *Centaurus* 5 [1958], pp. 192–208). Therefore, $\lambda_{\text{Babylonian}} = \lambda_{\text{tropical}} + \Delta\lambda$, where $\Delta\lambda = 3.08° + 1.3825° \times$ (year date number). $3.08°$ is the correction factor for the year 0 and $1.3825°$ is the constant of precession per 100 years.

21. ACT 310 for Mercury, 654 and 655 for Jupiter.

22. The subdivision of the synodic arc of Jupiter in ACT Nos. 654 and 655 produces a retrograde arc of 8;47°. This is smaller than any of the known Babylonian Jupiter tables, such as System A or A', and 1° less than the actual value of the retrogradation for the date of the table (9.8° in −163), as pointed out by Aaboe, unpublished ms V, p. 17. In his excursus (ibid. V, p. 19) on the third-order scheme for Jupiter's daily motion, Aaboe said, "The motivation of this fine scheme cannot, then, be based in practical astronomy, for if the observational acumen of the Babylonian astronomers was dull enough to allow them to tolerate an error of one degree in the length of the retrograde arc, it was also too dull to enable them to detect the need for so elaborate a scheme to account satisfactorily for Jupiter's daily motion." I would like to thank Prof. Aaboe for allowing me to quote from his unpublished manuscript.

23. Ibid. V, p. 11.

24. Sachs, "A Classification of the Babylonian Astronomical Tablets of the Seleucid Period," JCS 2 (1948), p. 278.

25. The oppositions (computed by means of *na*) and conjunctions (computed by means of KUR), for which the length of the month would be needed, might be the data needed for computing the time of conception (D. Pingree, personal communication). Further research is needed before the use of the date of conception versus that of birth in Babylonian astrology is understood. It is clear though that omens for the date of conception were compiled, see for example, LBAT 1588 and 1589 (*šumma ...* LÚ.TUR *re-ḫi*).

26. See texts 10 and 16a and 16b of my forthcoming edition.

27. See Neugebauer, *ACT*, vol. I, pp. 186–211, and Aaboe-Henderson, "The Babylonian Theory of Lunar Latitude," p. 208–211.

28. Aaboe, "The Babylonian Theory of Lunar Latitude," p. 209.

29. Texts 3, 4, 13, 14, 19, 20, 21, 22, 23, 24, 25, and 26 of my forthcoming edition.

30. See Sachs-Hunger, *Diaries I*, p. 23.

31. Th. von Oppolzer, *Canon der Finsternisse* (Math.-Naturwiss. Cl. d.... Akad. d. Wiss. Denkschriften 52, Vienna, 1887).

32. See my *Aspects of Babylonian Celestial Divination* (AfO Beiheft 22, 1988), pp. 36–57.

33. Such as in *LBAT* 1413, *1414, 1415 + 1416 + 1417, *1419, *1420, 1421, 1426, 1427, 1437–*1450.

34. For the literature on the scheme, see Neugebauer, "A Table of Solstices from Uruk," *JCS* 1 (1947), pp. 143–148; Neugebauer, "Solstices and Equinoxes in Babylonian Astronomy," *JCS* 2 (1948), pp. 209–222; Neugebauer, HAMA, pp. 357–363; and A. Slotsky, "The Uruk Solstice Scheme Revisited," in H. Galter, ed., *Die Rolle der Astronomie in den Kulturen Mesopotamiens* (Beitrage zum 3. Grazer Morgenländischen Symposion, 1993), pp. 359–366.

35. Texts 2, 4, 6, 7, 8, 13, 14, 15, and 17 of my edition.

36. For the form of a diary entry, see Sachs, *JCS* 2: 285–6, and idem, "Babylonian Observational Astronomy," in Hodson ed., *The Place of Astronomy in the Ancient World* (= Phil. Trans. R. Soc. Lond. A. 276, 1974): 43–50. For texts, see Sachs-Hunger, *Diaries*, vols. I–III.

37. For details, see my "Babylonian Horoscopes and Their Sources," OrNS 58 (1989), pp. 102–123.

38. Neugebauer, ACT.

39. Neugebauer and H. B. van Hoesen, *Greek Horoscopes* (Philadelphia, American Philosophical Society, 1959), p. 185.

40. Ibid.

BABYLONIAN OBSERVATIONS OF SATURN DURING THE REIGN OF KANDALANU

C. B. F. Walker

BM 76738 + 76813 (AH 83-1-18, 2109 + 2185) (joined by Dr. I. L. Finkel) is a fragment of a tablet containing a collection of observations of the planet Saturn made during the reign of Kandalanu (647–627 B.C.).[1] As such it is the first collection of planetary data since the Venus observations made during the reign of Ammiṣaduqa (1702–1682 B.C., according to Huber's chronology) in the Old Babylonian period. It was presumably excerpted from the Astronomical Diaries, or a similar source. One line includes the comment, ḫe-pí, "broken," indicating that at this point the scribe was copying from a broken text. The earliest surviving Astronomical Diary dates from just five years earlier, −651/650 (652/651 B.C.), and includes an observation of the last visibility of Saturn. As in the few surviving Astronomical Diaries dating from before the fourth century B.C., the names given to the various constellations and stars used by the astronomers as reference points for their observations of Saturn differ somewhat from the later standard list of so-called "Normal Stars." However, the general terminology is similar to that of the Diaries.

The registration AH 83-1-18 might at first sight seem to suggest that the tablet came from Sippar (Abu Habba), but as Reade (1986), xxxiv, has pointed out, the collection plainly also included material from Rassam's excavations at Babylon and Borsippa. Not wishing to revive the old discussion about whether there might have been astronomical archives at Sippar, I assume that Babylon or Borsippa is the most likely source of the present tablet.

The text lists the dates of successive first and last visibilities of Saturn in terms of the regnal year, month and day in the lunar calendar. In two cases, lines 7′ and 10′, the date is given only as "at the end of month 4/5." In some cases it is noted that the first or last visibility was not actually observed, and in such cases we must assume that the date was estimated, or as stated in line 23′, muš-šúḫ, "calculated." In two cases of first visibility, lines 6′ and 8′, the expression NIM, "high," indicates that the astronomers themselves were aware that the planet was higher in the sky than usual for

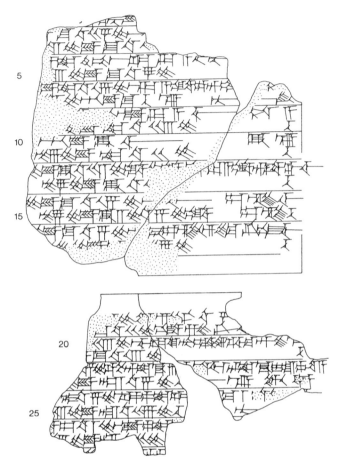

Figure 1
BM 76738 + 76813.

first visibility, implying that under better conditions it would have been visible a day or two earlier.

The location of the planet in the sky is also given in all cases of first visibility except in line 10′ (for which also a precise date was lacking); but for last visibility the location is only given in two cases (lines 15′ and 19′). The planet is nowhere named on the tablet, but the intervals between the various observations (both chronologically and in terms of ecliptic longitude) fit the movements of Saturn and of no other planet. Modern theory calculates the last and first visibility of planets as a pair by reference to the time and location of the conjunction of the planet with the sun that occurs between last and first visibility. That is also the pairing that we find on the Venus Tablet of Ammiṣaduqa a thousand years earlier, where the concern is with the total length of invisibility. By contrast, on this tablet the Babylonian scribe presents the observations in the sequence of first and last visibility of the planet, each group of two observations being marked off by a ruling with the second line being indented; the conjunction is not yet a point of reference.

The tablet may perhaps be seen as part of the process of collecting and analyzing data whereby the Babylonian astronomers eventually discovered satisfactory means of mathematically describing and predicting planetary movements in terms of time and longitude. Paleographically it seems likely that the tablet was written within a few decades of the date of the latest observation recorded on it. It is of course obvious from the circumstance that the tablet is incomplete, that the chronological range of observations being examined was wider than the reign of Kandalanu.

The name of the planet Saturn is not given on the tablet, and the name of Kandalanu is to be restored from only a few traces in the first line. It is, however, certain that we are dealing with Saturn and Kandalanu. Saturn is the slowest moving of the visible planets, and only Saturn would move the distances indicated between successive first visibilities. A complete cycle of Saturn phenomena in relation to the stars takes 59 years. But when that cycle has to be fitted to the lunar calendar of 29 or 30 days then identical cycles recur at intervals of rather more than 17 centuries. Thus there is no difficulty in determining the date of the present text.

TRANSLITERATION

1′   [MU 1-KAM *kan-d*]*a-*⌜*la-nu* ITU⌝-[x U$_4$ x-KAM ŠÚ]

2′   [MU 1-KAM IT]U-ŠU U$_4$ 24-KAM *ina* I[GI … ALLA … IGI]

3′   [M]U 2-KAM ITU-ŠU U$_4$ 10+[x ŠÚ]

4′   [MU 2-KAM IT]U-NE *ḫe-pí ina* SAG ⌜UR-A IGI NU⌝ [ŠEŠ? ŠE/KIN DIRI]

5′   [M]U 3-KAM ITU-ŠU U$_4$ 7-KAM [ŠÚ]

6′   [MU 3-KAM] ITU-NE U$_4$ 16-KAM *ina* UR-A EGIR MUL-LUGAL [IGI] NI[M?-A]

7′   [MU] ⌜4⌝-KAM *ina* TIL ITU-ŠU ŠÚ DIR NU ŠEŠ

8′   [MU 4-KAM ITU-KIN?] ⌜U$_4$⌝ [x]-⌜KAM⌝ *ina* MURUB$_4$ UR-A IGI NIM-A

9'   [MU 5]-⌜KAM ITU-NE⌝ U$_4$ 23-KAM ŠÚ

10'   [MU 5-KAM] ⌜ina TIL⌝ ITU-KIN IGI KIN DIRI

11'   MU 6-KAM ITU-NE U$_4$ 20-KAM ŠÚ

12'   [MU 6-KA]M ITU-KIN U$_4$ 22-KAM EGIR ⌜GÌR EGIR *ša*? UR⌝-A EGIR AN.GÚ.ME.MAR IGI

13'   MU 7-KAM ITU-KIN U$_4$ 10+[x-KAM] ŠÚ

14'   [MU 7]-KAM ITU-DU$_6$ U$_4$ 15-KAM ⌜ina IGI⌝ AB.SÍN IGI

15'   MU 8-KAM ITU-KIN 2-KAM U$_4$ 5-KAM EGIR AB.SÍN ŠÚ

16'   [MU 8-KA]M ITU-DU$_6$ U$_4$ 5-KAM *ina* D[AL]-BAN AB.SÍN *u zi-ba-ni-tum* IGI

17'   [M]U ⌜9-KAM ITU-KIN⌝ [U$_4$] ⌜20⌝+⌜7/8⌝-KAM ŠÚ

Reverse

18'   [MU 9-KAM ITU-APIN U$_4$ x]+1-KAM ⌜ŠÀ?⌝⌜*zi*⌝-*bānī*(DÙ)-*tú* ⌜*šá*⌝ [... IGI ...]

19'   [MU 10-KAM ITU]-⌜DU$_6$⌝ U$_4$ 20-KAM EGIR *zi-bānī*(DÙ)-*t*[*ú* ŠÚ]

20'   [MU 10-KAM ITU-APIN U$_4$] 23-KAM *ina* IGI SAG-KI GÍR-TAB Á IM-SI IGI *ba-ìl* ŠE D[IRI?]

21'   [MU 11-KAM IT]U-DU$_6$ U$_4$ 13-KAM ŠÚ

22'   [MU 11-KAM ITU-APIN U$_4$] ⌜15⌝-KAM *e-lat* ^{d}*li$_9$-si$_4$* ⌜6$\frac{1}{2}$?⌝ UŠ⌝ IGI *ana* ^{d}*li$_9$-si$_4$ i-ṣi pa-na-a*

23'   [MU 12-KA]M ITU-APIN U$_4$ 5-KAM ⌜ŠÚ⌝⌜*ina*⌝ DIR *muš*-⌜*šúḫ*⌝

24'   [MU 12-KAM ITU]-GAN U$_4$ 5-KAM *ina* SAG PA-BIL-SA[G IGI?] x *ma* 1 ⌜UŠ?⌝ x [x] x

25'   [MU] 13-KAM ITU-APIN U$_4$ 26-KAM ŠÚ [DIR?] NU [ŠEŠ]

26'   [MU 13-KAM IT]U-AB U$_4$ 1-KAM *ina* MURUB$_4$ PA-BIL-S[AG IGI ...]

27'   [M]U ⌜14-KAM⌝ ITU-⌜APIN U$_4$ 20⌝-KAM [ŠÚ ...]

28'   [MU 14-KAM ITU-GAN U$_4$] ⌜20⌝+[?-KAM x] x ⌜MUL?-x⌝ [... IGI ...]

TRANSLATION

1'   [Year 1 of Kand]alanu, ⌜month⌝ [..., day ..., last appearance.]

2'   [Year 1, mont]h 4, day 24, in fr[ont of ... the Crab, first appearance.]

3'   [Ye]ar 2, month 4, day 10+[x, ..., last appearance.]

4'   [Year 2, mon]th 5, broken, in the head of the Lion, first appearance; not [observed?.]

5′   [Ye]ar 3, month 4, day 7, [last appearance.]

6′ [Year 3], month 5, day 16, in the Lion behind the King (= α Leonis), [first appearance]; ⌜high⌝.

7′   [Year] ⌜4⌝, at the end of month 4, last appearance; (because of) cloud not observed.

8′ [Year 4, month 6?], day [x], in the middle of the Lion, first appearance; high.

9′   [Year 5], month 5, day 23, last appearance.

10′ [Year 5], at the end of month 6, first appearance; intercalary Ululu.

11′   Year 6, month 5, day 20, last appearance.

12′ [Year 6], month 6, day 22, behind ⌜the rear foot of⌝ the Lion (= β Virginis), behind AN.GÚ.ME.MAR, first appearance.

13′   Year 7, month 6, day 10+[x], last appearance.

14′ [Year 7], month 7, day 15, ⌜in front of⌝ the Furrow (α+ Virginis), first appearance.

15′   Year 8, month 6, day 5, behind the Furrow (α+ Virginis), last appearance.

16′ [Year 8], month 7, day 5, ⌜between⌝ the Furrow (α+ Virginis) and the Balance (Libra), first appearance.

17′   [Year] ⌜9, month 6⌝, [day] ⌜27/28?⌝, last appearance.

Reverse

18′ [Year 9, month 8, day x]+1, ⌜within?⌝ the Balance,... [..., first appearance].

19′   Year 10, month] ⌜7⌝, day 20, behind the Balance, [last appearance].

20′ [Year 10, month 8, day] 23, in front of the Forehead of the Scorpion, on the north side, first appearance; it was bright; intercalary Addaru.

21′   [Year 11, month] 7, day 13, last appearance.

22′ [Year 11, month 8, day] ⌜15⌝, above Lisi (= α Scorpii) ⌜$6\frac{1}{2}$ degrees⌝, first appearance; with reference to Lisi a little in front(?).

23′   [Year 12], month 8, day 5, last appearance; ⌜because of⌝ cloud computed.

24′ [Year 12, month] 9, day 5, at the beginning of Pabilsag (= Sagittarius + part of Ophiuchus), [first appearance?];... 1 degree?...

25′   [Year] 13, month 8, day 26, last appearance; [cloud?], not [observed?].

26′ [Year 13, month] 10, day 1, in the middle of Pabilsag, [first appearance;...]

27′   [Year] ⌜14⌝, month ⌜8⌝, ⌜day 20⌝, [last appearance;...]

28′ [Year 14, month 9, day] ⌜20⌝[+ ?,...] ... [..., first appearance;...]

a	b	c	d
Conjunction	Long.		LE/FM
−646/6/28	89°	−15	e6/13
		+19	m7/17
−645/7/13	102°	−15	e6/28
		+18	m7/31
−644/7/26	116°	−16	e7/10
		+16	m8/11
−643/8/8	128°	−16	e7/23
		+16	m8/24
−642/8/21	141°	−16	e8/5
		+15	m9/5
−641/9/3	153°	−17	e8/17
		+15	m9/18
−640/9/15	166°	−17	e8/29
		+14	m9/29
−639/9/27	178°	−17	e9/10
		+13	m10/10
−638/10/8	189°	−17	e9/21
		+13	m10/21
−637/10/20	200°	−17	e10/3
		+13	m11/2
−636/10/30	211°	−17	e10/13
		+14	m11/13
−635/11/10	222°	−16	e10/25
		+14	m11/24
−634/11/21	233°	−16	e11/5
		+15	m12/6
−633/12/2	244°	−15	e11/17
		+16	m12/18

e	f	g	h	i
NM	Day	Bab. Y/M/D	−	+
e5/25	20	[1/.../...]	?	
e6/24	23	[1]/4/24		+1
6/13	16	2/4/10+[x]	?	
7/13	18	2/5/*ḫe-pī*		?
7/1	10	3/4/7	−3	
7/31	11	3/5/16 high		+5
7/20	4	4/TIL ITU-ŠU	?	
8/19	5	[4/6?/...] high		?
7/10	27	[5]/5/23	−4	
8/8	28	[5]/TIL ITU-KIN		?
7/28	21	6/5/20	−1	
8/27	22	[6]/6/22		0
8/15	15	7/6/10+[?]	?	
9/14	15	[7]/7/15		0
9/3	8	8/6/5	−3	
10/3	7	[8]/7/5**		−2
8/23	30	9/6/20 + 7/8	−2?	
9/22	29			
or 10/22		[9/8]/[x]+⌜1⌝		+2?
9/11	23	10/7/20	−3	
10/11	22	[10/8]/23		+1
9/29	15	[11]/7/13	−2	
10/29	15	[11/8]/15		0
10/18	8	[12]/8/5	−3	
11/17	7	[12]/9/5**		−2
10/8	29	13/8/26	−3	
11/6	30			
or 12/6		[13]/10/1		+1
10/27	22	14/8/⌜20⌝	−2	
11/25	23	[14/9]/⌜20⌝+[x]		?

THE VISIBILITY OF SATURN

A procedure for establishing the dates and place in the ecliptic of first and last visibility of Saturn is given by Schoch (1928), 109 (with pl. 14); his Table C (Conjunction) for the time intervals between conjunction and first and last visibility has been revised by van der Waerden (1943). The procedure can now be simplified by the use of the published tables of Hunger and Dvorak (1981). For the years −647 to −633 the table on pp. 66–67 gives:

(a) the date (year, month, day) of conjunction between Saturn and the sun,

(b) the approximate longitude of conjunction (to the nearest degree),

(c) the number of days before (−) and after (+) conjunction at which last evening and first morning visibility may be expected for a conjunction at that longitude,

(d) the resulting date in the Julian calendar (month, day) for theoretical last evening (LE) or first morning (FM) visibility (by modern, not Babylonian, calculations),

(e) the date (month, day) in the Julian calendar of the preceding new moon (NM),

(f) the day in the Babylonian lunar month on which last/first visibility may be expected [N.B. Babylonian observations of first morning visibility are made on the Babylonian day, which began on the previous evening],

(g) the Babylonian date (year/month/day, using the lunar month) on which last/first visibility is recorded,

(h) comparison of (g) with (f) for last visibility,

(i) comparison of (g) with (f) for first visibility.

(a) and (b) are derived by interpolation from the tables of Hunger and Dvorak (1981); (c) is taken from van der Waerden (1943); (d) is taken from a computer program by P. Huber, but an approximate date can be obtained by taking the dates of "new moon" given in Goldstine (1973), and adding 2 or 3 days.

On the assumptions that the text is not corrupt and that the modern theory is good for ancient Babylonian observations (i.e., that the Babylonians would not have seen Saturn after the calculated date for last visibility, nor before the calculated date for first visibility), where the figure in column (c) is negative, the day number in column (g) should be equal to or lower than the figure in column (f), and this should result in a 0 or a minus quantity in column (h); where the figure in column (c) is positive, the day number in column (g) should be not less than the figure in column (f) −1 (this allows for the possibility of the Babylonian lunar month starting one day later than calculated), and this should result in a figure of −1 or higher in column (i). One must also make the appropriate adjustments at the month boundaries. In practice all the recorded observations fit this pattern well, except for the two cases marked **, where the Babylonians appear to have seen the planet for the first time at least one day earlier than expected (according to van der Waerden's visibility tables) even allowing for the possibility that they had previously seen the new moon one day late. These two cases are perhaps to be explained as scribal errors, the miscopying of 5 for 8.

The Babylonian Calendar

The Babylonian calendar was luni-solar with an additional "intercalary" month being added on average 7 times in 19 years to bring the lunar and solar cycles into line. In the seventh century B.C. the later "Metonic" pattern of regular intercalations was not yet in place and it is a matter of interest to establish in which years the intercalary months were inserted.

The synodic period of Saturn is 378.09 days. Hence phenomena recur about 24 days later in the Babylonian calendar than in the previous year (Schoch [1928], 109). In consequence the present text allows us to establish the approximate, or sometimes the precise, position of intercalary months in the years covered by this text (647–634 B.C.). The textual indications, line 10′ (MU 5-KAM ... KIN DIRI), line 15′ (MU 8-KAM ITU-KIN 2-KAM), line 20′ (MU 10-KAM ... ŠE D[IRI]), fit the astronomical requirements. The intercalary months in years 5 and 10 are also confirmed by contemporary economic texts (Brinkman and Kennedy [1983], 40–42). Additionally an intercalary month (Ululu or Addaru) is required in year 2, and either an intercalary Addaru in year 13 or intercalary Ululu in year 14. If we assume, on the basis of the record which the scribe gives of intercalary months in years 5, 8, and 10, that he was deliberately recording these in order to assess the synodic period of Saturn, then we may also note that there is room to restore KIN/ŠE DIRI at the end of line 4′ (for year 2), and to restore either ŠE DIRI at the end of line 26′ (for year 13) or KIN DIRI at the end of line 27′ (for year 14).

The exact dates that can be derived from the present text allow us to give the following table for years 2–14 of Kandalanu, using dates for first lunar visibility computed by P. Huber, and following the same principles as the tables in Parker and Dubberstein (1956); that is that the Babylonian day is equated with the Julian day that began at the midnight following the evening beginning of the Babylonian day (this is in contrast with the previous table of Saturn visibility). The last part of year 2, the end of year 13, and the first part of year 14 are italicized because of the uncertainty over whether the relevant intercalary months were Ululu or Addaru.

Kandalanu

YR	B.C.	Nis	Aia	Sim	Duz	Abu	Ulu
2	646	3/18	4/17	5/16	6/14	7/14	8/12
3	645	4/5	5/5	6/3	7/2	8/1	8/30
4	644	3/25	4/24	5/23	6/22	7/21	8/20
5	643	3/14	4/13	5/12	6/11	7/11	8/9
6	642	4/2	5/1	5/31	6/30	7/29	8/28
7	641	3/21	4/20	5/19	6/18	7/17	8/16
8	640	3/11	4/9	5/9	6/7	7/7	8/5
9	639	3/30	4/28	5/28	6/26	7/25	8/24
10	638	3/20	4/18	5/17	6/16	7/15	8/13
11	637	4/7	5/6	6/4	7/4	8/2	9/1
12	636	3/27	4/25	5/25	6/23	7/23	8/21
13	635	3/16	4/15	5/14	6/13	7/12	8/11
14	634	4/3	5/3	6/2	7/2	7/31	8/30

We are getting close to the point at which we could extend Parker and
Dubberstein's tables for Babylonian chronology back in a continuous
series from the reign of Nabopolassar to the beginning of the reign of
Kandalanu. An intercalary Ululu is attested for Kandalanu's 19th year
(Brinkman and Kennedy [1992], 47–48). It remains to identify the intercalary month expected in Kandalanu's 16th year, and to identify the
missing intercalary month between Kandalanu year 19 and Nabopolassar
year 2. This may be the intercalary Addar in Nabopolassar's first year
recorded by Kennedy (1986), 179, but the reading of the date is disputed.

THE BABYLONIAN STARS AND CONSTELLATIONS

The following list gives those Babylonian observations in which the
planet's approximate longitude is indicated, the Julian date of calculated
first/last visibility, the planet's ecliptic longitude and latitude at that date
(derived by interpolation from the tables of Hunger and Dvorak [1981]),
and the ecliptic longitude and latitude of the nearest relevant star(s); all
figures are rounded to one decimal place. Wherever possible this corresponds to a star from the list of Normal Stars. For a list of Normal Stars
and their approximate ecliptic coordinates in −600 see Sachs and Hunger
(1988), 17–19.

2′   [MU 1-KAM IT]U-ŠU U₄ 24-KAM *ina* I[GI … ALLA … IGI]
−646/7/17   91.1°, 0.0°   θ Cnc, 89.0°, −0.7°

Ulu 2	Tas	Ara	Kis	Teb	Sha	Add	Add 2
	9/10	*10/10*	*11/8*	*12/8*	*1/7*	*2/6*	*3/6*
	9/29	10/28	11/26	12/26	1/25	2/23	
	9/18	10/18	11/16	12/16	1/14	2/13	
9/8	10/7	11/6	12/5	1/4	2/2	3/4	
	9/27	10/26	11/25	12/24	1/23	2/21	
	9/15	10/14	11/13	12/13	1/11	2/10	
9/4	10/3	11/2	12/2	1/1	1/30	2/1	
	9/23	10/22	11/21	12/21	1/20	2/18	
	9/12	10/11	11/10	12/10	1/9	2/7	3/8
	9/30	10/29	11/28	12/28	1/26	2/25	
	9/20	10/19	11/17	12/17	1/16	2/14	
	9/9	10/9	11/7	12/6	1/5	2/3	3/5
	9/28	10/28	11/26	12/26	1/24	2/22	

4′ [MU 2-KAM IT]U-NE *ḫe-pí ina* SAG ⌜UR-A IGI NU⌝ [ŠEŠ?]
−645/7/31   104.6°, 0.5°   ε Leo, 104.0°, 9.8°

6′ [MU 3-KAM] ITU-NE U$_4$ 16-KAM *ina* UR-A EGIR MUL-LUGAL [IGI] NI[M?-A]
−644/8/11   117.6°, 1.0°   α Leo, 113.1°, 0.6°

8′ [MU 4-KAM ITU-KIN?] ⌜U$_4$⌝ [x]-⌜KAM⌝ *ina* MURUB$_4$ UR-A IGI NIM-A
−643/8/24   130.5°, 1.5°   · ρ Leo, 119.7°, 0.3°
                                     θ Leo, 126.7°, 9.9°
                                     χ Leo, 127.8°, 1.6°

12′ [MU 6-KA]M ITU-KIN U$_4$ 22-KAM EGIR ⌜GÌR EGIR *ša*? UR⌝-A EGIR AN.GÚ.ME.MAR IGI
−641/9/18   155.4°, 2.1°   β Vir, 140.4°, 1.1°
                                     η Vir, 147.9°, 2.3°
                                     γ Vir, 153.5°, 3.0°
                                     Jupiter, 151.5°, 1.12°

14′ [MU 7]-KAM ITU-DU6 U$_4$ 15-KAM ⌜*ina* IGI⌝ AB.SÍN IGI
−640/9/29   167.4°, 2.3°   α Vir, 167.2°, −1.8°

15′ MU 8-KAM ITU-KIN 2-KAM U$_4$ 5-KAM EGIR AB.SÍN ŠÚ
−639/9/10   175.5°, 2.3°   α Vir, 167.2°, −1.8°

16′ [MU 8-KA]M ITU-DU6 U$_4$ 5-KAM *ina* D[AL]-BAN AB.SÍN *u zi-ba-ni-tum* IGI
−639/10/10   179.0°, 2.3°   α Vir 167.2°, −1.8°
                                      α Lib 188.4, 0.6°

18′ [MU 9-KAM ITU-APIN U$_4$ x]+1-KAM ⌈ŠÀ?⌉ ⌈zi⌉-bānī(DÙ)-tú ⌈šá⌉
[... IGI ...]
−638/10/21   190.4°, 2.3°   α Lib 188.4°, 0.6°
                                          β Lib, 192.8°, 8.7°

19′ [MU 10-KAM ITU]-⌈DU6⌉ U$_4$ 20-KAM EGIR zi-bānī(DÙ)-t[ú ŠÚ]
−637/10/3   198.2°, 2.2°   β Lib, 192.8°, 8.7°

20′ [MU 10-KAM ITU-APIN U$_4$] 23-KAM ina IGI SAG-KI GÍR-TAB Á
IM-SI IGI ba-ìl ŠE D[IRI?]
−637/11/2   201.7°, 2.2°   δ Sco, 206.0°, −1.8°

22′ [MU 11-KAM ITU-APIN U$_4$] ⌈15⌉-KAM e-lat ^{d}li$_9$-si$_4$ ⌈6$\frac{1}{2}$?⌉ UŠ⌉ IGI ana
^{d}li$_9$-si$_4$ i-ṣi pa-na-a
−636/11/13   212.9°, 2.0°   α Sco. 213.2°, −4.4°

24′ [MU 12-KAM ITU]-GAN U$_4$ 5-KAM ina SAG PA-BIL-SA[G IGI?] x
ma 1 ⌈UŠ?⌉ x [x] x
−635/11/24   224.0°, 1.7°   β Oph, 224.8°, −1.7°

26′ [MU 13-KAM IT]U-AB U$_4$ 1-KAM ina MURUB$_4$ PA-BIL-S[AG IGI
...]
−634/12/6   235.2°, 1.4°   μ Sgr, 236.7°, 2.4°

28′ [MU 14-KAM ITU-GAN U$_4$] ⌈20⌉+[?-KAM x] x ⌈MUL?-x⌉ [... IGI
...]
−633/12/18   246.5°, 1.0°   σ Sgr, 245.8°, −3.4°
                                          ν Sgr, 245.9°, 0.2°
                                          ξ Sgr, 246.9°, 1.8°

STAR IDENTIFICATIONS

2′. The broken reference is probably to the constellation Cancer, but note
that Saturn is already to the west of the first Normal Star in Cancer, θ Cnc.
This star is first attested written as MUL IGI ša ALLA ša ULÚ in Diaries of
the 4th century B.C.

4′. SAG UR-A: in the Diaries from −380 onward SAG UR-A designates ε
Leonis.

6′. Interpretation of MUL-LUGAL as α Leo raises no problems.

8′. ina MURUB$_4$ UR-A: of the preceding Normal Stars in Leo ρ is described by the Babylonians as the small star 4 cubits behind Regulus, and θ is
described as the Lion's flank. The nearest bright star to Saturn seems to be χ
Leo, but in all probability ina MURUB$_4$ UR-A simply means within the
constellation Leo.

12′. EGIR ⌈GÌR EGIR ša? UR⌉-A EGIR AN.GÚ.ME.MAR: in the standard list of Normal Stars it is β Virginis, which is known as GÌR ár šá A (so

already in the Diary for −567, written GÌR *ár šá* UR-A). In fact the calculated location of the planet at first visibility is considerably to the east of *β*
Virginis, and just behind *γ* Virginis. But *γ* Virginis is known in the Diaries
as DELE *šá* IGI ABSIN. Why should our text refer at all to GÌR EGIR *šá*
UR-A, and what is AN.GÚ.ME.MAR? H. Hunger pointed out to me some
time ago that in the Diary for −567 GÌR *ár šá* UR-A (*β* Virginis) is written
by mistake for *γ* Virginis, and wondered whether the same mistake had happened here. He also pointed out that on this occasion Jupiter is also in the
vicinity, and wondered if there could be some way of bringing together
AN.GÚ.ME.MAR and the traditional writing of Jupiter as ^{d}SAG-ME-GAR.
It seems problematic, but I have no better solution.

14′. AB.SÍN: "the Furrow." In later Babylonian astronomical texts AB.SÍN
and ABSIN (KI.AŠ.AŠ) are used to designate the whole zodiac sign Virgo.
Since *γ* Vir, which Saturn passed a year before, is described as DELE *šá* IGI
ABSIN, it seems that AB.SÍN here must refer to *α* Virginis (Spica); but note
that Saturn has in fact just passed it. The reference to AB.SÍN in the next line
raises no such problems.

16′. The description of Saturn's position here raises no problems.

18′. ⌜ŠÀ?⌝$^r zi^1$-*bānī*(DÙ)-*tú* ⌜*šá*⌝ [...]: the reading ⌜ŠÀ⌝ seems assured from
the traces, and makes sense astronomically, but the commoner expression in
this context would be ina MURUB$_4$; cf. lines 8′ and 26′. The following *šá* is
suggestive of the Normal Star names RÍN *šá* ULÚ and RÍN *šá* SI, but it is
not obvious that either of these is appropriate here.

19′. The description of Saturn's position here raises no problems.

20′. SAG-KI GÍR-TAB: this expression is not attested elsewhere. In the
Diaries for −384, −375, −373 and −372 SAG GÍR-TAB = the head of the
Scorpion, and in Diaries from −382 onwards MÚL *e šá* SAG GÍR-TAB = *β*
Scorpii.

22′. ^{d}*li$_9$-si$_4$*: the Diaries for −651 and −567 use the abbreviation SI$_4$ for *α*
Scorpii. ⌜$6\frac{1}{2}$?⌝ UŠ⌝: the use of the UŠ, normally translated "time-degree" in
ACT, as a unit of angular measurement in observations (rather than calculations) here and apparently in line 24′, is of considerable interest. The normal
units of measured observation in Late Babylonian astronomical texts are the
cubit (KÙŠ) and the finger (SI). The only other evidence for this use of UŠ is
given by Pingree (1993), 271, with reference to observing the movement of
the rising-point of Venus on the horizon. *ana* ^{d}*li$_9$-si$_4$ i-ṣi pa-na-a*: cf. the regular use of *i-ṣi* in the Diaries in descriptions of the position of the moon and
planets relative to fixed stars.

24′. *ina* SAG PA-BIL-SA[G IGI?] x *ma* 1 ⌜UŠ?⌝ x [x] x: the constellation
Sagittarius falls in that part of the zodiac in which, for no reason apparent
to us, there are large gaps in the list of Normal Stars; thus there is no later

terminology comparable with *ina* SAG/MURUB₄ PA-BIL-SAG here and in line 26'. Calculation shows that Saturn will have become visible in the vicinity of θ Ophiuchi; the earliest references to this Normal Star in the Diaries give its name as MÚL KUR *šá* KIR₄ *šil* PA (diary for −381) or MÚL KUR *šá šil-taḫ* PA (diary for −380(B)). The translation of SAG as "beginning" assumes that the constellation was visualized approximately as in the star catalogue in Ptolemy's *Almagest*. The translation "head" would assume a significantly different configuration, and would also be at variance with the idea that θ Ophiuchus was part of Pabilsag's arrow.

26'. *ina* MURUB₄ PA-BIL-S[AG]: I give the co-ordinates of θ Sagittarii for reference, but the expression probably simply means in the middle of Pabilsag/Sagittarius.

28'. [x] x ⌜MUL?-x⌝ [...]: given the approximate ecliptic longitude of the planet at first visibility, it is possible that we have here a reference to the star group in Sagittarius known from the Normal Star Almanacs as MUL-4-ÁM *šá* PA *šá ár*. This star group has been discussed recently by Roughton and Canzoneri (1992).

ADDITIONAL COMMENTS

1'. The name Kandalanu is restored in the light of astronomical considerations, but the position in the line of the sign ITU already indicated that it was preceded by a royal name.

4'. *ḫe-pí*: broken: this would appear to indicate that in the original diary from which this entry is taken the day number is lost; however the fact that the planet's reappearance was apparently not observed is at variance with the idea that the original diary may have contained the date of an observation. Cf. also lines 23' and 25'. It is preferable to assume that in these cases the date of theoretical first or last visibility was deduced from the planet's position first or last actually seen (cf. lines 6' and 8', NIM, "high"). NU [ŠEŠ]: restored after line 7'; but note that unlike line 7' (and 25'?) there is no reference to clouds here.

6' & 8'. NIM, high: this term indicates that when first observed the planet was higher above the horizon than normal for first visibility, leading to the conclusion that theoretical first visibility had occurred a day or two earlier, but had not been observed. See Huber (1982), 12–13.

10'. *ina* TIL ITU-KIN: the omission of the day of the month suggests either that the (theoretical) first visibility was not observed (cf. line 7'), or that the day was not recorded in the sources available to the compiler of the present text. Note that the use of the expression *ina* TIL ITU is common in the summaries of planetary positions in the Diaries.

14′. In the Diaries from −440 onward α Virginis is SA₄ *šá* ABSIN, "the bright star of ABSIN." The Sumerian ABSIN means "furrow," and the constellation name is generally so translated. Note, however, that Sachs (1952), 146 n. 3, and Thureau-Dangin (1938), 36 n. 2, preferred the translation "Barley-stalk" (French épi) for AB.SÍN. The picture of the lady holding an ear of barley (Spica?), which appears on AO 6448 (Weidner [1967], pl. 10), may reflect the later influence of Greek ideas.

16′. DAL.BAN = *birīt*; elsewhere normally written DAL.BA.(AN).NA.

17′. Note the late form of 9 in MU 9-KAM; hence the day number here can only be 27 or 28.

18′. Astronomically it seems quite probable that one should read simply [. . . ITU-APIN U₄] 1-KAM; see above, on the dates of first and last visibility of Saturn. A restoration [. . . ITU-DU₆ 2]⌜8⌝-KAM would only be possible if the planet had been seen on the first possible day allowed by van der Waerden's visibility table.

20′. ŠE D[IRI]: the restoration takes into account the astronomical requirement for an intercalary month at the end of Kandalanu year 10 or in the middle of year 11, and is supported by contemporary economic texts; see below.

22′. [. . . U₄] 15-KAM: the reading 15, rather than ⌜2⌝5, is dictated by astronomical considerations; see above.

23′. *ina* DIR: the Diaries normally say only DIR.

muš-⌜*šuḫ*⌝, "measured": cf. P. J. Huber, Babylonian eclipse observations, 750 B.C. to 0 (privately circulated typescript, Harvard University, 1973), 6.

25′. Cf. the comment on line 4′.

NOTE

1. The text has been discussed or alluded to previously in Walker (1983), 20–21, Brinkman (1984), 105 n. 521, 118 n. 576, Brinkman (1991), 66 n. 472, and Sachs and Hunger (1988), 13.

BIBLIOGRAPHY

Brinkman, J. A. (1984). *Prelude to Empire: Babylonian Society and Politics, 747–626 B.C.* (Occasional Publications of the Babylonian Fund, 7; Philadelphia).

Brinkman, J. A. (1991). Babylonia in the Shadow of Assyria (747–626 B.C.). *Cambridge Ancient History* III/2, 1–70.

Brinkman, J. A., and Kennedy, D. A. (1983). Documentary Evidence for the Economic Base of Early Neo-Babylonian Society. *JCS* 35, 1–90.

Goldstine, H. H. (1973). New and Full Moons, 1001 B.C. to A.D. 1651 (MAPS 94)

Huber, P. J. (1982). *Astronomical Dating of Babylon I and Ur III* (Occasional Papers on the Near East 1/4)

Hunger, H., and Dvorak, R. (1981). *Ephemeriden von Sonne, Mond und hellen Planeten von −1000 bis −601.* (Österreichische Akademie der Wissenschaften, Vienna)

Kennedy, D. A. (1986). Documentary Evidence for the Economic Base of Early Neo-Babylonian Society: Part II. *JCS* 38, 172–244.

Parker, R. A., and Dubberstein, W. H. (1956). Babylonian Chronology, 626 B.C.–A.D. 75. (Brown University Press)

Pingree, D. (1993). Venus Phenomena in Enūma Anu Enlil, in H. D. Galter (ed.) *Die Rolle der Astronomie in den Kulturen Mesopotamiens.*

Reade, J. E. (1986). Rassam's Babylonian Collection: The Excavations and the Archives, in E. Leichty, *Catalogue of the Babylonian Tablets in the British Museum*, VI, xii–xxxvi.

Roughton, N. A., and Canzoneri, G. L. (1992). Babylonian Normal Stars in Sagittarius. *JHA* 23, 193–200.

Sachs A. J. (1952). A Late Babylonian Star Catalog. *JCS* 6, 146–150.

Sachs, A. J., and Hunger, H. (1988). Astronomical Diaries and Related Texts from Babylonia, I. Diaries from 652 B.C. to 262 B.C.

Schoch, C. (1928). Astronomical and Calendrical Tables, in Langdon, S. and Fotheringham, J. K., *The Venus Tablet of Ammizaduga*, 94–109 and Tables I–XVI.

Thureau-Dangin, F. (1938). *Textes Mathématiques Babyloniens.*

van der Waerden, B. L. (1943). Die Berechnung der ersten und letzten Sichtbarkeit von Mond und Planeten und die Venustafeln des Ammiṣaduqa. *Ber. d. Math.-Phys. Kl. d. Sächs. Ak. d. Wiss. zu Leipzig* 94 (1943), 23–56.

Walker, C. B. F. (1983). Episodes in the History of Babylonian Astronomy. *Bulletin of the Society for Mesopotamian Studies* (Toronto) 5, 10–26.

4

Non-mathematical Astronomical Texts and Their Relationships
Hermann Hunger

One of the unsolved problems of the astronomical cuneiform texts of the late first millennium B.C. is the relation between the computational texts, edited mostly in Neugebauer's book ACT, and the others, called non-mathematical astronomical texts for lack of a better term; furthermore, the relations between the different types of these non-mathematical texts remain to be clarified. Since I am engaged in editing these texts, it seemed to me appropriate that I should try to get an idea if and how they depend on each other.

In an article on classification of these texts, Sachs (1948) did address the problem of relations among them only in general terms; he had to base his article on only 37 tablets. In spite of the fact that far more material became available later, his classification remained valid for the new material. To provide a background for further considerations, let me present a description of each group, as it can be seen now.

1. *Diaries* (Akkadian *naṣāru ša ginê*, "regular watching"): After Sachs had worked on these texts for a long time, they are now in the process of being edited (Sachs and Hunger 1988, 1989, 1996). You find in them: Lunar Six (see table 1); planetary phases, like first and last visibility, what Neugebauer called Greek-Letter phenomena (see table 2); monthly summaries of planetary positions; conjunctions between planets and the so-called Normal Stars (Sachs and Hunger 1988, 17ff.); conjunctions of the moon with Normal Stars; eclipses; solstices and equinoxes; phenomena of Sirius. Toward the end of the 3rd century B.C., Diaries begin to record the dates when a planet moved from one zodiacal sign into another. The rest of the Diaries' contents is non-astronomical.

Almost all of these items are observations. Exceptions are the solstices, equinoxes, and Sirius data, which were computed according to a scheme (HAMA, 357ff.); furthermore, in many instances when Lunar Sixes, lunar or solar eclipses, or planetary phases could not be observed, a date or time is nevertheless given, marked as not observed. Expected passings of Normal Stars by the moon are sometimes recorded as missed

Table 1
Lunar Six

	Name	Definition
1	NA	Time from sunset to moonset on the evening when the moon was visible for the first time after conjunction
2	ŠÚ	Time from moonset to sunrise when the moon set for the last time before sunset
3	NA	Time from sunrise to moonset when the moon set for the first time after sunrise
4	ME	Time from moonrise to sunset when the moon rose for the last time before sunset
5	GE$_6$	Time from sunset to moonrise when the moon rose for the first time after sunset
6	KUR	Date and time from moonrise to sunrise when the moon was visible for the last time before conjunction

Table 2
Greek-letter phenomena

Γ	Reappearance in the east
Φ	Stationary point in the east
Θ	Opposition
Σ	Disappearance in the east
Ξ	Reappearance in the west
Ψ	Stationary point in the west
Ω	Disappearance in the west

because of bad weather, but never is a distance between moon and Normal Star given as computed.

2. *Goal year texts* (Akkadian UD-1-KAM *tāmarātu etēqu u antalû*, literally, "first days, visibilities, passing and eclipses"): contain Lunar Sixes; planetary phases; conjunctions of planets and Normal Stars; eclipses. These data refer to a year which is preceding the "Goal year" by one of the following periods: 71 or 83 years in case of Jupiter, 8 years for Venus, 46 years for Mercury, 59 years for Saturn, 47 or 79 years for Mars, 18 years for the moon. The Goal year texts, like the Diaries, contain mostly observations, as can be seen from remarks about weather conditions which interfered with observation. Like the Diaries, too, Goal year texts substitute computed values for certain missed observations, especially for the moon. This again makes it probable that they were excerpts from Diaries. As emphasized by Sachs (1948), the excerpting must have been a

laborious process, because more than a dozen Diary tablets had to be consulted to produce one Goal year text; also, one had to read through the whole of the relevant Diaries to sift out the data for the planet in question.

3. *Almanacs* (Akkadian *mešḫī ša kašādi ša bibbī*, "measurements of the reaching of the planets"): Lunar Three (which is a group similar to the Lunar Six, viz. a) the length of the preceding month, 29 or 30 days, b) the date in the middle of the month when the moon set for the first time after sunrise, c) the date when the moon was visible for the last time before conjunction with the sun); planetary phases; entries of planets into zodiacal signs; monthly summaries of planetary positions; solstices and equinoxes, Sirius phases. Some expected eclipses are mentioned.

4. *Normal Star Almanacs* (Akkadian *mešḫī* "measurements"): Planetary phases; conjunctions of planets with Normal Stars; sometimes Lunar Three or Lunar Six; solstices and equinoxes, Sirius phases. Some expected eclipses are mentioned.

5. *Planetary observations*: no common format can be described for these compilations. In some of them the observations are arranged in groups according to the periods appropriate for the planet. When complete, such tables could have served as basic material for the construction of planetary tables of the kind contained in ACT. Aaboe (1980) has shown a possible way of doing this. A collection of Jupiter observations (LBAT 1394(+)1395(+)1399+1400) explicitly mentions Diaries written on tablets and wooden boards as its source. One tablet (LBAT 1374+1375+1376) combines data for Mars from years 50 to 52 of the Seleucid era with those of Mercury from the years 83 to 85; both are by one period, that is, 79 or 46 years, earlier than 129 to 131 S.E.; this is therefore the same arrangement as is found in Goal year texts.

Similar texts are also preserved for lunar phenomena. Especially eclipse collections have attracted attention, and several of these were the subject of recently published articles. But there are also lists of Lunar Sixes, which possibly were used in constructing lunar ephemerides.

Looking at the tabulation of these contents we see that in principle everything in Goal year texts, Almanacs and Normal Star Almanacs could have been excerpted from Diaries. Even the entries of planets into zodiacal signs are attested in Diaries at an earlier time than in Almanacs, although the older Diaries do not yet have them. It is conceivable that the zodiacal entries were added to the Diaries because they were needed for the Almanacs; of course this cannot be proven, but Rochberg (Rochberg-Halton 1989:108ff.) has argued that horoscopes were one of the uses of

Table 3

	Diaries	Goal year texts	N.-S. almanacs	Almanacs
Lunar Six	x	x	(x)	
Lunar Three			(x)	x
Planetary phenomena	x	x	x	x
Entrances into zod. signs	(x)			x
Summary of planets' positions	x			x
Solstices, equinoxes, Sirius	x		x	x
Eclipses	x	x	(x)	(x)
Planets and Normal stars	x	x	x	
Moon and Normal stars	x			

the Almanacs; this in turn would then be the motive for introducing into the Diaries the date when a planet entered a zodiacal sign.

Otherwise, the Diaries changed little in design although they were compiled over a rather long time. So the purposes for which they were produced must have been clear early and be served reasonably well by the Diaries that came to be in the collection.

In the sources preserved, Goal year texts and Almanacs begin rather late as compared to Diaries. It is tempting to say that it took some time to collect enough material before one could produce the Goal year texts, which could then serve as auxiliaries for the predictive texts. We may, however, simply not be in the possession of enough texts. After all, 300 years is a long time to wait for results. Which will bring about a lot of questions: how was such an observation program kept going? who financed it? who could hope to use it?

We know very little about this aspect of the texts. There are two pieces of evidence from the 2d century B.C. (CT 49 144 and BOR 4 132ff., see McEwan 1981:17ff.), that is, from a rather late period. Both are official documents of the temple council of the Marduk temple in Babylon. One confirms the position of a certain man as a "scribe of Enūma Anu Enlil" (Enūma Anu Enlil is the traditional collection of celestial omens) in succession of his father; he is to receive a yearly amount of 1 mina of silver, in addition to some unspecified arable land. For some reason not mentioned, this land had in the time after the father did his work been used by someone else, who now agrees to return it to the newly appointed son. The duties of the son include making observations (*naṣāru*), preparing tables (*tersītu*) and Normal Star Almanacs (at least the same word, *mešḫī*, is used), and deliver them, working together with 5 other named "scribes of Enūma Anu Enlil." The second document is similar in character, except that it concerns two men who also are to succeed their father. "Scribes of Enūma Anu Enlil" are also mentioned in the rather fragmentary piece CT 49 186. As noted by Rochberg (1993:42ff.), some of the "scribes of Enūma Anu Enlil" belonged to the priestly staff of the temples, as can be seen from their titles. They wrote not only what we call astronomical texts, but omens as well. Both kinds of texts were part of the same scribal tradition. It is still largely unknown to us in what way the astronomers served the needs of the temples which employed them. In the second century B.C. the kings, be they Seleucid or Parthian, did not reside in Babylonia and did not use the services of "scribes of Enūma Anu Enlil." In earlier times the situation had been different. In the royal archives of Nineveh we have the correspondence of scribes who were engaged in observations for the purpose of detecting and interpreting omens. These people, as described by Oppenheim (1969), depended on the king for their livelihood, and they sometimes speak of their material needs in their letters. It is obvious that they were employed for a service that was of vital importance for the king and the whole country (Parpola 1983, Hunger 1992). No Diaries have been found in Nineveh, but Parpola (1983, 92) has drawn attention to a fragment which may have been something similar. In any case, the Assyrian scribes and their Babylonian colleagues were clearly interested in the astronomical part of their studies, as can be seen from the existence and copying of Mul-Apin (Hunger and Pingree, 1989), and from other astronomical texts found in the Assyrian cities (e.g., Pingree and Reiner, 1975). After the end of the Assyrian empire, the kings of the Neo-Babylonian dynasty should have had the same interest in an expert handling of astral omens, and therefore continued to need the services of "scribes of Enūma Anu Enlil". In this way royal support may have been available to them. Whatever their fates were during the Achaemenid period, they made great progress in analysing and computing astronomical phenomena. We can guess that it was then that the temples began to provide an income for the astronomers. Beaulieu and Britton (1994) have found indications that lunar eclipse predictions were made in 6th century Uruk so that the temple administration could have the appropriate rituals performed to avert the evil announced by the eclipse.

While we have a large collection of nonmathematical astronomical texts, not all of it needs to be from the same ancient archive. It is not certain that all these texts formed a coherent and meaningfully correlated whole. Some seem to be related to each other, but that does not explain every detail of the arrangements. Still one more problem is the relation, if there was any, to the computed astronomical texts, which come from the same source and should again be part of the same archive, if there was such a thing.

As I mentioned before, a preliminary evaluation of nonmathematical astronomical texts was given by Sachs (1948 and 1974). While there is little that I could add to his description, I have tried to find passages in the texts that deal with the same celestial events, whether observed or predicted. Now contrary to what one might expect from an archive of ca. 1500 tablets, such passages are not available in abundance.

To give you an idea how much was originally contained in that archive, and how much is still preserved, I made a few rough estimates. From well preserved Diaries, I found that in each month about 15 lunar and 5 planetary positions, both in relation to Normal Stars, are reported. Also, every month the so-called lunar Six are recorded. Each year will in addition contain 3 Sirius phases, 2 solstices and 2 equinoxes, at least 4 eclipse possibilities or eclipses, and about 25 planetary phases. Together, this results in about 350 astronomical observations per year. In 600 years, 210,000 observations are accumulated. Now I do not know whether the archive was ever complete to this extent. Sometimes copies of older Diaries indicate that things were missing in the original. But on the whole, this is the order of magnitude. By counting the number of reasonably (i.e., not completely, but more than half) preserved months, I arrived at ca. 400 months preserved in dated Diaries (undated fragments do not help for the purposes of this lecture). If we compare this to a duration of 600 years for the archive, we see that we have preserved ca. 5% of the months in Diaries. Now if the Goal year texts and Almanacs are preserved to a similar extent, you can imagine that rather few instances will be available where a diary and a Goal year text or almanac refer to the same observed event. The following remarks are based on only 425 cases, many of which are partly broken. The results can therefore only be preliminary. I have not been able to check all the originals; so far only the dated Diaries could be collated.

I compared data for the same events among the following groups:

A1. Diaries to Goal year texts (93 cases);

A2. Diaries to Almanacs and Normal Star Almanacs (75 cases);

A3. Goal year texts to Almanacs and Normal Star Almanacs (96 cases).

I also compared data in Goal year texts that were earlier by one period (in the sense in which such periods are used in the Goal year texts) than the corresponding data in:

B1. Diaries (17 cases);

B2. Almanacs (33 cases);

B3. Normal Star Almanacs (111 cases).

Depending on what groups are compared, different conjectures can be made from agreement or disagreement about data. I proceed according to the list just given.

A1. Since it seems that Goal year texts were excerpts from Diaries, data for the same event in a Goal year text should be identical to those in a Diary. Following the example of Sachs, I sometimes even have restored broken passages in the edition of the Diaries (Sachs and Hunger 1988 and 1989) from a corresponding Goal year text. After comparing about 90 cases, I am not so sure about the certainty of restoring anymore. While most of the passages agreed, in 20 cases there were differences. The easiest explanation is that these Goal year texts were excerpted from other Diaries than the ones which happen to be preserved. Similarly, sometimes variant data are attested in parallel Diaries, and they are probably due to different observers. Such differences are to be expected if two people independently observe the same phenomenon.

A2. While Diaries and Goal year texts seem to contain mostly observations, Almanacs and Normal Star Almanacs clearly are predictive texts. It is not very surprising then that only about half of the events available in parallel show the same calendar date and/or position.

A3. When comparing Goal year texts on the one hand and Almanacs or Normal Star Almanacs on the other, the majority of parallel passages shows deviations. It is not clear to me why Almanacs and Normal Star Almanacs should deviate from Goal year texts more frequently than from Diaries, but this is how it looks on the basis of our admittedly rather restricted evidence.

Comparing data in any text with data which are earlier by one period as collected in the Goal year texts is of course another matter. It presupposes the applicability of these periods. In order to have an idea of their usefulness, I compared them to data computed by modern theories expressed in the Babylonian calendar. The resulting average corrections to be applied to the Goal year text periods are:

Table 4

Planet	Period in years	Δt in days
Jupiter	71	0
Venus	8	−4
Mercury	46	−1
Saturn	59	−6
Mars	47	ca. +12
	79	ca. +4

Some of these values agree with those attested in astronomical procedure texts (Neugebauer and Sachs 1971:206f.).

B1. The data from Diaries and Goal year texts, being both essentially observations, show in general the same differences as those derived from modern computations in table 2.

B2 and B3. This view tries to evaluate the quality of the predictions contained in Almanacs and Normal Star Almanacs. I arranged the data according to planets, and there were really few for each planet.

DIFFERENCES: GOAL YEAR TEXTS VS. ALMANACS AND NORMAL STAR ALMANACS

This table compares data collected in Goal year texts (i.e., data earlier than the goal year by one period) with data for the goal year found in Almanacs and Normal Star Almanacs. In the column "Planet," numbers with Jupiter and Mars indicate which of the two periods used for these planets is applicable. "comp." means that the entry in the Goal year text is labeled as not observed.

Planet	Goal year text LBAT	Δt (in days)	(Normal Star) Almanac LBAT
Jupiter 71	1339:1	−1	1059:15
Jupiter 71 (comp.)	1285:1	+1	1057:2
Jupiter 71	1285:1f.	0	1057:14
Jupiter 71	1280:2f.	0	1049f. r.7
Jupiter 71 (comp.)	1275:3	0	1047:13
Jupiter 71 (comp.)	1296:2	+1	1174:4
Jupiter 71	1339:1	−1	1151:7
Jupiter 71 (comp.)	1263:2	0	1129 r.6
Venus	1285:8	−2	1057:5f.

Venus	1285:9	−1	1057:9
Venus	1285:10	−1	1057:13
Venus	1285:10	−1	1057:13f.
Venus	1285:11	−1	1057:15f.
Venus	1285:12	−1	1057:18
Venus	1285:13	−1	1057:20
Venus	1284:8	−2	1056:5
Venus	1284:10	−2	1056 r.5
Venus	1284:22	−2	1056 r.7f.
Venus	1308:10	−4	1051 r.20
Venus	1275f.:19	−2	1048:7
Venus	1225:8	−1	1008:4
Venus	1225:18	−2	1008 r.7
Venus	1225:19	2	1008 r.9
Venus (comp.)	1296f.:12	−5	1174:3
Venus (comp.)	1296f.:12f.	−2	1174:5
Venus (comp.)	1296f.:13	−2	1174:7
Venus (comp.)	1296f.:13	−3	1174:9
Venus (comp.)	1296f.:14	−2	1174:11
Venus (comp.)	1296f.:16	−2	1174:14
Venus (comp.)	1296f.:18	0	1174 r.5
Venus (comp.)	1296f.:19	−4	1174 r.7
Mercury	1288	−8	1059:2
Mercury	1288	+4	1059:10
Mercury	1288	+5	1059:20
Mercury	1288	0	1059:35
Mercury	1288	+1	1059:30
Mercury (comp.)	1288	+6	1059:54
Mercury	1285:22	0	1057:7f.
Mercury	1285:22f.	0	1057:8
Mercury	1285:24	−3	1057:18
Mercury	1285:25	0	1057:21
Mercury	1285:28	−2	1057 r.5
Mercury	1285:29	−2	1057 r.8
Mercury	1285:30	−1	1057 r.8f.
Mercury	1285:31	−1	1057 r.9f.
Mercury (comp.)	1285:31f.	−3	1057 r.14

Mercury	1284:15	−5	1056 r.4
Mercury	1284:16	−1	1056 r.8f.
Mercury (comp.)	1296f.:20	+8	1174:3
Mercury (comp.)	1296f.:20f.	0	1174:5
Mercury	1296f.:21	−4	1174:6
Mercury (comp.)	1296f.:22	0	1174:9
Mercury	1296f.:22	−3	1174:11
Mercury (comp.)	1296f.:23	0	1174 r.1
Mercury (comp.)	1296f.:23f.	0	1174 r.2
Mercury (comp.)	1296f.:24	0	1174 r.3
Mercury (comp.)	1296f.:24	+10	1174 r.6
Mercury	1296f.:25	−1	1174 r.8
Mercury (comp.)	1296f.:25	−2	1174 r.8
Mercury (comp.)	1296f.:26	+10	1174 r.9
Saturn	1320:3	−4	1059:27
Saturn	1285:35	−8	1057:19
Saturn (comp.)	1285 e. 1	−5	1057 r.5
Saturn	1285 e. 2	−1	1057 r. 13
Saturn	1284:22	−5	1056 r.7
Saturn (comp.)	1236:5	−5	1016 r.3
Saturn	1300:17	−11	1179f.
Saturn (comp.)	1296f.:27	−7	1174:1
Saturn	1296f.:27	−7	1174:8
Saturn (comp.)	1296f.:28	−13	1174:14
Mars 47	1339 r.9	+16	1059:61
Mars 47	1339 r.10	+16	1059:68
Mars 47	1285 e. 4f.	+17	1057:17
Mars 47	1285 r.2	+16	1057 r.2
Mars 47	1229:4	+16	1011:1
Mars 47	1229:5	+17	1011:2
Mars 47	1225 r. 7	+16	1008:8
Mars 47	1225 r. 7f.	+16	1008:14
Mars 79 (comp.)	1285 e. 3	0	1057 r. 12
Mars 79 (comp.)	1300:19	+9	1179f.
Mars 79	1300:19f.	+8	1179f.
Mars 79 (comp.)	1296f.:38	+4	1174 r.3
Mars 79	1296f. r. 13	+6	1174 r.5f.

On the whole, the differences agree with the expected values. Especially for Jupiter (8 cases) and Saturn (10 cases) this is visible. Mercury shows the expected results in many of the ca. 30 cases, but there is quite a number of widely varying differences between data for its phenomena. Differences for Venus are generally smaller than expected. Data for Mars are few (13 cases); those for the 79-year period fit, those for the 47-year period are a little too high. No agreement at all is found between data for the Lunar Six, but this is not surprising: the so-called "Saros" period of 18 years is more precisely 6585 $\frac{1}{3}$ days, and the fraction of a day makes the return of the values for the Lunar Six impossible. I gave up an attempt to compute the Lunar Six by Babylonian methods from the ACT procedure texts; too many assumptions had to be made so that the results could not be meaningfully compared to Lunar Six given as predicted in the nonmathematical astronomical texts.

In conclusion, then, the predictions of the Almanacs and Normal Star Almanacs are certainly not simply values taken over from Goal year texts. If they were derived from them at all, they were modified. However, our material is far too small to determine how these modifications were made. It seems that in many cases they were made in the right direction, but a quantitative description is not possible at present.

BIBLIOGRAPHY

Aaboe, A. 1980. Observation and Theory in Babylonian Astronomy: Centaurus 24, 14–35.

ACT. O. Neugebauer, Astronomical Cuneiform Texts. London 1955.

Beaulieu, P.-A., and Britton, J. P. 1994. Rituals for an Eclipse Possibility in the 8th Year of Cyrus: JCS 46, 73–86.

BOR. The Babylonian and Oriental Record. London.

CT. Cuneiform Texts from Babylonian Tablets etc. London 1896ff.

Hunger, H. 1992. Astrological Reports to Assyrian Kings. State Archives of Assyria, vol. 8. Helsinki.

Hunger, H., and Pingree, D. 1989. MUL.APIN, An Astronomical Compendium in Cuneiform. Horn.

McEwan, G. J. P. 1981. Priest and Temple in Hellenistic Babylonia. Wiesbaden, Franz Steiner.

Neugebauer, O., and Sachs, A. J. 1967. Some Atypical Astronomical Cuneiform Texts, I: JCS 21, 183–218.

Oppenheim, A. L. 1969. Celestial Divination in the Neo-Assyrian Empire: Centaurus 14, 97–135.

Parpola, S. 1983. Letters from Assyrian Scholars to the Kings Esarhaddon and Assurbanipal. Part II: Commentary. Kevelaer and Neukirchen-Vluyn.

Pingree, D., and Reiner, E. 1975. Observational Texts Concerning the Planet Mercury: Revue d'Assyriologie 69, 175–180.

Rochberg, F. 1993. The Cultural Locus of Astronomy in Late Babylonia: Die Rolle der Astronomie in den Kulturen Mesopotamiens (H. D. Galter, ed.), 31–45.

Rochberg-Halton, F. 1989. Babylonian Horoscopes and Their Sources: Orientalia 58, 102–123.

Sachs, A. J. 1948. A Classification of the Babylonian Astronomical Tablets of the Seleucid period: JCS 2, 271–290.

Sachs, A. J. 1974. Babylonian Observational Astronomy: The Place of Astronomy in the Ancient World (F. R. Hodgson, ed.), 43–50.

Sachs, A. J., and Hunger, H. 1988, 1989, 1996. Astronomical Diaries and Related Texts from Babylonia, vols. I–III. Vienna.

APPENDIX

While the datable Diaries are available in a recent edition, Almanacs and Goal year texts will be edited only in several years. I therefore present here a specimen of each category in translation. Babylonian month names are represented by Roman numerals.

LBAT 1251 + 1252: Goal Year Text for Year 140 of the Seleucid Era

Obverse

1 [Year 69, Seleucus] king. II 3, Jupiter's first appearance in Aries; it was small, rising of Jupiter to sunrise: 12° 30′.

2 [....] stationary point. VIII 3, Jupiter's acronychal rising. X, until the 1st, Jupiter's [stationary point to the west ....]

3 [Year 56,] intercalary month VI. Year 57, Antiochus king. III, night of the 15th, last part of the night, Jupiter [....]

4 [....] while moving back to the west, was $2\frac{2}{3}$ cubits below η Tauri; oral information. XI, night of the 21st, [....]

5 [Year 1]32, Seleucus king. I 15, Venus's last appearance in the east in Aries: from the 14th on when I watched I did not see it. IV [....]

6 [....] (ideal) first appearance on the 1st. VI, night of the 7th, first part of the night, Venus was 1 cubit above α Virginis. Night of the 25th, first part [of the night, ...]

7 [Night of the 2]9th, first part of the night, Venus was $3\frac{1}{2}$ cubits below β Librae. VI$_2$, night of the 10th, first part of the night, Venus was [....] above δ Scorpii,

8 having passed 4 fingers to the east. Night of the 16th, first part of the night, Venus was 2 cubits above α Scorpii. Night of the 25th, first part of the night, Venus was [....] above θ [Ophiuchi.]

9 VIII, night of the 1st, first part of the night, Venus was $2\frac{1}{2}$ cubits below β Capricorni. Night of the 15th, first part of the night, Venus was [....] above γ Capricorni.

10 Night of the 18th, first part of the night, Venus was 1 finger above δ Capricorni, Venus having passed 6 fingers to the east. XI 23, 11° [sunset to setting of Venus; ....]

11 Venus's last appearance in the west in the end of Pisces. The 30th, Venus's first appearance in the east in Pisces; it was bright (and) high, rising of Venus to sunrise: 8°, (ideal) first appearance on the 29th.

12 Year 94, Antiochus king. I 16, Mercury's last appearance in the east in Aries; clouds, I did not watch. II 19, Mercury's [first appearance] in the west in [....]

13 sunset to setting of Mercury: 16°, (ideal) first appearance on the 17th. III, night of the 1st, first part of the night, Mercury was $3\frac{1}{2}$ cubits below α Geminorum. Night [...., Mercury was]

14 $2\frac{1}{2}$ cubits below β Geminorum. Night of the 17th, first part of the night, Mercury was 2 fingers above δ Cancri. IV [....]

15 The 26th, Mercury's first appearance in the east in Cancer; rising of Mercury to sunrise: 15°, (ideal) first appearance on the 25th. V 12, rising of Mercury to sunrise: 16°. [.... Mercury's]

16 last appearance in the east in the beginning of Leo. VI$_2$ 9, Mercury's first appearance in the west ...., omitted. The 23rd, Mercury's last appearance in the west ...., [omitted.]

17 VII 12, Mercury's first appearance in the east in Libra, $1\frac{1}{2}$ cubits behind α Librae to the east; rising of Mercury to sunrise: 17°, (ideal) first appearance on the 10th. [....]

18 Mercury was $2\frac{2}{3}$ cubits below β Librae. VIII, night of the 1st, last part of the night, Mercury was 10 fingers above β Scorpii. Ni[ght ....,]

19 Mercury was $3\frac{1}{2}$ cubits above α Scorpii. The 21st, Mercury's last appearance in the east in Sagittarius; I did not watch. IX 27, Mercury's first appearance in the west in Aquarius [....]

20 X 15, sunset to setting of Mercury: 18°. Around the 18th, Mercury's last appearance in the west in Aquarius. XI 7, Mercury's first appearance in the east in Aquarius, 2 cubits [. . . .]

21 having passed $\frac{2}{3}$ cubit to the east; rising of Mercury to sunrise: 14° 30′. XII 7, rising of Mercury to sunrise: 14°. Around the 10th, Mercury's [last appearance] in the east in Pisces.

22 Year 80, XII$_2$, until the 25th, when Saturn became stationary to the west, it became stationary $1\frac{1}{2}$ cubits in front of ϑ Leonis. Year 81, [. . . .]

23 III, night of the 22nd, first part of the night, Saturn, while moving back to the east, was 4 cubits below ϑ Leonis. IV 15, sunset to setting of Saturn: 16°. Around the 18th, [Saturn's last appearance in . . . .]

24 V 25, Saturn's first appearance in the end of Leo; rising of Saturn to sunrise: 16° 30′, (ideal) first appearance on the 23rd. VI, night of the 20th, last part of the night, Saturn was [. . . .] above β Virginis.

25 IX, until the 19th, when Saturn became stationary to the east, it became stationary 2 cubits 8 fingers behind β Virginis. XI 16, Saturn's acronychal rising [. . . .]

26 Night of the 29th, first part of the night, Saturn, while moving back to the west, was 14 fingers above β Virginis.

27 Year 61, Antiochus king. VII, until the 13th, when Mars became stationary to the east, it became stationary $\frac{2}{3}$ cubit above η [Geminorum, being . . . .]

28 back to the west. VIII 26, Mars's acronychal rising. IX 27, Mars became stationary in Taurus [. . . .]

29 [Year] 93, Antiochus king. II, night of the 17th, last part of the night, Mars was $3\frac{1}{2}$ cubits below η Piscium [. . . .]

30 4 cubits below β Arietis. III, night of the 6th, last part of the night, Mars was 4 [cubits] below α Arietis [. . . .]

31 $2\frac{2}{3}$ cubits below η Tauri. Night of the 26th, last part of the night, Mars was [. . . .]

32 $2\frac{1}{2}$ cubits below β Tauri. Night of the 19th, last part of the night, Mars [. . . .]

33 while moving back to the west, was $1\frac{1}{2}$ cubits above ζ Tauri . . . . [. . . .]

34 X, night of the 21st, first part of the night, Mars, while moving back to the east, [. . . .]

35 2 cubits above ζ Tauri. Night of the 20 + [xth, ...]

36 1 cubit 8 fingers [. . . .] μ Geminorum(?) [. . . .]

Edge

1 α Geminorum .... [....]

Reverse

Col. I

1   Year 121, Antiochus king.

2   VII   moonset to sunrise plus sunrise to moonset: 15°

3   moonrise to sunset plus sunset to moonrise: 6°

4   VIII moonset to sunrise plus sunrise to moonset: 13°

5   moonrise to sunset plus sunset to moonrise: 8° 40′

6   [IX] moonset to sunrise plus sunrise to moonset: [....]

7   moonrise to sunset plus sunset to moonrise: [....]

8   [X   moonset to sunrise plus sunrise to moonset: ....]

9   moonrise to sunset plus sunset to moonrise: [....]

10 XI   moonset to sunrise plus sunrise to moonset: ....

11   moonrise to sunset plus sunset to moonrise: 14° 10′

12 XII moonset to sunrise plus sunrise to moonset: 6° 30′

13   moonrise to sunset plus sunset to moonrise: 13° 20′

14 Year 122, Antiochus king.

15 V, night of the 14th, moonrise to sunset: 1°, measured (despite) clouds.

16 When the moon came out from a cloud, 2 fingers on the [....] side

17 remained to clearing. At 1 bēru

18 before sunset. The 28th, [solar] eclipse;

19 when I watched I did not see it. At 28°

20 before sunset. X 28,

21 solar eclipse, (after) 5 months, ... omitted.

22 At 78° before sunset.

23 XI, night of the 14th, moonrise to sunset: 8°, measured (despite) mist.

24 Lunar eclipse. When it began on the east side,

25 in 15° of night all was covered.

26 It set eclipsed. At 34° before sunrise.

Col. II

1   Year 121, An[tiochus king.]

2   XII$_2$ the 1st (of which followed the 30th of the preceding month), sunset to moonset: 21°; it was bright, earthshine,

3 it was low to the sun.

4 14th moonset to sunrise: 5° 30′, measured

5 15th moonrise to sunset: [. . . .], measured

6 15th sunrise to moonset: [. . . .], measured

7 16th sunset to moonrise: 12°, measured

8 27th moonrise to sunrise: 16° 40′; clouds, I did not watch

9 Year 122, Antiochus king.

10 I (the 1st of which was identical to) the 30th (of the preceding month); sunset to moonset: 14° 10′; clouds, I did not watch.

11 15th moonrise to sunset: 11°, measured

12 15th moonset to sunrise: 4° 40′, measured

13 16th sunset to moonrise: 6° 40′, measured

14 16th sunrise to moonset: 5° 30′, measured

15 28th moonrise to sunrise: 9° 40′; clouds, I did not watch

16 [II] the 1st (of which followed the 30th of the preceding month), sunset to moonset: 21°; it was bright, measured (despite) clouds; it was high to the sun

17 [. . . .] moonrise to sunset: 17° 40′

18 [. . . .] moonset to sunrise: . . . ., measured

19 15th sunset to moonrise: 1° 30′, measured

20 15th sunrise to moonset: 2°, measured

21 28th moonrise to sunrise: 9°; mist, when I watched I did not see it

22 III (the 1st of which was identical with) the 30th (of the preceding month), sunset to moonset: 13°; it was faint, measured; it was high to the sun

23 15th moonrise to sunset: 2° 30′, measured

24 15th moonset to sunrise: 1°, measured (despite) clouds

25 16th sunset to moonrise: 13°, measured

26 16th sunrise to moonset: 15°, measured

27 28th moonrise to sunrise: 13°, measured

28 IV the 1st (of which followed the 30th of the preceding month), sunset to moonset: 15°; it was bright, measured.

29 14th [. . . .] . . . ., measured

30 [. . . ., measu]red

31 [. . . ., measu]red

Col. III

1   V   the 1st (of which followed the 30th of the preceding month), sunset to moonset: .... [....]

2   13th [....] 6°

3   14th moonrise to sunset: 1° [....]

4   14th sunrise to moonset: 9° 30', clouds [....]

5   15th sunset to moonrise: 7°, measured (despite) clouds

6   27th moonrise to sunrise: 16°, measured

7   VI   the 1st (of which followed the 30th of the preceding month), sunset to moonset: 11°, measured; it was low to the sun

8   12th moonset to sunrise: 10° 40', measured (despite) clouds

9   13th moonrise to sunset: 8° 50', measured (despite) clouds

10 13th sunrise to moonset: 2° 30', measured

11 14th sunset to moonrise: 4° 40', measured (despite) mist

12 27th moonrise to sunrise: 13°, measured (despite) clouds

13 VII the 1st (of which followed the 30th of the preceding month), sunset to moonset: 12°; it was bright, measured; it was low to the sun

14 12th moonset to sunrise: 2° 30', measured (despite) clouds

15 13th moonrise to sunset: 2° 30', measured (despite) mist

16 13th sunrise to moonset: 13°, measured

17 14th sunset to moonrise: 4°, measured (despite) mist

18 26th moonrise to sunrise: 22°, measured (despite) clouds

19 VIII (the 1st of which was identical with) the 30th (of the preceding month), sunset to moonset: 8° 50'; dense mist, I did not see the moon

20 12th moonset to sunrise: 8° 30', measured (despite) clouds

21 13th sunrise to moonset: 6°, measured (despite) clouds

22 14th moonrise to sunset: 40', measured (despite) clouds

23 15th sunset to moonrise: 9°, measured (despite) clouds

24 27th moonrise to sunrise: 13° 30', measured (despite) clouds

Col. IV

1'   .... [....]

2'   X   (the 1st of which was identical with) the 30th (of the preceding month), sunset to moonset: ....

3'   it could be seen while (the sun) stood there

4'   12th [....] 7°

5'   13th sunrise to moonset(?): 3° [....]

6'   13th moonrise to sunset(?): 30' [....]

7'   14th sunset to moonrise: 15° [....]

8'   27th moonrise to sunrise: 12° [....]

9'   XI   (the 1st of which was identical with) the 30th (of the preceding
month), sunset to moonset: 14°; it was bright [....]

10' 13th [....] 4° 30'

11' 14th moonrise to sunset: 8° [....]

12' 14th [....] 2°

13' 15th [....] 5°

14' 27th [moonrise to sunrise:] 17°

15' XII   (the 1st of which was identical with) the 30th (of the preceding
month), [....]

16' 14th [....]

17' 15th [....]

    (Rest broken)

LBAT 1174 (Restored from Duplicates): Almanac for Seleucid Era 236

Edge

1 At the command of Bel and Beltiya may it go well.

Obverse

Month I, the 1st (of which followed the 30th of the preceding month). Jupiter in Taurus, Saturn in Scorpius, Mars in Pisces. The 15th, first sunrise before moonset. The 19th, Jupiter's last appearance in Taurus. The 26th, Mars reaches Aries. The 27th, Saturn's acronychal rising. The 27th, last visibility of the moon. The 29th, first appearance of Pleiades in the east.

Month II, (the 1st of which is identical with) the 30th (of the preceding month). Saturn in Scorpius, Mars in Aries. The 2nd, Sirius's last appearance. The 6th, Venus's first appearance in the west in Gemini. The 6th, Mercury's first appearance in the east ...., omitted. The 15th, first sunrise before moonset. The 20th, Jupiter's first appearance in the end of Taurus. The 20th, Mercury's last appearance in the east, omitted. The 26th, Venus reaches Cancer. The 28th, last visibility of the moon. The 29th, Jupiter reaches Gemini.

Month III, the 1st (of which follows the 30th of the preceding month). Jupiter in Gemini, Venus in Cancer, Saturn in Scorpius, Mars in Aries. The

9th, Mars reaches Taurus. The 14th, first sunrise before moonset. The 16th, solstice. The 15th, Mercury's first appearance in the west in Cancer. The 21st, Venus reaches Leo. The 21st, Mercury reaches Leo. The 27th, last visibility of the moon. Night of the 29th, solar eclipse, which is omitted.

Month IV, the 1st (of which follows the 30th of the preceding month). Jupiter in Gemini, Venus and Mercury in Leo, Saturn in Scorpius, Mars in Taurus. The 1st, Saturn stationary in Scorpius. The 7th, Sirius' first appearance. Night of the 13th, 8° before sunrise, lunar eclipse, it makes one-fourth of the disk; it sets eclipsed. The 13th, first sunrise before moonset. The 15th, Venus reaches Virgo. The 23rd, Mercury's last appearance in the west in the end of Leo. The 25th, Mars reaches Gemini. The 27th, last visibility of the moon.

Month V, the 1st (of which follows the 30th of the preceding month). Jupiter and Mars in Gemini, Venus in Virgo, Saturn in Scorpius. The 11th, Venus reaches Libra. The 13th, first sunrise before moonset. The 21st, Mercury's first appearance in the east in Leo. The 26th, last visibility of the moon.

Month VI, (the 1st of which is identical with) the 30th (of the preceding month) .... Jupiter and Mars in Gemini, Venus in Libra, Mercury in Leo, Saturn in Scorpius. The 3rd, Mercury reaches Virgo. The 8th, Venus reaches Scorpius. The 12th, Mercury's last appearance in the east in Virgo. The 13th, first sunrise before moonset. The 19th, equinox. The 22nd, Jupiter stationary in Gemini. The 28th, last visibility of the moon. The 29th, Mars reaches Cancer.

Month VII, the 1st (of which follows the 30th of the preceding month). Jupiter in Gemini, Venus and Saturn in Scorpius, Mars in Cancer. The 2nd, Venus reaches Sagittarius. The 13th, first sunrise before moonset. The 21st, Saturn's last appearance in Scorpius. The 25th, Mars stationary in Cancer. The 27th, last visibility of the moon. The 28th, Venus reaches Capricorn.

Month VIII, (the 1st of which is identical with) the 30th (of the preceding month). Jupiter in Gemini, Venus in Capricorn, Mars in Cancer. The 4th, Pleiades's setting in the morning in the west. The 8th, Mercury's first appearance in the east[sic] in Sagittarius. The 13th, first sunrise before moonset. The 23rd, Mercury's last appearance in the west in Sagittarius. The 24th, Jupiter's acronychal rising. The 25th, Mars reaches Gemini. The 26th, Venus reaches Aquarius. The 27th, last visibility of the moon. The 28th, [Saturn's first appearance in Scorpius.]

Month IX, (the 1st of which is identical with) the 30th (of the preceding month). [Jupiter and] Mars in Gemini, Venus in Aquarius, Saturn in Scorpius. The 4th, Saturn reaches Sagittarius. The 6th, Mars' acronychal rising. The 7th, Mercury's first appearance in the east in Sagittarius. Night of the 14th, lunar eclipse (after) 5 months, ... omitted. The 14th, first sunrise before moonset. The 22nd, solstice. The 27th, Sirius's acronychal rising. The 28th, last visibility of the moon. Night of the 29th, solar eclipse which is omitted.

Month X, the 1st (of which follows the 30th of the preceding month). Jupiter and Mars in Gemini, Venus in Aquarius, Mercury and Saturn in Sagittarius. The 5th, Venus reaches Pisces. The 8th, Mercury reaches Capricorn. The 14th, first sunrise before moonset. The 16th, Mars stationary in Gemini. The 19th, Jupiter stationary in Gemini. The 23rd, Mercury's last appearance in the east in Capricorn. The 24th, Venus reaches Aquarius. The 27th, last visibility of the moon.

Month XI, (the 1st of which is identical with) the 30th (of the preceding month). Jupiter and Mars in Gemini, Venus in Aquarius, Saturn in Sagittarius. The 6th, Venus's last appearance in the west in Aquarius. The 8th, Venus's first appearance in the east in Aquarius. The 15th, first sunrise before moonset. The 23rd, Mercury's first appearance in the west in Pisces. The 27th, last visibility of the moon. The 28th, Mercury reaches Aries.

Month XII, (the 1st of which is identical with) the 30th (of the preceding month). Jupiter and Mars in Gemini, Venus in Aquarius, Mercury in Aries, Saturn in Sagittarius. The 14th, Mars reaches Cancer. The 16th, first sunrise before moonset. The 27th, Mercury's last appearance in the west in Aries. The 25th, equinox. The 28th, last visibility of the moon. The 29th, Saturn stationary in Sagittarius. The 30th, [Venus] reaches Pisces.

Month I, the 1st (of which follows the 30th of the preceding month). Measurements of the reachings of the planets of year 172, which is year 236, Arsaces king and Ispubarza, his sister, queen.

5

NORMAL STAR OBSERVATIONS IN LATE ASTRONOMICAL BABYLONIAN DIARIES
Gerd Graßhoff

1 INTRODUCTION

Several types of astronomical texts attest the earliest elaborate form of astronomy in the first millennium B.C.E. in Ancient Mesopotamia. Most of the texts were found around Babylon and Uruk, where they were imprinted into wet clay tablets with wooden stylus. The resulting patterns of cuneiform engraving gave the script its name. After the tablets were baked and fined, the texts were well preserved for centuries.

Early astronomy dealt with *events* such as eclipses and the first and last visibility of the moon, planets and stars after (or before) a period of time during which their close proximity to the sun prevented them from being seen. Babylonian theories accurately predicted these events. Their underlying parameters are of such good quality that they merged into Greek astronomy and continued to be the backbone of astronomical knowledge up to the time of Kepler. Although we have a clear picture of the precision and predictive scope of Babylonian astronomical theories, their empirical basis has remained unclear. It is neither established which types of observations provided the information to construe those theories, nor on which documents one might find such information.

2 ASTRONOMICAL DIARIES

The recently published *Astronomical Diaries* is a collection of astronomical cuneiform texts in Akkadian, the language used by Babylonian astronomy during the first millennium. These texts report a variety of astronomical events over a certain period including eclipses and the visibility phenomena of the moon and planets.[1] Yet the majority of the reports describe topographical relations between two celestial objects of the type "Object O_1 is [above/below/in front of/behind] object O_2 at a distance of L cubits (KÙŠ)." Since those reports are by far the majority in the *Diaries*, they also should describe observations, if the texts are to be taken as observational diaries at all.

Any interpretation of the topographical statements starts with many uncertain assumptions: it begins with the usual errors introduced by the ancient scribe or the contemporary editor. Furthermore, the usage of the Babylonian lunar calendar involves uncertainties about the beginning of the month. Together with the unknown observational hour of the night plus a unit of measurement which is only approximately known, one quickly obtains so many variables in each interpretation that the finding of a consistent solution seems a hopeless enterprise.

Despite a careful analysis of possibly false assumptions, all attempted reconstructions have failed to provide a consistent interpretation of the *Diaries* as observational reports. Consequently, Otto Neugebauer drew the conclusion that the *Diaries* are not related in a direct way to the Babylonian theoretical astronomy:

> Serious difficulties are caused by the expressions "above" and "below" a Normal Star, always combined with an integer number of cubits, or occasionally with simple fractions of cubits, at greater distances, fingers at smaller intervals. Since modern tables provide us with the ecliptic coordinates of star and planet one can graphically determine the direction from the star to the planet at the date in question.
>
> If "above" and "below" referred to a sidereally fixed direction (e.g. perpendicular to the ecliptic or to the equator) the directions from a given star to any planet above or below it should always coincide. In fact, however, this is not the case.[2]
>
> For example Mars is said to be below θ Leonis in S.E. 114 VII 22 = −197 Oct. 29. The star was at $\lambda = 132.86$ and $\beta = 9.65$, the planet at $\lambda = 135.4$ and $\beta = 1.8$. The line connecting these two points meets the ecliptic at a longitude about $3°$ greater than the orthogonal from the star. Similarly Mars in S.E. 81 VI 28 = −230 Oct. 8 (at $\lambda = 221.4$, $\beta = −0, 6$) is "above" α Scorpii (at $\lambda = 218.74$, $\beta = −4.28$). The line from the star to the planet again points to a longitude about $3°$ greater than the star's longitude. This, and many similar cases, rule out any system of spherical coordinates in fixed relation to the ecliptic or the equator. What remains as a possibility seems to be orthogonality to the horizon but this would involve rules as to the time of the night at which the observation had to be made.[3]

Such inconsistencies lead to drastic consequences for the understanding of the *Diaries*:

To summarize the situation: I think one has to admit that the texts which operate with Normal Stars are farther removed from spherical geometry than even the arithmetical patterns used in the ephemerides for the oblique ascensions and their applications. This being the case it must have been difficult to make full use of the empirical data accumulated in the Normal Stars texts. For example a "conjunction" of a planet and a star on an altitude circle cannot be easily used as information about ecliptic coordinates. The Normal Stars texts and the ephemerides appear to be two unrelated approaches to planetary theory.[4]

It is indeed conceivable that, as in Greece, a sort of astrological literature distinct from scientific texts could have flourished. Such a solution would immediately lead to another dilemma. First of all, the tablets are explicitly observational texts: at the end or on the edges of the tablets the expression for "regular watching" is engraved and there are at least two reports about the professional observers who were employed for that purpose.[5] Secondly, if the *Diaries* do not report observations on the level of accuracy of the theoretical texts, then we have to attest two different approaches to astronomy, producing entirely different traditions even on the observational level. Thirdly, no text would attest the empirical input for the theoretical astronomy. If one wants to avoid these consequences, one has to find an interpretation of the observational reports in the *Diaries* which in accuracy and methodology matches that of the well-documented theoretical astronomy—a goal which has not been achieved so far.

3   SEARCH FOR OBSERVATIONS

The only obvious candidate for observational texts in Babylon—the *Astronomical Diaries*—contain entries of observable astronomical events in chronological order on a nearly daily basis. An example of a transcribed and annotated cuneiform tablet is shown in figure 1.

The apparent difficulties in interpreting the *Diaries* required a completely new approach. It should start from the complete set of Akkadian tokens of the *Diaries*, construe from that in a systematic way all possible astronomical statements one could think of, and evaluate the underlying interpretation hypothesis on the basis of all published tablets.

We transliterated the entire corpus of cuneiform texts up to 165 B.C.E. into a computer readable form. A set of grammar rules transformed the Akkadian expressions into a representation of more than three

No. -361

BM 37073(=80-6-17,818)   Artaxerxes II year 43
Listed as LBAT *187   X [XI XII XII₂]
Photo: Pl. 22
Upper edge
1: EN-NUN *šá gi-né-e šá* TA A[B ....]
2: [*ina a-mat* ᵈEN] *u* ᵈGAŠAN-*iá liš-lim*
Obv.'
1: [.... MU-43]-KÁM ᴵ*Ar-šú šá* ᴵ*Ar-tak-šat-su* LUGAL [....]
2: [....] AN *ana* NIM LAL GE₆ 3 3 AN UTAH   [....]
3: [.... GE₆ 5 SAG] GE₆ *sin ár* GENNA 2 KÙŠ 8 SI *ana* NI[M GUB ....]
4: [....] ⌈KÙŠ⌉ GE₆ 9 SAG GE₆ *sin ár* M[ÚL-BABBAR ....]
5: [.... *sin* nn mm *ana* NI]M ⌈DIB⌉ *ina* ⌈IGI AN 1 KÙŠ⌉ [*ana* ŠÚ GUB ....]
'Rev.
1': [....] *Ár-šú šá* ᴵ⌈*Ár-tak-šat-su* LUGAL⌉ [....]
2': [EN-NUN *šá gi-né-e*] *šá* TA AB EN TIL DIR-ŠE MU-⌈43⌉-[KÁM ....]
3': [....] ŠUᴵᴵ ᴵTIN-*su* -ᵈEN DUMU *šá* ᴵ·ᵈEN-A-[....]

Figure 1
Transcription of an astronomical cuneiform tablet.

thousand astronomical events. The range of possible interpretations is huge, despite the relatively small number of terms involved. In subsequent tests such interpretations can be sorted out which obviously do not refer to any astronomical event.

A new interpretation technique had to be developed, in which models of interpretation were constructed and subsequently refined according to novel search strategies. A typical, structurally simple example of an assumed observational report reads in its English version:

23 year Seleucic Era, 4th day of month 7 , moon above θ Ophiuchi $1\frac{1}{2}$ KÙŠ .

Besides the numerical signs and proper names the sentence contains critical words with unclear meanings marked by a box: (1) the observational time is not fully determined, since the beginning of the month varies with the observation of the first visible moon after new moon, (2) the meaning of the topographical relation is completely pending, and (3) the unit of measure is not determined. The general form of the topographical sentences is

D: At t object O_1 is [under/above/in front of/behind] object O_2 with L KÙŠ.

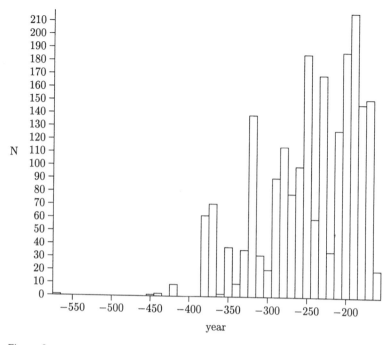

Figure 2
Number of topographical observations per decade.

Such sentences seem to report the relative position of two celestial objects O_1 and O_2 with an (unknown) topographical relation "below," "above," "in front of," or "behind." This relation is followed by a quantitative measure L of a unit KÙŠ. There are more complex variations of this type of sentence, which will be discussed after the simple versions are solved.

If we interpret the first part of the sentence as a reference to the time of observation, we find a distribution (figure 2) of recorded observations over several centuries. The distribution does not necessarily reflect observational activities in Mesopotamia. Nearly all tablets come from Babylon, where they were found accidentally and not by a systematic archaeological excavation.[6] Villagers around the ancient site of Babylon recycled the tablets as tiles, and they certainly found other functions for them as well. Therefore it is difficult to assign these texts to a specific archive with a precise scientific function. On the other hand, since the tablets come from almost one single area, they might be highly selected and not a representative sample of all observational records.

4   TRANSLITERATION AND REPRESENTATION OF TEXTS

4.1   Sentence Form

The first step of interpretation forms sentences from a sequence of Akkadian signs. Akkadian, as it is used for astronomical texts, made use of the originally Sumerian symbolic system. Over the centuries it remained largely unchanged even though major cultural, political and ethnical transformations took place.[7] Especially the transmission from Sumerian to Akkadian introduced a complicated correlation between phonetic and semantic values of signs. For example, one sign can be read equally as *DU* or *RA*. Conversely, for one phoneme one can have different symbolic expressions.

In the case of the *Diaries*, which belong to the later Neo-Babylonian period, the syntactic structures of sentences are not difficult to establish. Each date is followed by a description of an event, which is terminated by the date of the next event. Full dates can be identified by names of the king or the Seleucid period, the month and day of the month, occasionally followed by a qualification for the part of the night at which the event occurred. Subsequent dates omit redundant information relative to the previous sentence. It is also easy to identify the very frequent names of the celestial objects. Even if the names for the celestial objects are not certain in all cases, these expressions define the logical structure of all event descriptions in the *Diaries*: at a given date an event takes place in which one or more celestial objects are involved. In case of the topographical events two other groups of symbols complete the sentences. Expressions of the first group stand between the object names and define a relation in the widest sense, the second contains numbers and an expression, which could be a unit of measurement.

4.2   Interpretations of the Relational Expressions

The relational expressions stand between names of celestial objects. There are mainly four different relational expressions, which in other contexts of language mean "above," "below," "in front of," and "behind." Searching for an adequate interpretation does not require to start with the most plausible assumption. On the contrary, it is advisable to include even the "wildest" hypotheses and try to foster them as far as possible. From the literature many different readings of the topographical relations could be assembled:

Ecliptic coordinates.   Kugler assumed the Zodiac as the basis for the Babylonian coordinate system. "Above" and "below" then describe differences in latitude and "in front of" and "behind" in longitude.[8]

Equatorial coordinates. In this case the relational expressions denote differences in declination and right ascensions.[9]

Horizontal systems. As in older astronomical texts the relations could refer to the horizontal system in various ways.[10]

We have tested all those possibilities and many more, especially in the case of horizontal coordinates. For them, the suggested quantities vary strongly with the assumed time of the night. The arcs between a celestial object and points of the horizon change within a short period of time. The texts, on the other hand, do not accurately specify the time of observation. Hence, if one wants to explore the interpretation of horizontal coordinates, one has to make strong assumptions about the exact moment of measurement, for which there is no direct evidence in the texts. In order to proceed these various paths of interpretation, a huge amount of combinational variations of the used assumptions had to be tested, for which the use of a computer had been essential. It had been even more important to use the computer, because the tests of the interpretation should be comprehensive: each proposal should be tested with all available events published so far. Such a test is far too laborious to be exercised manually.

4.3 Computer Representation

In order to be able to vary all interpretative assumptions, the computer representation of the Babylonian reports starts from the Akkadian script. Hermann Hunger kindly provided a disk with the text of the first two volumes of his edition of the *Diaries*. The codes of these texts were converted into a computer representation: all linguistic tokens are stored in small letters, and accents and indices are represented by suitable codes (or omitted). All linguistic entities—words—are separated by commas. An Akkadian sentence is typically transcribed like the following:

GE_6 4 SAG GE_6 *sin e* MÚL KUR *šá* KIR_4 *šil* PA $1\frac{1}{2}$ KÙŠ.

It describes an event with the moon "sin" and a star of a name starting with "MÚL." The word 'e' stands for an unknown topographic relation between those celestial objects, which is canonically translated as "above." These linguistic tokens are represented in the computer as

ge6,4,sag,ge6,sin,e,mul,kur,sa,kir4,sil,pa,$1,\frac{1}{2}$,kus.

The aim of the first step of interpretation is the formation of sentences from the sequence of words. One can easily identify the beginning of events by an expression for the date because of its schematic formulation.

In most of the cases the intermediate linguistic tokens then belong to the description of specific events at that date. Therefore we finally represent the events in the following way:

```
event(time(23, seleucid_era, 7, 4, beginning_of_night),
  configuration(moon, above, th_ophiuchi, distance((1, 1/2),
    kus))).
```

The form is typical for the computer language PROLOG, by which all astronomical events are represented and the interpretation models are programmed.[11] The basic data structures of this powerful modern programming language are propositions, which include rules as conditionals and grammar parsing capabilities. Propositions are represented as *terms* with the predicate name (here "event") preceding the parameters in parentheses. The first parameter is again a term ("time") with five parameters for the segments of a Babylonian date. The second parameter ("configuration") is a term that can have different forms according to the described astronomical configuration. The meanings of these terms still have to be determined. The occurrence of a term "above" in the complex expression does not yet imply any interpretation. PROLOG provides inference mechanisms, with which one can hypothetically define the meanings of the linguistic entities and perform hermeneutic tests on its consequences.

According to a first tentative interpretation, the sentence reports an event at the 4th day of the 7th month of the 23rd year of the Seleucid Period. On the basis of the preliminary assumptions the moon is observed "above" θ Ophiuchi with $1\frac{1}{2}$ units of an unknown standard called "KÙŠ." Of course, the interpretation make some assumptions—the identification of objects, the Babylonian date and that there is a kind of quantitative relation between these objects. Yet in principle these can be retracted and substituted by alternatives.

4.4   From Tokens to Akkadian Sentences

Figure 3 shows a typical example of a list of tokens read from one cuneiform tablet. The first tablet of the year -288 is labeled according to Hunger's notation as "pl288A." The transcribed words are separated by commas in the list of tokens, which is delimited by square brackets. The sign "x" indicates an unreadable character.

The list is processed sequentially by a parser, which builds sentences from the list of tokens according to the defined grammatical rules. Figure 4 shows a parsing tree of the first sentence. The lowest level of the tree contains the words in their sequence on the tablet. Depending on the

```
keilschriftplatte(pl288A,[
[obv,x,#],
mu_23_kam,i,se_lu_ku,u,i,an_ti_u_uk_su,lugal,mes,du6,30,16,na,ge6,2,sin,ina,
igi,sag,gir_tab,2,x,x,x,x,
x,x,x,x,ge6,4,sag,ge6,sin,e,mul,kur,sa,kir4,sil,pa,1,1/2,kus,5,lal_tim,nu,pap,
ge6,6,x,x,x,x,
x,x,x,x,ge6,7,sag,ge6,sin,ar,si,mas,2,kus,ge6,8,sag,ge6,sin,sig,
mul,ar,sa,suhur,mas,6,u,
4,x,x,x,x,ge6,14,5,!,30,ge6,nu,pap,14,12,na,nu,pap,ge6,15,ina,zalag,
sin,ar,mul_mul,1,kus,15,dele_bat,ina,su,
ina,gir_tab,igi,x,x,x,x,kus,17,gu4_ud,ina,nim,ina,rin,igi,nu,pap,ge6,18,
murub4,sin,sig,mul,ar,sa,se_pit,mas_mas,8,u,ge6,21,
x,x,x,x,kus,in,en,21,mul_babbar,ana,nim,ki,us_u,ina,alla,us,ge6,22,ina,zalag,sin,
ina,igi,lugal,2,kus,
x,x,x,x,ge6,23,ina,zalag,sin,ar,mul,tur,sa,4,kus,ar,lugal,1,kus,ge6,24,
ina,zalag,sin,
ina,igi,gir,ar,sa,a,x,1,kus,ge6,25,ina,zalag,sin,ina,igi,genna,2,kus,ana,su,gub,
x,x,x,x,kus,ge6,27,ina,zalag,sin,ina,igi,mul_babbar,2,kus,ana,su,gub,27,24,kur,
itu,bi,ki,x,lam,se_im,0,
x,x,x,x,x,1,pi,se_gis,2,b,sig,ha,3,ma_na,i_nu_su,mul_babbar,ina,alla,dele_bat,
u,gu4_ud,ina,rin,ina,til,wr,x,as_til,itu,ina,gir_tab,
genna,ina,absin,an,ina,pa,ina,til,itu,ina,mas,itu,bi,illu,4,si,lal,pap,37,
na,
[#], . . .
```

Figure 3
List of tokens.

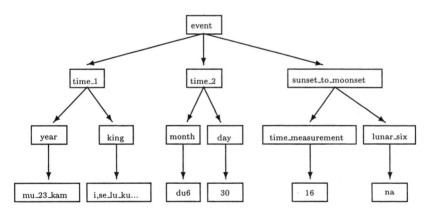

Figure 4
Parser tree.

```
babylonian_events( p1288A, [
event(time(mu_23_kam, an_ti_u_uk_su, du6, 0, G6251),
    sunset_to_moonset((16, 0), none, standard)),
event(time(mu_23_kam, an_ti_u_uk_su, du6, 4, sag),
    configuration(sin,e,(mul,kur,sa,kir4,sil,pa),distance((1,1 / 2),kus))),
event(time(mu_23_kam, an_ti_u_uk_su, du6, 7, sag),
    configuration(sin, ar, (si, mas), distance(2, kus))),
event(time(mu_23_kam, an_ti_u_uk_su, du6, 8, sag),
    configuration(sin, sig, (mul, ar, sa, suhur, mas), distance(6, u))),
event(time(mu_23_kam, an_ti_u_uk_su, du6, 14, G6335),
    sunset_to_moonrise((5, 30), none, calculated)),
event(time(mu_23_kam, an_ti_u_uk_su, du6, 14, G6347),
    sunrise_to_moonset((12, 0), none, calculated)),
event(time(mu_23_kam, an_ti_u_uk_su, du6, 15, zalag),
    configuration(sin, ar, mul_mul, distance(1, kus))),
event(time(mu_23_kam, an_ti_u_uk_su, du6, 18, murub4),
    configuration(sin, sig, (mul, ar, sa, se_pit, mas_mas), distance(8, u))),
event(time(mu_23_kam, an_ti_u_uk_su, du6, 22, zalag),
    configuration(sin, igi, lugal, distance(2, kus))),
event(time(mu_23_kam, an_ti_u_uk_su, du6, 23, zalag),
    configuration(sin, ar, (mul, tur, sa, 4, kus, ar, lugal), distance(1, kus))),
event(time(mu_23_kam, an_ti_u_uk_su, du6, 25, zalag),
    configuration(sin, igi, genna, distance(2, kus), to(ana_su_gub))),
event(time(mu_23_kam, an_ti_u_uk_su, du6, 27, zalag),
    configuration(sin, igi, mul_babbar, distance(2, kus), to(ana_su_gub))),
event(time(mu_23_kam, an_ti_u_uk_su, du6, 27, G6485),
    moonrise_to_sunrise((24, 0), none, standard)),
event(time(mu_23_kam, an_ti_u_uk_su, du6, 27, G6485),
    configuration(mul_babbar, ina, alla, distance(0))),
event(time(mu_23_kam, an_ti_u_uk_su, du6, 27, G6485),
    configuration(gu4_ud, ina, rin, distance(0))),
event(time(mu_23_kam, an_ti_u_uk_su, du6, 27, G6485),
    configuration(genna, ina, absin, distance(0))),
event(time(mu_23_kam, an_ti_u_uk_su, du6, 27, G6485),
    configuration(an, ina, pa, distance(0))),
```

Figure 5
Event descriptions after parsing the token list.

context of their neighboring expressions these words are classified syntactically. The classifications are construed by going though the tree bottom up. If they can be concatenated to a full event description, the parser has successfully assembled a sentence from the token list; otherwise it fails and leaves the list of tokens as it is. Since the parser forms sentences purely syntactically, the meanings of the sentence phrases are not yet fixed. The parser could also insert possible completions for unreadable expressions or breaks in the tablets, which are then subsequently tested.

After the first parsing step the token list is converted to a set of event descriptions, as in figure 5 for the first tablet of the year −288. The next step (figure 6) is of a more cosmetic nature, when the Akkadian expressions are preliminarily substituted by English phrases. It facilitates the reading of the lists of events.

```
translated_events( p1288A, [
event(time(23, seleucid_era, 7, 0, G613),
    sunset_to_moonset((16, 0), none, standard))),
event(time(23, seleucid_era, 7, 4, beginning_of_night),
    configuration(moon, above, th_ophiuchi, distance((1, 1 / 2), kus)))),
event(time(23, seleucid_era, 7, 7, beginning_of_night),
    configuration(moon, behind, b_capricorni, distance(2, kus)))),
event(time(23, seleucid_era, 7, 8, beginning_of_night),
    configuration(moon, below, d_capricorni, distance(6, u)))),
event(time(23, seleucid_era, 7, 14, G728),
    sunset_to_moonrise((5, 30), none, calculated))),
event(time(23, seleucid_era, 7, 14, G747),
    sunrise_to_moonset((12, 0), none, calculated))),
event(time(23, seleucid_era, 7, 15, last_part_of_night),
    configuration(moon, behind, eta_tauri, distance(1, kus)))),
event(time(23, seleucid_era, 7, 18, middle_part_of_night),
    configuration(moon, below, m_geminorum, distance(8, u)))),
event(time(23, seleucid_era, 7, 22, last_part_of_night),
    configuration(moon, in_front_of, a_leonis, distance(2, kus)))),
event(time(23, seleucid_era, 7, 23, last_part_of_night),
    configuration(moon, behind, r_leonis, distance(1, kus)))),
event(time(23, seleucid_era, 7, 25, last_part_of_night),
    configuration(moon, in_front_of, saturn, distance(2, kus), to(to_the_west)))),
event(time(23, seleucid_era, 7, 27, last_part_of_night),
    configuration(moon, in_front_of, jupiter, distance(2, kus), to(to_the_west)))),
event(time(23, seleucid_era, 7, 27, G922),
    moonrise_to_sunrise((24, 0), none, standard))),
event(time(23, seleucid_era, 7, 27, G922),
    configuration(jupiter, is_standing_in, cancer, distance(0)))),
event(time(23, seleucid_era, 7, 27, G922),
    configuration(mercury, is_standing_in, libra, distance(0)))),
event(time(23, seleucid_era, 7, 27, G922),
    configuration(saturn, is_standing_in, virgo, distance(0)))),
event(time(23, seleucid_era, 7, 27, G922),
    configuration(mars, is_standing_in, sagittarius, distance(0)))),
```

Figure 6
List of events in English rendering.

After the parsing of all cuneiform tablets in the first two volumes of the *Diaries* I obtained descriptions of 3285 events, of which 2781 are complete without unreadable words or broken plates. Out of those are 1882 topographical events, 604 are lunar observations called *Lunar Six* (which will be characterized below) and 295 are locations of a celestial object in a constellation. For the purpose of interpreting the *Diaries* the reported heliacal risings and settings are omitted, while the *Lunar Six* observations are needed as auxiliary information.

5   Sky over Babylon

Interpretations of observational statements require accurate calculations of the described historical events. The algorithms used for the calculation of the planetary and lunar positions are based on Montenbruck and Pfleger (1989). After some initial tests an error in the lunar theory was corrected. The ephemerides were compared with Tuckerman's tables and in partic- ular with the very accurate programs of Bretagnon and Simon (1986) for planets and Chapront-Touzé and Chapront (1991) for the moon. The last are based on new relativistic models of celestial mechanics. A critical parameter of historical ephemerides is the conversion from ephemerides time to world time.[12] While the ephemerides time is a uniform measure of time, world time is defined by the passing of the mean sun through the meridian of Greenwich at noon. Because of tidal friction the rotation of the earth is non-uniformly slowed down over the centuries. The conver- sions in our calculation are done according to Stephenson and Morrison (1984), like Chapront-Touzé and Chapront (1991), p. 6. All coordinates are apparent coordinates of true equinox and true equator at the time of observation. All coordinates include refraction, and parallax in the case of the moon.[13] The lunar coordinates were compared with independent programs of Graeme Waddington (Oxford). The error of planetary posi- tions should be less than a few hundredths of a degree and the lunar positions better than a few tenths of a degree.

6   FIRST PARTIAL INTERPRETATION

The following topographical report describes an event involving the moon and the star θ Ophiuchi:

23 year Seleucid Era, month VII, night of the 4th, beginning of the night, moon above θ Ophiuchi $1\frac{1}{2}$ KÙŠ.

It is the goal to find a consistent interpretation of the topographical rela- tion (here "above") and the measured quantity, which is $1\frac{1}{2}$ KÙŠ in this example. The set of grammatical rules defining the meaning of the involved terms exceeds by far the number of Akkadian terms. Which of those possible interpretations should be tested first and compared with alternative interpretations? According to the strategy one should prefer those elements which can be most securely tested with suitable sets of observation reports, independently of the interpretation of the other terms. Some aspects need to be considered before this rule is applied.

1. The Akkadian names can be securely translated. Especially the names of the celestial objects and the names of kings and months are well established.

2. Despite that the meanings of the calendaric expressions are well known, the corresponding Julian date is not certain. The Babylonian month starts with the first visibility of the moon in the western sky. Some tables of lunar phases, e.g., Goldstine's, tabulate the astronomical new moon. A schematic addition of one day to that date gives only a rough estimate of the first lunar visibility. Even if one estimates the first visibility by an *arcus visionis* model,[14] the difference between observation and calculation could always differ by one day. After having completed the successful interpretation of the observation reports, the analysis shows that 90% of the beginnings of the months are correctly predicted with an *arcus visionis* model, the rest differs only by one day. Schematic additions of one day to the date of astronomical new moon are much less accurate and should not be used at all.

3. While the well-known and mostly secure meanings of the Akkadian terms are easily transliterated, the meanings of the remaining terms can only be determined by a comprehensive testing of a large variation of different interpretation hypotheses. For each of those the assumed astronomical configuration has to be calculated and compared with a suitable set of observation reports. Such comparison requires a precise and complete information of the date of observation. Since only the beginning of the tables specify the full date, while the following dates comprise only the discriminating information, broken tablets pose a particular problem. Their dates have to be determined internally by the events mentioned on the plate.

As shown schematically in table 1, one usually finds several different types of topographical observations in one month. The strategy suggests starting with a minimal interpretation model. Among those one should prefer such models that can be tested most sharply. Tests are performed by comparing the interpreted event description with the computed celestial configuration. In table 1 a schematic table of events has been composed for illustrative purposes. Several different types of topographical relations

Table 1
Schematic pattern of observation records

Event	king	Y;M	D	object1	dist	relation	object2
e_7	k_1	3;1	1	sunset to moonset	12°		
e_1	k_1	3;1	2	moon	3k	above	Venus
e_2	k_1	3;1	5	moon	2k	behind	star x
e_3	k_1	3;1	6	Mars	2k	below	star y
e_4	k_1	3;1	8	moon	3k	before	Jupiter
e_5	k_1	3;1	12	moon	1k	above	star z
e_6	k_1	3;1	25	moon	4k	behind	star u

are reported between different celestial objects, as well as one instance of a
Lunar Six observation. The first step of interpretation determines a suitable
subset of events against which a minimal interpretation model can be
tested most sharply. We again mark the critical expressions in the example

23 year Seleucid Era, $\boxed{\text{month}}$ VII, night of the 4th, beginning of the
night, moon $\boxed{\text{above}}$ θ Ophiuchi $1\frac{1}{2}$ $\boxed{\text{KÙŠ}}$.

In the beginning the interpretation focuses on the following points.

1. One has to take into account that the beginning of the month is not
exactly known.

2. The exact observational time in the night is not known, although in this
case the date is followed by the qualification "beginning of the night."

3. The topographical relation is unclear.

4. The unit of measurement is not precisely established.

5. In addition either the transcription could be erroneous or the original
recording could be corrupt.

Initially, the interpretation of the topographical relations is the most difficult to determine. The many different possible readings depend highly
on assumptions about other astronomical circumstances, which prevent
clear–cut tests. But we can test two other elements of event descriptions,
which are not so sensitive to variations of the others. These are, for
example, such events that would not change their appearance if the date
had been shifted by one or two days. Such events would be *insensitive* to
variations of the time dimension, including the only vaguely specified
observational time. On the other hand, if we could accurately determine
the beginning of the month and hence eliminate the insecurity in the time
dimension, we could use events sensitive to time shifts for the determination of other interpretative aspects.

6.1 Initial Models and Expansion Strategies

In more abstract terms, in principle we start the interpretation with two
basic alternative readings of the *Diaries*, represented as the two most general classes of models.

M 1 The sentences of the *Diaries* report true observations, unless it is
explicitly marked otherwise.[15]

M 2 The sentences of the *Diaries* are speculative by nature or express
astronomical (or astrological) theoretical knowledge of unknown
origin.

The second interpretation $\mathcal{M}2$ contradicts $\mathcal{M}1$, yet cannot be avoided if all possible interpretations in the line of $\mathcal{M}1$ are exhausted. Since we pursue the interpretation of the *Diaries* as observation texts, we will expand this class of models. The most secure next expansion of the model includes the expressions for time and the celestial objects.

$\mathcal{M}1.1$ The Akkadian terms for calendaric expressions and the canonical names of celestial objects are assumed.

The expanded model $\mathcal{M}1.1$ is neither certain beyond doubt, nor is it sufficiently rich for a test. The assignment of names to celestial objects is certain in the case of the planets and the moon, but as it turned out, the canonical identification of η Tauri standing for the measured star in the Pleiades needs further inquiry. Preferably one should not include this constellation in the test set of observations.

Expansion heuristics for interpretations   An interpretation model is to be expanded by such a minimal set of elements, which is (a) constructive and (b) maximally efficient.

The expansion by one element is constructive, if it increases the testability of the hypothesis. It should add information to the hypothesis, which implies new testable consequences. Furthermore it is only then constructive, if it is consistent with the other statements of a model. Among the selection of constructive expansions we should prefer those that are maximally efficient. In other words, they lead to the most direct and sharp tests of the model. This is reasonable, because otherwise the interpretation model might carry untested false assumptions, which would be much harder to isolate and detect in a model with many statements than in a less complex model. It is not crucial whether the expanded new assumptions are plausible. In fact, an interpreter initially should not put too much weight on his judgments of whether one or the other interpretation element is plausible. Tests will bear out wrong assumptions soon enough. For a comprehensive interpretation attempt it would be more important to include as many variations as possible. In case of clear falsifications, models are corrected according to the *contraction heuristics*.

Contraction heuristics   If an interpretation model is clearly falsified, elements of the model are retracted until the model is again consistent with the data.

Usually, the last expanded elements are the critical components that lead to a refuted hypothesis. In general, any element of a model could be responsible for a falsification. In case of a falsification, the interpreter must ensure that all variations of the elimination of elements are tried. Strategically, the interplay of expansion and contraction heuristics allows a gradual construction of hypotheses with most of the used assumptions tested as early as possible. Technically speaking, with a minimal effort of testing this method allows the evalutation of a rather complex combinational variation of assumptions within an interpretation hypothesis.

The first expansions of model $\mathcal{M}1.1$ could be either the fixing of the beginnings of the months or a preliminary determination of the KÙŠ. According to the expansion heuristics, such modifications would be preferable which could be tested largely independently of all possible variations in the other components (this leads to the sharpest tests), in particular the topographical relations. Indeed, in the case of the beginnings of the months this condition is perfectly met. The *Diaries* contain many reports of *Lunar Six* data. Most of them seem to be observed, while some rely on theoretical information, when Babylonian astronomers marked them as *nu pap* or *not observed*. Because of the speed of the moon the data should clearly signify the julian date to which they refer, independent of whether they were observed or theoretically derived. However, the method requires an accurate algorithm for the lunar motion, which was initially not at my disposal, so the second possible expansion—the determination of the KÙŠ—was preferred.

6.2   First Approximations to the KÙŠ

All topographical observations in the *Diaries* include quantitative measures with a unit called KÙŠ and sometimes a smaller unit SI. The English canonical translation is "cubit" and "finger."

In Old-Babylonian times the astronomical units arose from measures of distances, with a plausible speculation that they indicate the distances of grades on a scaled measuring device. In the astronomical context this could quantify angular distances. Units are divided by simple ratios. So we find that 6 *še* (barleycorn) yield one SI (finger), and 30 finger make at least since the pre-Sargonic era one KÙŠ (cubit). 12 KÙŠ then yield one NINDA.[16]

In Neo-Babylonian time one finds an alternative division of one KÙŠ into 24 SI or fingers. According to Kugler, both norms for the division of KÙŠ were simultaneously used in the Persian and Seleucid eras, even with a more frequent usage of the 30-finger division in later

times. Yet, this result is based on a very limited analysis of few planetary tablets, which apparently are Babylonian excerpts of *Diaries*. As far as
I can see, all references to the division of KÙŠ and fingers trace back
to Kugler's statement,[17] and one should seek for an independent analysis.
If converted to angular distances, the cubit with a 30 finger division
should be equivalent to 2.5 angular degrees, while the later variation
amounts to 2 degrees. This assumes a constant length of a finger for both
systems.[18]

Astronomical event descriptions use only two basic types of measure: time and angular distances. Since we also find a well established
cluster of time concepts in the *Diaries*, the unit KÙŠ should be a unit for
angular distances.

With this background there are good reasons to assume that KÙŠ
as a measure of arc corresponds to $2°$ to $2°.5$. This interval must be
narrowed, for further inquiries. A Babylonian measurement of few KÙŠ
would be wrong by some degrees of arc, if the assumed unit of measure is
half degree wrong.

M 1.1.1   Topographical relations measure quantities of angular distances.

At this point of the construction of hypothesis we have alternative
readings for the size of the unit and we should seek to diminish the range
of possible values. However, tests require assumptions about the meaning
of the topographical relations, which seems to be a vicious circle: a specified value for the KÙŠ should be the basis for resolving the topographical
relations; conversely the determination of the KÙŠ requires assumptions
about unknown topographical relations. How can one escape this circle?
According to the hermeneutic strategy one should search for such events,
where variations in the meaning of the topographical relations have hardly
any bearing on the test. Applying the idea of falsifying possible values for
the KÙŠ, we should exclude such values, for which we find falsifying
instances for *every* different version of topographical relation. We choose
from the four relations—"above," "in front of," "below," and "behind"—
such relations, which measures are most insensitive to changes in the other
parameters. It is obvious that the relations "behind" and "in front of"
exhibit a much larger variation then the other couple, which means that
they should be oriented in the direction of movement of planets and the
moon. Any small temporal variation like the shift from evening hours to
morning hours, effects relations in the direction of motion much more
than roughly perpendicular orientations. For the initial estimation of the

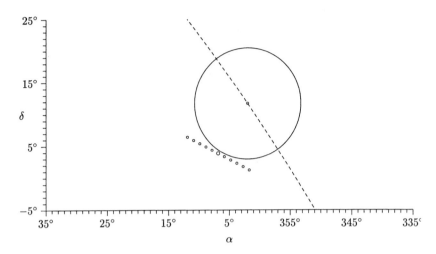

Figure 7
79^y Seleucid era 11^m 20^d, first part of night, Venus below α Arietis, $3\frac{1}{2}$ KÙŠ.

size of KÙŠ only events with "above" and "below" relations were chosen
and the moon was excluded. Its motion is so fast that a slightly wrong
assumption concerning the time of measurement might have drastic effects
on the outcome. Events with planets in "above"/"below" relations remain
as the best type of events for the initial determination of the unit of
measurement.

Figure 7 shows the motion of Venus in the equatorial coordinate
system. This sidereal coordinate system moves with the apparent rotation
of the sky. The plotted motion of Venus in these coordinates abstracts
from the daily apparent motion. The figure shows the equatorial position
of Venus at midnight of successive days for Babylonian times. The slightly
bigger dot in the middle marks the position of Venus for the recorded day
of the *Diaries*. Venus moves from the right side along the dotted line five
days before the recorded date to the left side five days after the recorded
date. The middle of the circle is the position of the star α Arietis (Hamal).
The circle has the radius of the measured distance on the basis of one
KÙŠ equals $2°.5$. Venus reaches the closest proximity to the star at the day
of observation. The circle increases its size, if one assumes a larger value
for the KÙŠ, and it decreases if the KÙŠ is closer to $2°$ or even smaller.
Which value fits the data best? In this case we consider three forms of
relations of sidereal coordinate system consistent with our metrological
assumption:

1. declination differences,
2. angular distances,
3. differences in latitude.

Although we do not know the exact observational time, we see that the measures in all three possible interpretations are almost the same. There still might be an error in the assumed day, but that can be securely controlled by the observations of the *Lunar Six*.

The next step excludes those unit values for the KÙŠ, which are not compatible with either of the three readings, for example, the apparent angular distance between the two objects. For the selected set of "above" and "below" events we thereby exclude values larger than $2°7$ and smaller than $1°8$. Given just an alternative between $2°$ and $2°5$, the latter is clearly preferable. Certainly this is a preliminary result so far, and it has to prove its value—and face possible modifications—when all other events can be included into the testing set. For the time being we use the preliminary hypothesis $\mathcal{M}\,1.1.1.1$.

$\mathcal{M}\,1.1.1.1$ The angular unit is $1\ \text{KÙŠ} = 2°5$.

This intermediate hypothesis has to be rechecked every time the interpretation could be extended or improved. We will summarize these results at the end.

7   Elimination of Day Errors

Further refinement of the interpretation requires the moon. Especially the fixing of the beginning of the month could proceed, after the algorithms for the lunar position were integrated into the PROLOG environment carrying the statements of the Babylonian tablets. When one looks at a number of topographical observations involving the moon, one can easily spot a day error for the assumed date. This error is due to a misidentification of the beginning of the month.

Figure 8 shows an event of the moon "behind" the star α Scorpii according to the *Diaries*. In this case the positions are plotted in the ecliptic coordinate system. Like the equatorial coordinate system it rotates together with the daily rotation of the sky around the observer. The plane of the lunar orbit is strongly inclined to the ecliptic (about 5°). The planetary motion occurs near the ecliptic with increasing longitude, in our figure from right to left, except for the retrograde motion. The lunar position is shown for one night from sunset to sunrise. When the moon

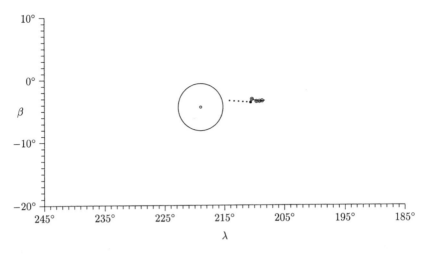

Figure 8
102^y Seleucid era 5^m 7^d, beginning of night, moon behind α Scorpii, $1\frac{1}{2}$ KÙŠ.

stands above the horizon, its position is plotted with billed circles. If it is
below the horizon during the night, it's position is marked with small
dots. Each dot is separated by one hour. In this case the moon is well
above the horizon in the evening. About four hours after sunset the moon
sets in the west and the lunar position for the rest of the night is plotted by
small dots.

According to the *Diaries* the moon is "behind" the star. In most
other such cases the moon has a larger longitude than the second celestial
object. In this case it should stand on the left side of the object. Because of
the rapid lunar movement it would stand there in the following night, as
is shown in figure 9. Could the assumed day of observation be wrong by
one day? Since the date is given in the Babylonian calendar, a conversion
to a modern calendar requires the exact knowledge of the beginning of
the Babylonian month. I calculated these dates with various models[19] and
it turned out that in this case the beginning of the month has to be shifted
by one day. In about 10% of the cases the actually observed first visibility
differs from the calculation. Luckily, there is an independent method to
determine the factually observed new moon in Babylon.

7.1 *Lunar Six* and the Beginnings of the Months
The time difference between rising and setting of moon and sun, both at
the eastern and western horizon, is of special interest for Babylonian

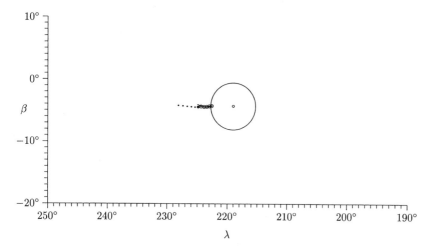

Figure 9
102^y Seleucid era 5^m 7^d, beginning of night, moon behind α Scorpii, 1$\frac{1}{2}$ KÙŠ (computed beginning of month plus one day).

astronomy. These times are measured around new or full moon. There are six different configurations (sunset to moonset, sunset to moonrise, etc.), because of which these time quantities are called *Lunar Six*. At the beginning of the month the Babylonian scribes reported the time difference between sunset and moonset as first entry in the *Diaries*. Not all of those entries were observed. Unlike the topographical observations, we find qualifications of the visibility condition by the astronomer. Sometimes he adds that he did not observe the time difference (hence it must be computed), sometimes bad weather conditions are noted. Are these indications that all values were computed? There are explicit statements of the Babylonian astronomer that he did indeed measure a time difference. Most of the entries do not have any additional qualification. Since the additions are mostly reported for days where we also find remarks about bad weather conditions, one should interpret the regular *Lunar Six* values without further qualifications as genuine measurements, which are only supplemented by theoretical values if circumstances prevent the sight of the new moon. In the following these regular statements are therefore called "normal" observations. Tables 2 and 3 summarize the frequency of all *Lunar Six* statements with their additional qualifications. Both tables show an equal distribution of all types of phenomena, hence the attention of the astronomer did not concentrate on the setting times of the new

Table 2
Frequencies of additional specifications in the first three *Lunar Six* observations

Observation	Comment	State	Number
sunset to moonset	bright	—	21
sunset to moonset	bright	measured	8
sunset to moonset	—	—	47
sunset to moonset	—	measured	19
sunset to moonset	clouds	—	3
sunset to moonset	clouds	measured	6
sunset to moonset	clouds	calculated	13
sunset to moonset	faint	—	3
sunset to moonset	mist	—	3
sunset to moonset	mist	measured	3
sunset to moonset	mist	calculated	2
sunset to moonrise	—	—	56
sunset to moonrise	—	measured	17
sunset to moonrise	—	calculated	5
sunset to moonrise	clouds	—	2
sunset to moonrise	clouds	measured	11
sunset to moonrise	clouds	calculated	16
sunset to moonrise	mist	—	1
sunset to moonrise	mist	measured	3
sunrise to moonset	—	—	38
sunrise to moonset	—	measured	48
sunrise to moonset	—	calculated	2
sunrise to moonset	clouds	—	4
sunrise to moonset	clouds	measured	7
sunrise to moonset	clouds	calculated	21
sunrise to moonset	mist	—	1
sunrise to moonset	mist	calculated	2

moon at the beginning of the month. The *Lunar Six* phenomena have their own interest in Babylonian astronomy. There is a significant difference in the distribution of additional qualifications. Only for the times of sunset to moonset we find the occasional addition "bright."[20] The Babylonian astronomer noted in this case whether the moon at the day of first sighting had gained some larger distance from the sun than usual. However, since the numerical value of the sunset to moonset timing expresses this fact already, the particularly good sighting conditions must have been worth recording by the ancient astronomer. As noticed by

Table 3
Frequencies of additional specifications in the last three *Lunar Six* observations

Observation	Comment	State	Number
moonrise to sunrise	—	—	67
moonrise to sunrise	—	measured	21
moonrise to sunrise	—	calculated	7
moonrise to sunrise	clouds	—	6
moonrise to sunrise	clouds	measured	4
moonrise to sunrise	clouds	calculated	13
moonrise to sunrise	faint	—	1
moonrise to sunrise	mist	—	2
moonrise to sunrise	mist	measured	3
moonrise to sunset	—	—	34
moonrise to sunset	—	measured	40
moonrise to sunset	—	calculated	5
moonrise to sunset	clouds	—	1
moonrise to sunset	clouds	measured	16
moonrise to sunset	clouds	calculated	23
moonrise to sunset	mist	measured	5
moonrise to sunset	mist	calculated	2
moonset to sunrise	—	—	69
moonset to sunrise	—	measured	13
moonset to sunrise	—	calculated	7
moonset to sunrise	clouds	—	3
moonset to sunrise	clouds	measured	8
moonset to sunrise	clouds	calculated	22

Lis Brack–Bernsen, the Babylonian astronomers could derive a very good prediction for the *Lunar Six* times using observation records of one Saros period earlier.[21] They could perform these calculations without reference to the lunar theory and rely entirely on ephemerides compiled from the *Diaries*. The error of these computed values for the *Lunar Six* does not exceed the one for the observed values. Therefore, for our purpose we do not need to discriminate between *Lunar Six* observations and theoretical derivations for the determination of the beginning of the month.

The distribution of *Lunar Six* observations over the historical periods resembles the one of topographical observations (see figure 10). It confirms that the *Diaries* cover a very stable scientific practice that hardly changed over the centuries. This finding also supports the assumptions

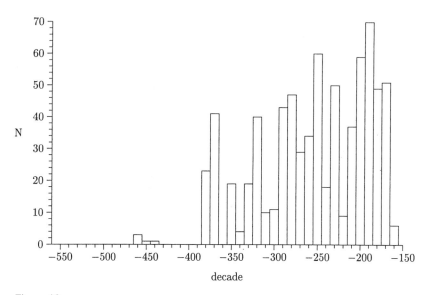

Figure 10
Total number of *Lunar Six* observations in the decades between −550 and −150.

that the meanings of the astronomical terms and the observational practice do not change significantly. It is remarkable indeed that the astronomy of the Persian empire could be maintained through the times of
Greek ruling.

The parser assembled 604 *Lunar Six* observations in total. Among
them are observed ones as well as calculated values. Since we assume that
the calculated values do indeed stem from observations of the previous
Saros cycle back, and the derivation does not add a significant error, we
can estimate the accuracy for both.[22] The second column of table 4 shows
the standard derivation of the time measurements in hours. We compare
the documented value with the time difference between the computed
setting for both the sun and moon including refraction and parallax. The
standard derivation is about a tenth of an hour or 6 minutes, while the
measurements are recorded with an accuracy of 4 minutes.

The obtained accuracy is excellent and far better than needed for the
determination of the beginning of the month. With a daily motion of the
moon of about 12° the *Lunar Six* clearly tag the day of measurement. In
the fourth column of table 4 we find the frequency of each type of *Lunar
Six* observation. We could determine the accurate beginnings of more
than a hundred months by comparing the *Lunar Six* quantities with those

Table 4
Standard derivation and mean of errors in hours, total number and number of outliers of *Lunar Six* observations, error threshold 24 minutes

Type	$\sigma[h]$	$\mu[h]$	N_{tot}	N_{out}
sunset to moonset	0.137	−0.065	103	1
sunset to moonrise	0.091	−0.098	93	2
moonrise to sunset	0.087	0.119	109	3
sunrise to moonset	0.083	0.105	106	4
moonrise to sunrise	0.151	−0.077	91	5
moonset to sunrise	0.095	−0.096	102	0
total	0.144	−0.014	604	15

computed for the assumed day of observation and some days around it. In most of the cases this led to an unambiguous decision. In the following we call these months *absolutely chronologically fixed*, in short *fixed* months. The next inquiries use only fixed months, because only for them large temporal errors in the observation reports can be excluded.

7.2   Scale Division

In extrapolation of the accurate *Lunar Six* measurements, one might speculate that the precision of the topographical quantities matches roughly the scale of measurement. Table 5 lists the frequency of measurements for particular quantities. Nearly equal distribution of 1 KÙŠ (302), $1\frac{1}{2}$ KÙŠ (290), and 2 KÙŠ (308) show that the main measurement resolution is $\frac{1}{2}$ KÙŠ. Large quantities are rarer. More precise measurements are less frequent, especially for those with small quantities less than 1 KÙŠ.

Generally it is assumed that there are divisions of one KÙŠ into either 24 or 30 fingers (SI). Do we find those divisions in the measured fractions of a KÙŠ? Probably, measurements with 8 fingers are done in a scale with 24 fingers in a cubit. On the other hand, measurements with 10 fingers express a division in the 30 fingers scale. Since we do not know anything about the mixture of measuring devices, we abstain from conclusions until the end of our inquiries.

We do find roughly the same number of 20 SI values (20 measurements) as those with $\frac{5}{6}$ KÙŠ (14 values). The second column shows the corresponding angles on the assumption of 1 KÙŠ = 2°5. We then see that the same angle would be once recorded as 20 SI as well as $\frac{5}{6}$ KÙŠ. One might resolve the ambiguity by correlating the measures with 20 SI to a different scale than the $\frac{5}{6}$ KÙŠ, for example, one that counts 30 SI in

Table 5
Frequencies of distances measured

Measurement	[°]	Number	Measurement	[°]	Number
1 SI	0.10	12	2 KÙŠ 2 SI	5.21	1
2 SI	0.21	15	2 KÙŠ 4 SI	5.42	4
3 SI	0.31	9	2 KÙŠ 6 SI	5.62	1
4 SI	0.42	22	2 KÙŠ 8 SI	5.83	23
5 SI	0.52	3	$2\frac{1}{2}$ KÙŠ	6.25	200
6 SI	0.62	16	$2\frac{2}{3}$ KÙŠ	6.67	39
8 SI	0.83	27	$2\frac{5}{6}$ KÙŠ	7.08	7
10 SI	1.04	8	2 KÙŠ 20 SI	7.08	1
$\frac{1}{2}$ KÙŠ	1.25	98	3 KÙŠ	7.50	163
14 SI	1.46	7	3 KÙŠ 5 SI	8.02	1
$\frac{2}{3}$ KÙŠ	1.67	104	3 KÙŠ 6 SI	8.12	1
18 SI	1.87	4	3 KÙŠ 10 SI	8.54	2
20 SI	2.08	20	$3\frac{1}{2}$ KÙŠ	8.75	58
$\frac{5}{6}$ KÙŠ	2.08	14	$3\frac{2}{3}$ KÙŠ	9.17	4
1 KÙŠ	2.50	302	4 KÙŠ	10.00	80
1 KÙŠ 4 SI	2.92	19	$4\frac{1}{2}$ KÙŠ	11.25	18
1 KÙŠ 6 SI	3.12	5	5 KÙŠ	12.50	51
1 KÙŠ 8 SI	3.33	54	$5\frac{1}{2}$ KÙŠ	13.75	5
$1\frac{1}{2}$ KÙŠ	3.75	290	6 KÙŠ	15.00	14
$1\frac{2}{3}$ KÙŠ	4.17	80	$6\frac{1}{2}$ KÙŠ	16.25	3
$1\frac{5}{6}$ KÙŠ	4.58	13	7 KÙŠ	17.50	1
1 KÙŠ 20 SI	4.58	4	8 KÙŠ	20.00	1
2 KÙŠ	5.00	308			

a KÙŠ. Only then we face a similar problem as before: 20 SI then mean the same as $\frac{2}{3}$ KÙŠ. Whatever turn one takes, we find variations in the recordings of fractions of a KÙŠ that might be due to different observation devices or different recording practices of possibly different observers or schools. These fine structures of the measurement data could supply further information of the measuring processes. For the purpose of the interpretation of the topographical relation we can plausibly conclude that the accuracy of the data should be in the dimension of one KÙŠ.

7.3 Frequencies of Events

Only a limited number of stars are used for topographical measurements of moon and planets. In the literature they are called "normal stars." Table 6 lists all those stars with their canonical identification. All stars are in the

Table 6
Ecliptical longitudes (epoch −128) of the stars referred to as second object and the number of their mentioning (for all relations except "is standing in")

Longitude [°]	Star	Number
4.8	β Arietis	46
8.2	α Arietis	59
29.9	η Tauri	101
40.2	α Tauri	87
53.0	β Tauri	63
55.2	ζ Tauri	59
63.9	η Geminorum	37
65.7	μ Geminorum	36
69.5	γ Geminorum	53
80.7	α Geminorum	52
84.0	β Geminorum	63
95.8	η Cancri	1
96.2	θ Cancri	2
98.0	γ Cancri	5
99.1	δ Cancri	41
111.1	ε Leonis	56
120.4	α Leonis	104
126.8	ρ Leonis	32
133.8	θ Leonis	52
147.1	β Virginis	93
160.8	γ Virginis	80
174.2	α Virginis	101
195.6	α Librae	90
199.9	β Librae	43
213.0	δ Scorpii	14
213.4	π Scorpii	5
213.6	β Scorpii	29
220.2	α Scorpii	90
231.8	θ Ophiuchi	89
274.4	β Capricorni	108
292.1	γ Capricorni	38
293.8	δ Capricorni	39
357.2	η Piscium	53

Table 7

Frequencies of references to the moon or a planet as second object (for all relations except "is standing in")

Object	Number
Moon	1
Mercury	24
Venus	62
Mars	57
Jupiter	80
Saturn	70

Table 8

Total number of relations mentioned

Relation	Number
above	367
balanced above	4
below	490
in front of	688
balanced in front of	2
behind	564
is standing in	436

vicinity of the ecliptic, hence they are selected because of their close proximities to the course of the planets or the moon. Observing close approaches to normal stars is a simple way of measuring sidereal coordinates without sophisticated instrumentation. Table 7 shows the second objects, relative to which the position of a moving celestial object is determined.

The moon is the most frequent object recorded on the first position of the topographical relation. Only once it is recorded as second object, which is clearly an exception of the rule. The general ordering of objects places the fastest objects first, which is natural if we want to determine the position of the moving object relative to slowly moving or sidereally fixed objects.

The frequency of the specific topographical relations is shown in table 8. The relations "before" and "behind" are the most frequent ones, also because they are typically used in conjunction with lunar observations. It does not mean that these types of relations were more interesting

Table 9
Frequencies for "above"

	Moon	Mercury	Venus	Mars	Jupiter	Saturn	Star
Moon	0	0	4	1	7	6	99
Mercury	0	0	3	2	1	2	18
Venus	0	1	0	2	3	3	75
Mars	0	0	0	0	0	1	38
Jupiter	0	0	1	0	0	0	14
Saturn	0	0	1	0	0	0	3

Table 10
Frequencies for "below"

	Moon	Mercury	Venus	Mars	Jupiter	Saturn	Star
Moon	0	2	7	8	4	5	189
Mercury	0	0	0	0	0	0	13
Venus	0	1	0	0	5	2	72
Mars	0	0	0	0	0	0	33
Jupiter	0	0	0	0	0	0	5
Saturn	0	0	0	1	0	0	4

to the ancient astronomer, but that they simply occur more frequent with the fast moving moon. Among the opposite topographical relations ("above" versus "below" and "in front of" versus "behind") "below" and "in front of" are slightly more frequent than their counterparts. Again, this need not reflect a bias toward one relation; an uneven distribution of stars around the ecliptic or preferred observational times could easily induce such differences. However, the differences are small enough to maintain that all topographical relations were of equal interest to the Babylonian astronomer.

Tables 9 and 10 sum up the combinations, in which the celestial objects occur in the relation "above" and "below." The first column of the matrix contains the first object in the relation and the other columns show the frequency by which this object is observed together with the second object of that column. A star never occurs as the first object, and in the case of these two relations the moon is never the second object. With few exceptions we find all events in the upper right side of the matrix. It confirms the view that the order of objects is strictly defined by the speed of the specific combination of objects. In the case of "above" we find only three exceptions with Venus as the second object and one with Mercury. The relation "below" has only one exception each in the case of Mercury and Saturn. The intermediate result is:

1. The relations "above" and "below," and "in front of" and "behind" are pairwise related, probably contrary concepts.

2. All four relations are recorded with equal interest.

3. The accuracy of measurement is in the dimension of one KÙŠ, decreasing for larger distances.

4. The accurate measurements of the *Lunar Six* allow to fix the beginnings of the Babylonian months. In the following, topographical observations in these months are suitable for further interpretative tests, since they do not suffer from dating errors.

8 MEASURING

Without any further clue it is difficult to work out the applied measuring procedure. Although Ptolemy describes in the *Almagest* the use of a spherical astrolab for a direct measurement of ecliptic coordinates, this measure seems much too delicate to be in use already hundreds of years earlier. The advantage of this instrument lies in its generality: one can measure any celestial position once it is properly aligned to the horizontal position of the zodiac. But this instrument needs to be permanently realigned with the daily rotation of the sky. The simplest position measurement, also attested in the Almagest, is the measurement of altitude at the meridian, which is the culmination height. Together with the known altitude of the observation site this value immediately yields the declination of the object. I investigated this possibility in earlier stages of the project up to the point where many "above"/"below"–cases clearly cannot suit the hypothesis of measuring declination differences. One should bear in mind that numerous measurements of that kind are reported for the young and old moon as well as for Mercury and Venus, which are invisible at their meridian transit. This excludes meridian measurements as basis of the *Diaries*.

At least three other possibilities are left for the direct measurement of ecliptic coordinates:

1. Differences in ecliptic coordinates are measured locally, that is, by an instrument that is held up to the two objects while aligned to the ecliptic. Then the coordinates differences are read off by scales or by a frame of strings, which form a grid and thereby allow an easy counting of distance lines.

2. The critical factor for the measurement of ecliptic coordinates is the alignment to the ecliptic, which always changes its horizontal position. How could one properly align an instrument to the ecliptic at the moment of observation? One might observe the positions right after sunset or before sunrise. The rising and setting points of the sun are prominent points in ancient astronomy. At dawn this point also marks the position of the ecliptic. From there one only needs to direct a measurement device to a planet or the moon (assuming they are on the ecliptic), and the connection line perfectly approximates the plane of the ecliptic. A frame of strings or a Jacob's staff could then allow the readings of the coordinates.

3. One could also estimate the distances without instrument. This again would involve knowledge of the position of the ecliptic. However, since we find an extended use of fractions of the measuring unit (KÙŠ) in the data, it seems implausible that such good data could be maintained by an estimation process alone.

9   Interpretative Decisions

The hypotheses concerning the topographical relation can be divided into two major classes: sidereal coordinates, which are synchronized to the daily rotation of the earth, and horizontal coordinates, in which the apparent position of a celestial object is measured. Horizontal positions change quickly with time, while the planets move daily only few degrees in their sidereal coordinates. The main historically known coordinate systems are:

Sidereal   1. Equatorial coordinates,

2. ecliptic coordinates,

3. angular distances,

4. polar distances.[23]

Horizontal   1. Differences of altitude and azimuth.

2. Arcs between celestial object and one point on the horizon, e.g., the intersection between ecliptic and horizon.

This is only a fraction of all conceivable coordinate systems. For testing purposes one should not restrict the imagined coordinate systems to the historically plausible alone. Especially since one can test all variations by a computer.

Other remaining critical components are the observational time and the measurement procedure. Coordinates relative to the horizon would be extremely sensitive to the exact time of observation. Since the *Diaries* record the observational time only as accurate as the night watch,[24] one could only think of various forms of procedural fixations, for example, a measurement at the moment of closest proximity of two celestial bodies or the beginning of a night watch. One should also take into account systematic errors that might be introduced by a method of measurement, about which we can only speculate up to this point. However complex these alternative interpretations become, with the advantage of a computer representation they can be comprehensively tested.

The investigations of the "below" and "above"–relations could not decide between any of the hypotheses. Another class of observation is much more promising to fulfill that task. Because of the large velocity of the moon and the quickly changing celestial configurations lunar observations especially of the "in front of" and "behind"–relations might be decisive.

9.1 First tests of Hypotheses

Early astronomy was very much concerned about astronomical phenomena at the horizon. Setting and rising phenomena of the sun, moon, planets and stars were intensely watched. Their observation formed the basis for all time concepts and the fundamental relations of astronomy. Since Babylonian astronomical theory predicts the occurrence of these horizontal phenonema, it would be a natural interpretation of the observations recorded in the diaries that they are related to the horizon as well. Independent of what angular distance is actually measured, one might expect the celestial objects always near the horizon.

$\mathcal{M}A.1$ The observed events take place near the horizon.[25]

Figure 11 shows a lunar observation, which occurred right after the beginning of the night according to the ancient report. Because of the *Lunar Six* we have narrowed the temporal uncertainty of the moment of observations to less than few hours. This is a very high accuracy, if one considers that the reported event took place more than two thousand years ago! The positions of the moon and the star are plotted in ecliptic coordinates, and the thick line for the position of the visible moon starts at the right side at the beginning of the night. Judging from the length of the dotted thick line, the moon was three hours before setting well above the horizon when it was observed. This and many other similar observations exclude the possibility that all topographical observations took place near the horizon. Yet, it remains a possibility that the objects stand somewhere in the sky and that the topographical relations were measured in some sort of horizontal system by, for example, differences of altitudes.

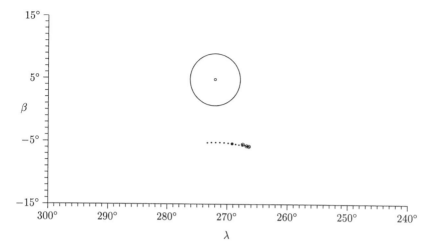

Figure 11
9^y Seleucid era $8^m\ 4^d$, beginning of night, moon in front of β Capricorni, $1\frac{2}{3}$ KÙŠ.

$\mathcal{M}A.2$ The coordinates were measured in a horizon related coordinate system.

Such an interpretation implies that the values of the measurement change rapidly in the course of the night. The most accurate time attributions of the recorded observation are either the beginning or end of the night and the night watches. This is far too inaccurate for a precise horizontal coordinate. Initial estimations of the accuracy of the measurements give values in the dimension of the scales. If the observations had varied within the permissible slot of the reported observational time period, one would get much less accurate measurements. One can escape this argumentation by assuming that implicitly the Babylonian astronomers used a well defined, small period of time for their observations. For example, if they report an event in the second watch, it had been measured at the beginning of the watch and not any time later. These considerations do not falsify model $\mathcal{M}A.2$ in general, but it proved impossible to reconcile the data with any specific horizontal coordinate system I could think of. This then leads to a refutation of the models $\mathcal{M}A.1$ and $\mathcal{M}A.2$.

9.2   "In Front of" and "Behind" Are Not Distances
Another typical group of events (figure 11) leads to the exclusion of a initially rather plausible interpretation.

$\mathcal{M}A.3$ "In front" and "behind" quantities are angular distances between the objects.

Direct angular distances between two objects are the simplest to measure and therefore the most plausible to test. They can be measured by estimation without any tool, or by using the hand as measuring device. Its accuracy could be improved by instruments like a Jakob's staff, or wooden frames with a grid of strings, which are then pointed to both objects and the angular distance read off immediately. Yet, when one looks at figure 11, one sees the moon distant from the circle that corresponds to the angular distance. Since this is by far not a singular case, which could be explained away by transcription errors etc., one must refute hypothesis $\mathcal{M}A.3$ as well. These situations show that the reported quantities must be projections onto one coordinate. This one coordinate then is described by a topographical relation. We have now already several puzzles together.

1. Topographical relations should be sidereal coordinates.

2. We know the exact beginnings of the months for a considerable number of cases.

3. The quantities "in front" and "behind" cannot be angular distances between the objects. They must be single coordinates of a sidereal coordinate system.

9.3   Refutations and Confirmation

One might associate the interpretations of "in front" and "behind" with their meanings in common language. One could then relate the topographical directions to the direction of the daily celestial motion. Each object rises in the east, culminates in the south (in the northern hemisphere) and sets in the west. A second object in close proximity can rise a little later than the first one, or pass behind the first object through the meridian. These interpretations would assign the topographical relations a temporal meaning. The quantitative value associated with the relation could then measure either a spatial distance in a sidereal coordinate system or a temporal difference (although the metrological units do not fit the temporal ones which were in use at that time).

$\mathcal{M}A.4$ "In front" and "behind" express a temporal direction of the topographical relation in a sidereal coordinate system.

Figures 12 and 13 show the same event—the moon "in front" of star β Arietis—in (a) the equatorial coordinate system (with right ascen-

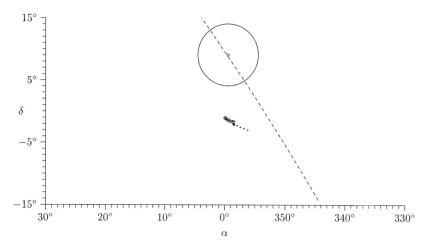

Figure 12
102^y Seleucid era 3^m 22^d, last part of night, moon in front of β Arietis, 2 KÙŠ (ecliptical coordinates).

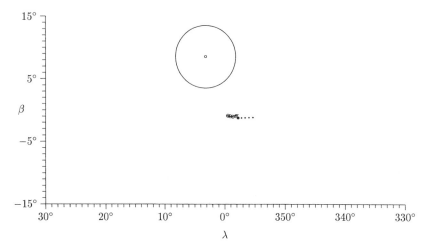

Figure 13
102^y Seleucid era 3^m 22^d, last part of night, moon in front of β Arietis, 2 KÙŠ (ecliptical coordinates).

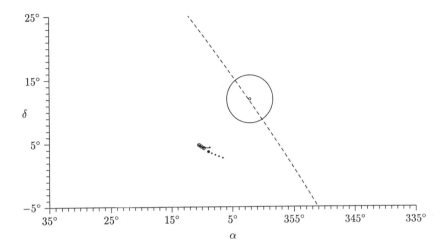

Figure 14
102^y Seleucid era 3^m 23^d, last part of night, moon behind β Arietis, 2 KÙŠ (ecliptical coordinates).

sion α and declination δ) and (b) in ecliptic coordinates with longitude λ and latitude β.

Again the night line of the moon is plotted as a dotted positional line with a circle around the second object for the measured distance according to the documented angular dimension and the working hypothesis of the metrology. In addition figure 12 shows the horizon as dotted line, as it was seen at the moment when the star rose in the east. Since the moon is observed at the 22nd day of the month, some days before it becomes invisible, one can see the moon only in the second half of the night. In figure 12 the dotted horizontal line moves up to the right, hence the moon rises shortly *after* (in the temporal sense) the star, although their relation in terms of the topographical relation is "before." Yet, in figure 14 practically the same situation is described as "behind." Both relations do not correlate with the temporal order of passing the horizon, and therefore I dismiss the interpretation of the topographical relations in temporal terms.

Nontemporal meanings of the relational expression lead to two basic interpretation models.

*M*A.5   "In front of" and "behind" denote differences in one coordinate of the equatorial coordinate system, the right ascension. One object stands "in front" of a second one, if its right ascension is smaller than the second one.[26]

In the same sense the next model defines the topographical coordinates by substituting equatorial coordinates by ecliptic coordinates.

MA.6   "In front of" and "behind" denote differences in longitudes. One object stands "in front of" of a second one, when its longitude is smaller than the second one.

Figure 12 shows an approximation of the moon to the star β Arietis in the equatorial coordinate system as demanded by *MA.5*. The observation took place in the morning shortly before sunrise. The moon is shown at the left end of its position line. We found already that the measured quantity should be one coordinate of a coordinate system, because the angular distance had been excluded as a viable interpretation. Unfortunately, there is no reasonable assignment of coordinates in this system for that particular day. If we try to match coordinates for the ecliptic coordinate system according to model *MA.6* (figure 13), we immediately see the solution! At the documented time—the end of the night—the moon stands at the left side of its position line roughly in southeast. The longitude fits well with the corresponding part of the measuring circle. The vertical line from the end of the lunar night line is nearly a tangent to the circle. It must be emphasized that this kind of observation occurs more often. This and many more similar cases strongly corroborate model *MA.6*. If one browses through the plots of all events, one realizes than almost all events of the "before"–relation find the moon on the left side of the figure (figure 13) and on the right side in case of the "in front of"–relation (figure 15). Hence the best candidate for an interpretation of the remaining topographical relation terms is

MA.6.1   "above" and "below" describe differences in latitude. One object stands "above" another, if its latitude is more northern, otherwise it is "below" the other object.

Figure 16 shows a typical case for the "above"–relation, which nicely plots how the moon approaches the star in the course of the night until they have the same longitude and when the latitudinal difference is nearly the same as the angular distance between the two objects.

10   Further Corroborating Findings

10.1   Dual Coordinates

Up to now we interpreted the standard form of the topographical relations. Moving to more complex observation reports allows even sharper tests of the favored interpretation hypothesis.

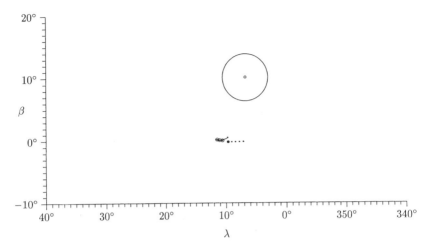

Figure 15
102^y Seleucid era 3^m 23^d, last part of night, moon behind α Arietis, $1\frac{1}{2}$ KÙŠ (ecliptical coordinates).

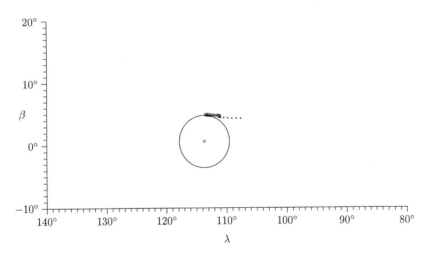

Figure 16
2^y Philip Arrhidaeus 6^m 23^d, last part of night, moon above Saturn, $1\frac{1}{2}$ KÙŠ).

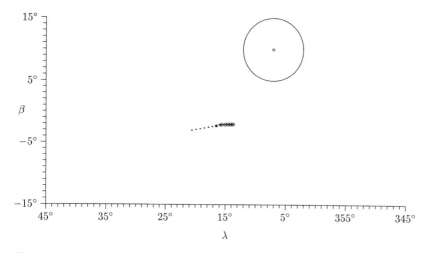

Figure 17
102^y Seleucid era 11^m 5^d, moon behind α Arietis, 2 KÙŠ, 5 KÙŠ to the south.

A number of observation reports mention more than one topographical relation and their quantity. In the translation these expressions are rendered as "low to the south," "high to the north," "back to the west," and "passed to the east" followed by another quantitative value. A schematic form of these expressions is:

at t : O_1 stands ['in front of'/'behind'] O_2 with D, low to the south with N.

What is the meaning of the additional term? The next plausible interpretation expands the current working model and assumes that the complementary ecliptic coordinate is recorded.

$\mathscr{M}A.6.1.1$ "Low to the south" and "high to the north" measures ecliptic differences. The first object stands in the north if its difference in latitudes with the second object is positive.

In figure 17 we find an interesting case of a dual coordinate observation with large quantitative values, which allow a very critical test of model $\mathscr{M}A.6.1.1$. Because both coordinates are tested in the same observation, a small rotation of the coordinate system would immediately show differences in the combined set of values.

In this case the moon is 2 KÙŠ behind the star α Arietis and 5 KÙŠ "low to the south." The position of the moon at the beginning of the night (on the right side of the night line) very nicely coincides with the

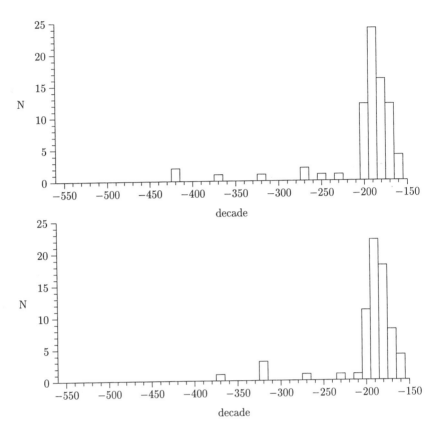

Figure 18
Frequencies *N* of relations "in front of" (top) and "behind" (bottom) with dual coordinates, accumulated for decades.

left side of the measurement circle. The circle has a radius equivalent to 2 KÙŠ. If one increases the size of the circle to 5 KÙŠ, the lower end of the circle touches the latitude of the moon! This is a fantastic confirmation of the interpretive model.

Over the entire historical period covered by the *Diaries*, there is no major revision of the observation records. The distribution of *Lunar Six*–observations remained the same as well as the standard form of the topographical relations. I could find more frequent observations of the mixed coordinate events in later periods. Figure 18 is a histogram of both types of relations, which shows many more observations from a period later than −200. This finding might reflect a more sophisticated observational technique, or a different measurement procedure that includes more arbitrary configurations of the celestial objects than the exact juxtaposition in one coordinate. It should be particularly interesting to see the development of this kind of observation in the yet unpublished later *Diaries*.

10.2 Unit of Measure Reconsidered

The unit of measurement—the KÙŠ—has been determined primarily on grounds of best fits with an intermediate hypothesis for one class of topographical relations ("above" and "below" observations). Therefore the best fitting values for the metrological unit must be redetermined.

Once the model for the topographical relations could be corroborated, one can conversely compute the appropriate values for the metrological units: for the given time of observation the differences of latitudes are computed and the corresponding value for one unit of KÙŠ follows directly. According to the standard hypothesis the measurement of the documented value D corresponds to a calculated quantity C. An estimate for the length of the unit is obtained by dividing C/D. The distribution of values for both topographical relations can be seen in figure 19 for 167 events in case of the "above"-relation and for all 260 events of the "below"-relation. In the first case one obtains a medium value of $\mu = 2°37 \pm 0°11$, and in the second $\mu = 2°43 \pm 0°09$. Both findings are compatible with the assumption that a KÙŠ has a value of 2°5. Yet, the distribution of values is not Gaussian. This fits the distribution of fractions of a KÙŠ, which shows both the fractions for a KÙŠ with 30 fingers and one with 24 fingers. As mentioned before, the Old Babylonian division has always been 30 fingers in a KÙŠ, without exception.[27] Metrologically the finger is linked with the night watch, 1 finger $= \frac{1}{12} \times \frac{1}{16}$ part of a watch. For equinoctial days this would yield 5'. 12 KÙŠ then make one NINDA, which in astronomical contexts corresponds to a full sign. There has been no variation in the link between a finger and a watch, so that it appears to be a fixed unit. In Late Babylonian times, especially in common life, the 24 finger variant becomes frequent.

There are at least two possible scenarios for the mixed usage of different angular units KÙŠ.

1. If the size of a SI (finger) was fixed, the unit of a KÙŠ had to vary and consequently loose its link to the NINDA (then 15 KÙŠ). Then the relations hold: $KÙŠ_{30} = 2°5$, while $KÙŠ_{24} = 2°$.

2. If the link of the KÙŠ to the NINDA $\frac{1}{12}$ stays fixed, the size of the KÙŠ would not vary despite the fact that both divisions into 24 or 30 fingers could occur. We find that in Late Babylonian the size of the KÙŠ remained

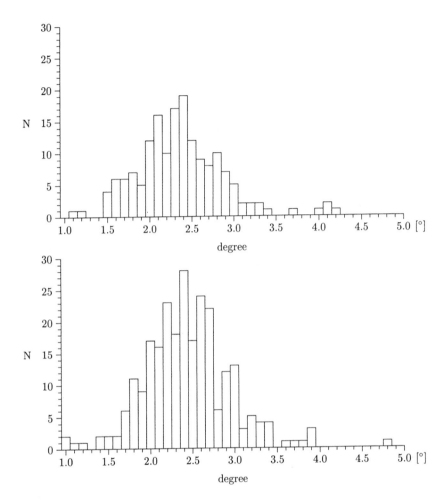

Figure 19
Possible values for a KÙŠ in degrees and the numbers of observations that support
them. In the first diagram 167 observations with the relation "above," in the second
260 with the relation "below" have been considered. To summarize, one KÙŠ best
corresponds to 2.37 ("above") or 2.43 ("below"), respectively.

Table 11
Standard derivation, total number and number of outliers of measurements of topographical relations

	σ	N_{tot}	N_{out}
above	1.086	29	2
below	1.497	108	19
in front of	1.021	54	9
behind	1.436	39	6

constant although it was divided differently.[28] In this case the value of one KÙŠ amounts to $2°5$ independently of the division into fingers. In this variant the dimension of a finger would change.

It is unclear to which extent each division was used for astronomical purposes. As the distribution of fractions shows, clearly both variations can be attested. Yet one must be careful in estimating the relative number of the variations. For small angular distances the astronomers might have used a different instrument with a different scale division than for larger distances. If the estimation is based on the small distances alone, one might get a biased result for the entire sample. The only certain outcome achieved so far is that the mixture of both scale systems can be attested and that the data fit $2°5$ for a KÙŠ best.

10.3   Accuracy of Measurement

With these results we can determine the accuracy of the Babylonian measurements again. The standard variation of the measurements for all four topographical relations is tabulated in table 11. With a general size of one degree and a resolution of the measurement of about half a KÙŠ both fit very well. The ratio between the resolution of measured values and the standard variation is similar to the time measurements of the *Lunar Six*.

11   Ancient Witness

Babylonian astronomy strongly influenced Ptolemaic astronomy especially through Hipparchus. In Hipparchus's *Commentary on Aratus* we find extensive usage of Babylonian terminology.[29] Ptolemy quotes two observations of Mercury with apparent Babylonian origin.[30] Ptolemy uses these observations for the derivation of his planetary model, and in these two cases he quotes the observations without full conversion to the Greek metrological system:

Table 12
Summary of the meaning of the relational expressions (lines) and the additional remarks (columns) used; for configurations between celestial bodies O_1 und O_2 with ecliptical coordinates λ_1, β_1 and λ_2, β_2. The measurement is denoted as D_1, accompanied by D_2 in case of dual coordinates.

	standard	back to the west	passed to the east	balanced	further specification
above	ecliptical difference of latitude, $\beta_1 > \beta_2$, $D_1 = \beta_1 - \beta_2$ small difference of longitude	difference of longitude $\lambda_1 < \lambda_2$, $D_2 = \lambda_2 - \lambda_1$	difference of longitude $\lambda_1 > \lambda_2$, $D_2 = \lambda_1 - \lambda_2$	very small difference of longitude	—
below	ecliptical difference of latitude, $\beta_1 > \beta_2$, $D_1 = \beta_2 - \beta_1$ small difference of longitude	difference of longitude $\lambda_1 < \lambda_2$, $D_2 = \lambda_2 - \lambda_1$	difference of longitude $\lambda_1 > \lambda_2$, $D_2 = \lambda_1 - \lambda_2$	very small difference of longitude	—
	standard	high to the north	low to the south	balanced	further specification
in front of	ecliptical difference of longitude, $\lambda_1 > \lambda_2$, $D_1 = \lambda_2 - \lambda_1$ undetermined difference of latitude	difference of latitude $\beta_1 < \beta_2$, $D_2 = \beta_1 - \beta_2$	difference of latitude $\beta_1 < \beta_2$, $D_2 = \beta_2 - \beta_1$	very small difference of latitude	with planets occasionally: to the west
behind	ecliptical difference of longitude, $\lambda_1 > \lambda_2$, $D_1 = \lambda_1 - \lambda_2$ undetermined difference of latitude	difference of latitude $\beta_1 < \beta_2$, $D_2 = \beta_1 - \beta_2$	difference of latitude $\beta_1 < \beta_2$, $D_2 = \beta_2 - \beta_1$	very small difference of latitude	with planets occasionally: to the east

In the 75th year in the Chaldean calendar, Dios 14, at dawn, [Mercury] was half a cubit above [the star on] the southern scale [of Libra]. Thus at that time it was in ♎ 14°10′, according to our coordinates. [...]

In the 67th year in the Chaldean calendar, Apellaios 5, at dawn, [Mercury] was a half a cubit above the northern [star in the] forehead of Scorpius [β]. Thus at that time it was in ♏ 21/3°, according to our coordinates.

Ptolemy paraphrases both observations in the form of topographical relations "object 1 above object 2 with X KÙŠ," which is a clear proof of their Babylonian origin. Even more interesting are the details of his evaluation. The measured quantities are differences of latitude between Mercury and the star. Ptolemy, however, uses this observation for the determination of the planetary longitude. How did he arrive there?

If he reduces the stellar longitudes for the epoch of the observations, according to his theory he has just to subtract the value for the precession. It amounts to 3°50′ for 373 years in case of the first observation and 4° for 381 years for the second one.[31] In the star catalogue of the Almagest β Scorpii has a longitude of ♏ $6\frac{1}{3}°$ and α Librae a longitude of ♎ 18°. In the second case the resulting longitude would be a little too large. It is plausible that Ptolemy actually did not reduce the longitudes from the star catalogue, but used Hipparchus's value or that of the Babylonian astronomers and added the precession constant to these values.

Independent of the exact derivation of the longitude, it is remarkable that Ptolemy assumes that Mercury and the stars have *the same longitude*. The measured coordinate of half a KÙŠ does not enter the calculation at all! This exhibits Ptolemy's understanding of the Babylonian report in two ways:

1. The measured topographical relation is a coordinate value, for example, either a longitude or a latitude.

2. Since he identifies the other coordinate with the longitude, the topographical relations must be understood in the framework of the ecliptic coordinate system!

Ptolemy quotes these excerpts from Hipparchus, who had extensive access to Babylonian ideas.[32] I have no doubt that Ptolemy understood the meaning of the Babylonian observation reports correctly. This is supported by the comparison of the reported observation with the calculated astronomical events.

In table 13 the ecliptic coordinates for Mercury and the stars are given and the difference in latitude is converted into units of KÙŠ. We

Table 13
Celestial positions of the Mercury observations

	Event 1	Event 2
Date (6 a.m.)	30.10.−236	19.11.−244
star	α Lib	β Sco
λ (Mercury)	193°.9	212°.3
λ (star)	193°.9	211°.9
β (Mercury)	2°.3	2°.5
β (star)	0°.6	1°.3
1 KÙŠ	3°.3	2°.4

find a perfect agreement with Ptolemy's report. Now it is also clear, why Neugebauer came to a different interpretation of these observations.

> In both cases one finds that the longitude of Mercury (as morning star) was almost exactly 1° greater than the longitude of the star (the latitudinal intervals are 1;35° and 1;5°, respectively). Since Ptolemy is only interested in the longitudinal component of the distance between the planet and the mean sun, the term ἐπάνω, literally "above" seems here to mean "ahead (in longitude)." This is reminiscent of the terminology of Theodosius, where, however, ἀνώτερον denotes the point ahead in the direction of the daily rotation.[33]

By comparing both observations with the plotted true configurations—as shown in figures 20 and 21—one sees that Neugebauer's calculations suffer from a day error. As we know from the *Diaries*, this can easily happen when the beginning of the month has only been schematically determined. My algorithm used for the determination of the beginning of the month yields the dates for which the plots of the close approaches of Mercury to the stars were drawn. Here in fact the longitudes of both objects coincide. Because of the day error Neugebauer had to believe that the topographical relation term "above" points to a direction in longitude. Since there is no noticeable difference in longitude at that day, but a difference in latitude of exactly the size which Ptolemy reports, the term "above" clearly describes the arrangement of the objects in latitudes. Since Ptolemy is solely interested in the longitudes, he can neglect the reported explicit measure of the latitudinal difference and make use of the implicit measure that both longitudes were the same at the day of observation.

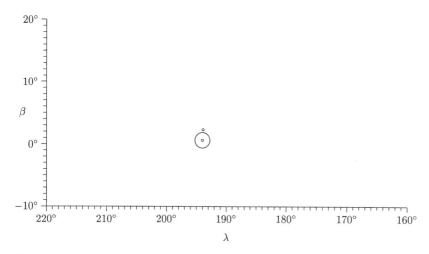

Figure 20
Map of the observation of Mercury 10/30/-236 A.M.

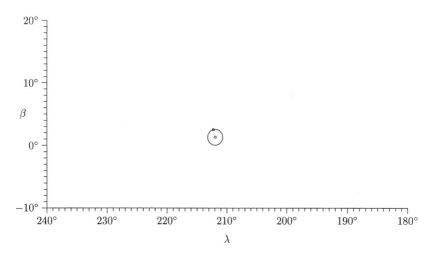

Figure 21
Map of the observation of Mercury on 11/19/-244, 6 A.M.

Therefore, in Ptolemy's report we find an independent confirmation of the interpretation hypothesis, which is tabulated in detail for the various variations of the topographical relation: *topographical relations in the Babylonian Diaries are accurately measured ecliptic coordinates.*

NOTES

1. A complete edition of these group of texts until −164 has been published in two volumes, Sachs and Hunger (1988) and Sachs and Hunger (1989).

2. Neugebauer (1975), p. 546f.

3. Ibid.

4. Ibid.

5. Sachs and Hunger (1988), p. 11.

6. Hunger (1993), p. 138.

7. An excellent introduction to the history of cuneiform script is in Neugebauer (1969), p. 40ff.

8. Kugler (1909–1924).

9. Roughton, according to Hunger (1993), p. 142.

10. For texts of the *Enūma Anu Enlil* tradition David Pingree (1993, p. 265) suggests, "For one planet to be 'in front of' another or a star means that it rises first of the two; therefore, 'in front of' near the eastern horizon signifies 'above,' near the western horizon 'below.'" Koch (1989, p. 143) has another variant.

11. The standard text book is Clocksin and Mellish (1981), advanced users consult O'Keefe (1990). Many example programs are discussed in Sterling and Shapiro (1986).

12. Since 1984 the ephemerides time is substituted by dynamic time, which is measured by the oscillation frequency of caesium.

13. Horizontal coordinates and parallax are calculated according to Meeus (1990).

14. This model assumes that the moon becomes visible when its altitudes exceeds a certain minimal altitude above the sun at the moment of geometrical setting. The minimal depression arc is called *arcus visionis*. The tables of P. V. Neugebauer are based on this model.

15. In case of the *Lunar Six* observations we occasionally find the additional remark *nu pap*—did not observe.

16. Hout (1990), p. 457ff.

17. E.g., Neugebauer (1955), p. 39.

18. Kugler (1909–1924), pp. 548–550.

19. My calculations are based on a *arcus visionis* model, which P. V. Neugebauer obtained from Schoch with improved parameters. It is also possible to construe models that derive the minimally required contrast of the waning moon in the twilight, which allows it to be perceived. Graeme Waddington (Oxford) suggested several variants and found out that the more complex models have hardly any differences compared to the much simpler Neugebauer models. Based on these evaluations, one should expect that an *arcus visionis* model for the lunar visibility is more or less the best one can get. Some improvement can be obtained by an *arcus visionis* value for the moon that decreases with the azimuth difference between sun and moon.

20. In one case we find a corresponding "faint" for moonrise to sunrise.

21. See the article in this volume.

22. No significant differences in accuracy could be found.

23. A coordinate used by Hipparchus, cf. Neugebauer (1975), p. 279.

24. Exceptions are the beginnings of the night.

25. The notation has been simplified for the models of $M\ 1.1.1.1$ to $M\ A$. Hence, model $M\ A.1$ is the first alternative for an expansion of the model.

26. One has to reverse the relation for cases in which the right ascension of one object is close to but smaller than 360 and the other larger than 0.

27. Cf. Friberg (1993), p. 387. I owe the metrological insights to discussions with Jöran Friberg.

28. Hout (1990), p. 467ff.

29. Cf. Neugebauer (1975), p. 279ff, 304, 544, 591ff.

30. Cf. Neugebauer (1975), p. 159.

31. Note that Ptolemy uses a precession constant of one degree per hundred years, which is much too small.

32. I carefully avoid reference to "sources," although it is highly probable that Hipparchus must have had a rather comprehensive access to either original or transcribed Babylonian sources, considering the wealth of Babylonian concepts utilized by Hipparchus. Cf. Toomer (1988).

33. Neugebauer (1975), p. 591.

References

Bretagnon, P., and Simon, J.-L.: 1986, *Planetary programs and tables from −4000 to +2800*, Willmann-Bell, Richmond.

Chapront-Touzé, M., and Chapront, J.: 1991, *Lunar Tables and Programs from 4000 B.C. to A.D. 8000*, Willmann-Bell, Richmond.

Clocksin, W. F., and Mellish, C. S.: 1981, *Programming in Prolog*, Springer, New York.

Friberg, J.: 1993, On the Structure of Cuneiform Metrological Table Texts from the −1st Millennium, *in* Galter (1993), 383−405.

Galter, H. D. (ed.): 1993, *Die Rolle der Astronomie in den Kulturen Mesopotamiens: Beiträge zum 3. Grazer Morgenländischen Symposion, 23.−27. September 1991*, Vol. 3 der Grazer Morgenländische Studien, Graz.

Graßhoff, G.: 1990, *The History of Ptolemy's Star Catalogue*, Springer, New York.

Hout, T.: 1990, Maße und Gewichte, *in* D. O. Edzard (ed.), *Reallexikon der Assyriologie und vorderasiatischen Archäologie*, Vol. 7, Berlin.

Hunger, H.: 1993, Astronomische Beobachtungen in Neubabylonischer Zeit, *in* Galter (1993), 139−147.

Koch, J.: 1989, *Neue Untersuchungen zur Topographie des babylonischen Fixsternhimmels*, Harrassowitz, Wiesbaden.

Kugler, F.: 1909−1924, *Sternkunde und Sterndienst in Babel, assyriologische, astronomische und astralmythologische Untersuchungen*, Vol. II, Münster.

Meeus, J.: 1990, *Astronomical Algorithms*, Willmann-Bell, Richmond.

Montenbruck, O., and Pfleger, T.: 1989, *Astronomie mit dem Personal-Computer*, Springer, Berlin.

Neugebauer, O.: 1955, *Astronomical Cuneiform Texts*, Lund Humphries, London. Reprint 1983.

Neugebauer, O.: 1969, *Vorlesungen über Geschichte der antiken mathematischen Wissenschaften*, Vol. 1, 2d ed., Springer, Berlin.

Neugebauer, O.: 1975, *A History of Ancient Mathematical Astronomy*, Springer, Berlin.

O'Keefe, R. A.: 1990, *The Craft of Prolog*, MIT Press, Cambridge.

Pingree, D.: 1993, Venus Phenomena in *Enūma Anu Enlil, in* Galter (1993), 259−273.

Ptolemy, C.: 1907, *Claudii Ptolemaei opera quae exstant omnia*, Vol II: Opera astronomica minora, Heiberg, Leipzig.

Sachs, A., and Hunger, H.: 1988, *Astronomical Diaries and Related Texts from Babylonia. Diaries from 652 B.C. to 262 B.C.*, Vol. I, Österreichische Akademie der Wissenschaften, phil. hist. Klasse 195, Vienna.

Sachs, A., and Hunger, H.: 1989, *Astronomical Diaries and Related Texts from Babylonia. Diaries from 261 B.C. to 165 B.C.*, Vol. II, Österreichische Akademie der Wissenschaften, phil. hist. Klasse 210, Vienna.

Sterling, L., and Shapiro, E.: 1986, *The Art of Prolog*, MIT Press, Cambridge.

Toomer, G.: 1984, *Ptolemy's Almagest*, Duckworth, London.

Toomer, G.: 1988, Hipparchus and Babylonian astronomy, *A scientific humanist: Studies in memory of Abraham Sachs. (Occasional Publications of the Samuel Noah Kramer Fund 9*, ed. E. Leichty et al., Philadelphia, 1988*)*, 353–362.

6

GOAL-YEAR TABLETS: LUNAR DATA AND PREDICTIONS
Lis Brack-Bernsen

This paper deals with lunar observations and predictions of the ancient Babylonians. Apart from lunar and solar eclipses (which are not dealt with here), the Babylonians regularly observed some characteristic time intervals in the days around conjunction and opposition, the so-called "Lunar Six." We elucidate these horizontal phenomena and mention the Babylonian sources in which they occur. We explain how the lunar data collected on the Goal-Year tablets can be used for predicting the Lunar Six, and we demonstrate through textual evidence that the Babylonians, indeed, predicted the phenomena in exactly this way. More precisely, we show that the "Lunar Four" time intervals around opposition can be predicted very accurately for an arbitrary full moon by a very simple calculation, using their values observed one Saros earlier than the full moon in question. In case of the new-moon phenomena KUR and NA, one has to know not only their values one Saros earlier but also the sums $SU + NA$ and $ME + GE$, respectively, occurring one Saros plus 6 months earlier. This fact explains for the first time why the Goal-Year text composed for a year Y contain not only all the Lunar Six of the year $Y - 18$, but also the sums $SU + NA$ and $ME + GE$ for the last 6 months of year $Y - 19$.

1   INTRODUCTION

The "Lunar Six" are some characteristic time intervals between sunrise or sunset and moonset or moonrise observed on the days around conjunction (KUR and NA) and opposition ($\check{S}\acute{U}$, NA, ME, and GE, which we call the "Lunar Four").[1] Although spectacular and easy to observe, these time intervals are very complicated quantities from a theoretical point of view. Neugebauer (1957), 107–109, describes in detail the factors that determine whether the new moon crescent can be seen after sunset on the evening after new moon (conjunction). The same factors also decide how long the new moon can be seen in the evening of its first visibility: The time interval between sunset and the first visible moonset after conjunction is called NA_N. The others of the Lunar Six are correspondingly determined by similar factors.

The Lunar Six time intervals were of great interest for the Babylonians. The Diaries [Sachs and Hunger (1988, 1989, 1996)] contain reports of some of these horizontal phenomena as early as 568 B.C., indicating that the Babylonians observed them month by month during a time span of at least 500 years (568–62 B.C.).

During the last three centuries B.C. (Seleucid time), ephemerides for the moon were calculated with the aim of determining the different Lunar Six, whose calculated magnitudes were recorded in the last column of the text. In the preceding columns leading to this goal, the parameters that determine this value were taken into account through periodically varying functions. The fundamental periods used here are the length $P_\odot$ of the solar year, the period $P_{\mathfrak{C}}$ of the lunar velocity, and the period P_Ω of the moon's movement in latitude. ($P_{\mathfrak{C}}$ and P_Ω are the periods of the moon velocity $v_{\mathfrak{C}}$ and of the moon latitude $\beta_{\mathfrak{C}}$, respectively, when these are computed once each synodic month, such as the Babylonians did in their ephemeris texts.) See ACT I, in which Neugebauer has published the lunar ephemerides, discussed the Lunar Six in detail, and explained how the Babylonians calculated them by a skillful combination of the relevant influences. See also HAMA, 474–555, and van der Waerden (1974), 205–249.

We have recently shown that the Lunar Four contain information on the fundamental periods mentioned above. For example, the period $(P_\Phi = P_{\mathfrak{C}})$ of the second column of a System A lunar ephemeris tablet can be determined empirically by the sum of all Lunar Four, whereas partial sums oscillate with the period $P_\odot$ (Brack-Bernsen 1990, 1994). In Brack-Bernsen and Schmidt (1994), we have explained why and how this works, through offering an astronomical interpretation and analysis of (partial) sums of the Lunar Four.

The present paper concentrates on the Babylonian treatment of these lunar observables. We shall see how they predicted the Lunar Six using collections of earlier lunar data recorded on the so-called Goal-Year tablets. Before doing so, we give a short overview of the variety of Babylonian texts concerned with the Lunar Six time intervals.

2 Lunar Six Data in Babylonian Texts

The interest of the Babylonians in the Lunar Six is documented through their occurrence in a variety of different types of texts (Sachs 1948; LBAT).

Over a period of more than 500 years, Lunar Six time intervals were regularly observed and recorded in the Diaries. There, however, we often find remarks about cloudy weather and that it was not possible to observe the moon. But still, the text gives the relevant lunar data. Clearly, such Lunar Six values must have been predicted somehow. Until now it has been a mystery how the Babylonians were able to predict such complicated phenomena, even as early as 568 B.C., and it has not been known how good their predictions were.

The *Almanacs* and *NS (normal star) Almanacs* record, among other information, Lunar Six data. The Babylonians even collected series of Lunar Six data over at least 60 consecutive months. For example, this is documented by a table (*LBAT 1431*)[2] containing a compilation of all Lunar Six data for a period of more than five years. It gives us a hint that they probably used these data for theoretical or empirical purposes.

Still another type of tablets contains compilations of Lunar Six data, excerpted from the Diaries, the so-called *Goal-Year tablets*. The Goal-Year tablets also contain observed characteristic phenomena for all the five planets known by the Babylonians. It is well known how the Goal-Year texts were used for predicting astronomical events for the planets, namely by using events of the same kind from some earlier characteristic time interval (see Sachs 1948 and *LBAT*, xxv).

Finally, the *Ephemerides* of the moon aimed at calculating the value of the different Lunar Six, whereas some *Procedure Texts* gave brief instructions on how to proceed. These lunar ephemerides attest to the most advanced part of the Babylonian astronomy. Throughout the columns leading to the calculated value of, for example, NA_N, all the variables influencing NA_N were correctly taken into account.

In this paper we shall present a detailed analysis of the lunar data collected on a Goal-Year tablet and explain how they could be used for predicting the Lunar Four. We will then demonstrate through textual evidence that the Babylonians predicted the phenomena in exactly this way. Indeed, a tablet from Uruk, TU 11, contains a short remark about finding *NA* and *GE*. This remark is so concentrated that it only became intelligible after we had reconstructed how the lunar data could possibly be used for predictions (Brack-Bernsen, 1994, section 4b). But then it reveals a procedure identical to the one we had reconstructed. In addition, the text TU 11 tells us how to calculate (and hence predict) the new moon phenomenon NA_N, another of the Lunar Six, by a very easy and precise method that we shall explain in detail. This text, also known as AO 6455, was published in cuneiform transcription by Thureau-Dangin (1922) and has recently been translated by Hunger. A small part of this text, namely, section 19 (rev. 8–15), was translated and published by Neugebauer (1947). Neugebauer just states that TU 11 contains a collection of rules for lunar and planetary phenomena and that it would lead far beyond the scope of his article to analyze all the relevant passages of the text. He writes that the translation of section 19 will show the general direction of these rules.

The rules demonstrated in section 19 through calculated examples are, indeed, very important. They reveal to us one (although rather primitive) method to extrapolate from one observed value of *KUR*. We therefore asked Hunger to translate the rest of the tablet TU 11.

Van der Waerden (1949) explained the rules of section 19 and connected them to older texts. Van der Waerden (1951), 29, must also have been in possession of at least a partial translation of other parts of the text TU 11. He surmises that in this text, some rules are given for calculating risings and settings of the Moon from observed values either a few days or 18, 36, or 54 years earlier. How these rules worked in detail, he does not tell; only that some indications about the methods might be drawn from TU 11. Our present result confirms his surmise and demonstrates how some of the rules worked.

Before presenting our systematic analysis and the new results from section 7 on, we shall give in sections 3–5 some basic knowledge and understanding of the Lunar Six phenomena. In section 6, we give a short survey of the Goal-Year tablets and summarize what so far has been known about them and their use.

3   THE PHENOMENON *KUR*

In figure 1 the phenomenon *KUR* is illustrated in detail. The horizontal (thin) great circle is the horizon, the (thick) oblique circle is the celestial equator (as seen from Babylon), and the dotted great circle is the ecliptic. We consider a morning shortly before new moon. The sun ☉ and the moon ☾$_{KUR}$ are placed somewhere on the ecliptic near the eastern horizon; we have neglected the latitude of the moon. The arc of the ecliptic between moon and sun may be around 20°; the moon has thus risen visibly about $1\frac{1}{2}$ hour before sunrise. On the next morning, however, the moon will be so close to the sun that the moonrise is invisible. The time difference between the last visible moonrise (before conjunction) and the sunrise is called *KUR*. It is the time it takes the elongation arc(☾$_{KUR}$, ☉) to rise.

The moon rises at the same time as point *B* of the equator, while the sun rises simultaneously with point *A*. Therefore, *KUR* is given by the length of arc(*A*, *B*) and depends on where the elongation arc(☾$_{KUR}$, ☉) is

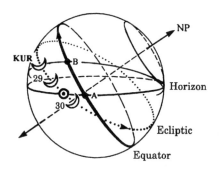

Figure 1
The celestial sphere for Babylon. Position of the moon near the eastern horizon at the moment of *sunrise*, shown on three consecutive mornings around conjunction (new moon). In the positions 29 and 30, the moon is too close to the sun to be observable.

placed on the ecliptic and also upon the length of $\mathrm{arc}(\mathbb{C}_{KUR}, \odot)$.[3] For measuring $\mathrm{arc}(A, B) = KUR$, we here use the Babylonian unit time degrees *uš*.[4] For further details, see Brack-Bernsen and Schmidt (1994).

Let us visualize the movement of the moon relative to the sun in the days around conjunction. On the following mornings, the rising moon will not be visible. Still, we have in figure 1 marked the position of the moon relative to the sun at the moment of sunrise. In our example we have assumed that the phenomenon *KUR* was measured on the 28th day of the Babylonian month. The next morning (of day 29), the invisible moon will be at the position $\mathbb{C}_{29}$ when the sun rises, while it will be at position $\mathbb{C}_{30}$ with the next sunrise. Sometimes between mornings 29 and 30, the moon will pass the sun: the conjunction has taken place.

Note that, if one knew the time (say, x *uš*) it takes $\mathrm{arc}(\mathbb{C}_{KUR}, \mathbb{C}_{29})$ to rise, it would be possible to determine the time difference between moonrise and sunrise on the morning of day 29. This nonobservable "KUR_{29}" is, of course, equal to $KUR - x$ *uš*. We will call this rising time, x *uš*, of $\mathrm{arc}(\mathbb{C}_{KUR}, \mathbb{C}_{29})$ the "daily change of KUR" and denote it by ΔKUR.

In this connection, we note that section 19 of TU 11, translated by Neugebauer as mentioned above, is very important. It shows us by some examples how the Babylonians, starting from the known value of *KUR*, found through extrapolation the estimated value of *KUR* for following mornings, when the moon was invisible. For ΔKUR, the daily change of *KUR*, they used four times the length of daylight measured in minas. In *Mul-Apin*, tablet II, ii 43–iii 15, *ŠÚ* (the setting of the moon) and *KUR* (the rising of the moon) are calculated as four times the length of the nighttime measured in minas. If, instead of $\check{SU}$ and KUR, we read $\Delta\check{SU}$ and ΔKUR, respectively, this scheme in *Mul-Apin* makes sense: It tabulates the values of the daily change of $\check{SU}$ and KUR to be used for extrapolations, as demonstrated in the examples of section 19 of text TU 11. Van der Waerden (1949) has exactly the same understanding of these texts.

4 LUNAR PHENOMENA AT THE EASTERN HORIZON: *ME*, *GE*, AND *ME* + *GE*

We now concentrate on the Lunar Four, the characteristic time intervals observable in the days around the opposition. As an example, we consider the opposition that occurs half a month later than the conjunction dealt with in figure 1. The time difference *ME* between the last moonrise before opposition and the sunset is based on the moon at the eastern horizon and the sun in the opposite direction (i.e., on the western horizon). In the following figures, we introduce for the sake of simplicity the symbol $\overline{\odot}$ for the "anti-sun," which we define as the point on the ecliptic situated directly opposite the sun. At the very moment when the sun rises, $\overline{\odot}$ sets, and vice versa. With this definition, *ME* is the time difference between the risings of the full moon and the anti-Sun $\overline{\odot}$. In figure 2 the phenomenon *ME* is illustrated analogously to *KUR* in figure 1. The horizontal (thin) great circle is the horizon, the (thick) oblique circle is the celestial equator (as seen from Babylon), and the dotted great circle is the

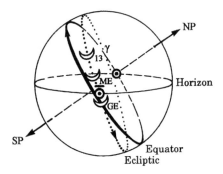

Figure 2
The celestial sphere for Babylon. The position of the moon near the eastern horizon at the moment of *sunset* is shown on three consecutive evenings around the opposition (full moon), which takes place half a month later than in figure 1.

ecliptic which in the present example stands very steep. We consider the evening just before opposition. The sun is about to set (i.e., the anti-sun is rising), while the moon already has risen. We have marked the position of the moon $\mathbb{C}_{ME}$ at the moment of anti-Sun rise. ME is the time it takes the elongation arc($\mathbb{C}_{ME}, \overline{\odot}$) to rise. This rising time is determined by the length of the arc and by its position on the ecliptic.

As in figure 1, we mark also here the positions of the moon at the moment of anti-sun rise on some consecutive days around opposition. Let us assume ME to be observed on the evening of the 14th day in the Babylonian month. On the previous evening, the moon was at the position $\mathbb{C}_{13}$ at the moment of anti-sun rise. Sometime during the day 14, at the moment of opposition, the moon will pass the anti-sun $\overline{\odot}$. On the following evening (15) at anti-sun rise, the moon will be in position $\mathbb{C}_{GE}$; still under the horizon. The time it takes the arc($\overline{\odot}, \mathbb{C}_{GE}$) to rise is the Babylonian observable GE. In short:

ME is the time from last moonrise to sunset before opposition and GE is the time from the first sunset after opposition to moonrise.

Both quantities denote actually the same time interval that changes sign between the two days around opposition; their definitions are chosen such that they always are positive. The sum $ME + GE$ can be interpreted as the decrease of ME during the day of opposition, or as the increase of GE during the same day. Now, $ME + GE$ is the rising time of arc($\mathbb{C}_{ME}, \mathbb{C}_{GE}$). It is determined by its length (1 day $\times$ ($v_{\mathbb{C}} - v_{\odot}$)) and by its position $\lambda_{\mathbb{C}}$ on the ecliptic. Here the velocities ($v_{\mathbb{C}}$ and $v_{\odot}$) are measured in degrees per day (°/day) and $\lambda_{\mathbb{C}}$ (the position of $\overline{\odot}$) is the longitude on the ecliptic at which the opposition takes place. The length (1 day $\times$ ($v_{\mathbb{C}} - v_{\odot}$)) of arc($\mathbb{C}_{ME}, \mathbb{C}_{GE}$) tells us how far the moon has moved relative to the sun during the day of opposition.

The sum $ME + GE$ is a rather simple quantity: It depends mainly on the two variables $v_{\mathbb{C}}$ and $\lambda_{\mathbb{C}}$:

$$ME + GE = (ME + GE)(v_{\mathbb{C}}, \lambda_{\mathbb{C}}).$$

Each single of the Lunar Four, however, is a much more complicated quantity, depending on four different astronomical variables. Let us, as an example, consider GE, which is the time it takes the arc($\overline{\odot}, \mathbb{C}_{GE}$) to rise. This rising time depends on $\lambda_{\mathbb{C}}$ (i.e., the position on the ecliptic at which the opposition takes place), on the lunar latitude $\beta_{\mathbb{C}}$ at opposition, and on the length of arc($\overline{\odot}, \mathbb{C}_{GE}$). This length equals the product of Δt and ($v_{\mathbb{C}} - v_{\odot}$), the lunar velocity relative to the sun. Here Δt is the time interval between opposition and anti-sun rise on the next morning. Hence *GE is also strongly depending on when the opposition takes place with respect to sunset.* In summary, we must consider it as function of all these four variables:

$$GE = GE(\Delta t, v_{\mathbb{C}}, \lambda_{\mathbb{C}}, \beta_{\mathbb{C}}).$$

The same holds for all the Lunar Four. The crucial point is now that through addition of *ME* and *GE*, the dependence on Δt and $\beta_{\mathbb{C}}$ is practically eliminated; for further details, see Brack-Bernsen and Schmidt (1994).

5   LUNAR PHENOMENA AT THE WESTERN HORIZON: ŠÚ, NA, AND ŠÚ + NA

We are now concerned with the *western* horizon in the days around the same opposition as in figure 2. The phenomena ŠÚ and NA are visualized in figure 3. Here we have drawn the celestial sphere (as seen from Babylon) at the moment of sunrise. In the same way as in figure 2, we have marked the positions of the (full) moon relative to the anti-sun—now at the moment of anti-sun set—in the mornings around opposition. The position of the ecliptic has the same inclination to the horizon as in figure 1.

ŠÚ is the setting time of arc($\mathbb{C}_{SU}$, $\overline{\odot}$), and NA is the setting time of arc($\overline{\odot}$, $\mathbb{C}_{NA}$). These observables are both strongly dependent upon the variable Δt, which equals the time from sunrise to opposition for ŠÚ, and

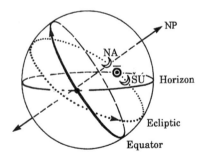

Figure 3
The celestial sphere for Babylon. Position of the moon near the western horizon at the moment of *sunrise* on two consecutive mornings around the same opposition (full moon) as in figure 2.

the time from opposition to next sunrise for NA. They also depend on $\beta_{\mathbb{C}}$, $\nu_{\mathbb{C}}$, and $\lambda_{\mathbb{C}}$:

$$NA = NA(\Delta t, \nu_{\mathbb{C}}, \lambda_{\mathbb{C}}, \beta_{\mathbb{C}}).$$

The sum $\check{S}\check{U} + NA$, finally, is the setting time of arc($\mathbb{C}_{SU}$, $\mathbb{C}_{NA}$) and depends only on $\nu_{\mathbb{C}}$ and $\lambda_{\mathbb{C}}$. The length of this arc is, of course, (1 day $\times$ $(\nu_{\mathbb{C}} - \nu_{\odot})$) as in the case of $ME + GE$, namely, the distance which the moon traveled relatively to the sun during the day of opposition.

6   THE GOAL-YEAR TABLETS

There exist about 150 Goal-Year texts, counting both fragments and whole tablets. They are registered, and most of them also published, as numbers 1213 through 1367 by Sachs in *LBAT*. They contain observations excerpted from the Diaries: raw material for the prediction of planetary and lunar phenomena for a given year Y, named the "goal year" by Sachs. (See also his discussion of the Goal-Year tablets in Sachs [1948], 282–285.) Of these tablets, 93 contain data concerning 56 different goal years in the time between 71 S.E. and 352 S.E.

The structure of all the Goal-Year tablets is the same. On the front side, planetary observations are recorded in a strict order while most of the reverse side is covered with lunar data, always written in a schematic and similar way. After the lunar section, a colophon-title usually follows. The title states the contents of the table and its purpose, and typically reads (in Sachs's translation): "the first day, appearances, passings, and eclipses which have been established for the year Y," where Y is the goal year. The tablet is covered with observations of appearances, passings, first days of visibility, etc., from different specific years prior to year Y.

For each of the five known planets, the characteristic phenomena, observed in a year that precedes the actual goal year Y by a number of years specific for each planet, are recorded in different sections. Obviously, the Babylonians were aware that a planet returned to the same characteristic appearance after the lapse of a relevant period of time; and obviously they used this knowledge for making predictions.

A concrete example may elucidate how this works in practice. In the first paragraph of the Jupiter section, the Greek-letter phenomena observed throughout year $Y - 71$ are recorded. The text utilizes the fact that after 71 years, namely in the year Y, these phenomena repeat themselves. However, Jupiter reaches the same position on the sky only after

83 years. Therefore, Jupiter's conjunctions with normal stars occurring during the year $Y - 83$ are collected in a second paragraph.

All this is well known, as well as the type of data that were put together in the lunar section: lunar and solar eclipses occurring one Saros (= 18 years) earlier than the goal year Y; the Lunar Six, month by month, during this whole year $Y - 18$; and their partial sums $\check{S}\acute{U} + NA$ and $ME + GE$ for the last 6 months of year $Y - 19$. We know that lunar eclipses repeat after one Saros; hence we understand how the Babylonians could utilize the recorded eclipse data. However, the way in which the Lunar Six data were used for predictions has not been investigated yet. (A part of our new results has been published in German in Brack-Bernsen (1994). Van der Waerden (1974, 110) made some conjectures about the use of these lunar data; we will comment on his remarks at the appropriate places.)

Let us therefore take a closer look at the Lunar Six data occurring on the Goal-Year tablets. Knowing that the structure is the same for all tablets, we take *LBAT* 1285 as an example.

7   The Goal-Year Tablet LBAT 1285

The colophon states the goal year to be 194 SE. The lunar data collected for predictions for this year are written in four columns on the reverse of the tablet. We know that year 194 SE had 13 months: I, II, ..., XII, and XII$_2$, and that the months one Saros = 223 synodic months prior to these were month XII$_2$ 175 SE and I–XII 176 SE. Correspondingly, the values of the Lunar Six as well as data from five (observed or expected) eclipses from this period of time were recorded. Table 1 contains a transcription of the text. Here we have, however, omitted observational remarks and only reproduced the values of the Lunar Six and their sums as well as the existing dates of the eclipses.

The first column records the values of the sums $\check{S}\acute{U} + NA$ and $ME + GE$ for months VII–XII of year 175 SE. Our reproduction is as short as that of Babylonians; lines 1 and 2 of column (1) state:

year 175 (month) VII 15 $\check{S}\acute{U}$ NA.

The meaning is the following: "In the year 175 (of the Seleucid era) for month VII the sum of $\check{S}\acute{U}$ and NA was 15 $u\check{s}$."

After the remarks on eclipses, the Lunar Six data for all the months from SE 175 XII$_2$ to SE 176 XII follow in thirteen entries, one for each month. The dates within the month on which the phenomenon was

Table 1
Transcription of the Goal-Year text LBAT 1285

year	175		year	176		year			year		
VII	15	ŠÚ NA	I 1			V 1	15 10	NA	IX 30	10 20	ŠÚ
VIII	8	ME GE₆	14.	4 10	ŠÚ	14.	4	ŠÚ	13.	7 30	
	15	ŠÚ NA	15.	3 20	NA	14.	2 20	NA		6 30	
IX	9 40*	ME GE₆	15.	4	ME	15.	5	ME	X 1		NA
	14 50	ŠÚ NA	16.	10 40	GE₆	15.	15	GE₆	11.	20	ŠÚ
	12 10	ME GE₆	27.	15	KUR	27.	20	KUR	12.	13 40	NA
X	14	ŠÚ NA	II 30	10	NA	VI 1	12	NA	13.	1 30	ME
XI	13	ME GE₆	15.	6	ŠÚ	13.	7	ŠÚ	14.	5 30	GE₆
	9 30	ŠÚ NA	15.	9	NA	14.	1 30	NA	27.	7 30	KUR
XII	13 30	ME GE₆	16.	9 50	ME			ME	XI 30	10 10	NA
	8 40	ŠÚ NA	16.	23	GE₆	27.	16	GE₆	12.	17	ŠÚ
	14 30	ME GE₆	28.	10	KUR	VII 1	11 30	KUR	13.	6	NA
...	...	...	III 1	9	NA	12.	13	NA	14.	2 50	ME
175 XII₂	night 15	☽ ecl.	14.	4 20	ŠÚ	13.	7	ŠÚ	15.	0*	GE₆
175 XII₂	day 29	⊙ ecl.	15.	5 30	NA	13.	3 40	ME	27.	15	KUR
176 VI	night 14	☽ ecl.	15.	20 30	ME	14.	1 10	GE₆	XII 30	11	NA
...	...	...	27.	13	KUR	27.	15 30	?	13.	13	ŠÚ
year 175			IV 30	12 20	NA	VIII 1	15	KUR	14.	4 30	ME
XII₂ 30	11	NA	14.	15	ŠÚ	12.	2	NA	14.	7	GE₆
14.	3 50	ŠÚ	14.	0 40	NA	13.	5	ŠÚ	15.	2 30	KUR
15.	7 40	NA	15.	2 20	ME	13.	13	ME	27.	6	NA
15.	3	ME	15.	14 20	GE₆	14.	4	NA	I 30	16	KUR
16.	6 20	GE₆	27.		KUR	26.	22	GE₆		10 40	NA
27.	16 30	KUR						KUR			

Between the dots (...) in the first column, the table contains three reports on eclipses of which we here only render the dates. An asterisk (*) signifies collation by Sachs (H. Hunger, private communication).

observed are written as ordinal numbers in the left side of the column, to the right is the name of the phenomenon, and in the middle of the column is its ascertained value.

As an example, we look closer at the first entry for the month following after the eclipse dates:

year 175 (month) XII$_2$ 30 11 NA.

Here we are told that "in the year SE 175 the first day of month XII$_2$ (which was identical with) the 30th (of the preceding month XII), NA_N was 11 uš." The next line states:

14 3 50 ŠÚ,

which means "(in the morning of) day 14 (of month XII$_2$ year 175 SE), ŠÚ (the time from moonset to sunrise) amounted to 3;50 uš."

Our task is now to find out if and how these lunar data can be used for predicting the Lunar Six for the 13 months of our goal year SE 194. Or more generally: How can the Lunar Six data collected on a Goal-Year tablet be used to predict Lunar Six phenomena of the goal year?

We know that the planetary phenomena and the eclipses collected on the table are expected to repeat themselves in year SE 194. Is the same true for the Lunar Six or for their sums? In order to answer this question we can, for example, compare corresponding values of the same phenomena for lunations 223 synodic months (= 1 Saros) apart.

8 SYSTEMATIC ANALYSIS OF LUNAR FOUR DATA

We know that the structure of each of the Lunar Six in principle must be the same. They are all either the rising or setting time of a little arc of the ecliptic: In case of *KUR* and NA_N this is the elongation arc between the moon and the sun, while for the Lunar Four (ŠÚ, NA, ME, and GE), it is an elongation arc between the moon and the anti-sun.

In our search for an eventual regular behavior of these quantities, it therefore suffices to examine one representative of the Lunar Six and the sums ŠÚ + NA and ME + GE. We look for regularities and for empirical information contained in the Lunar Four that the Babylonians might have found and used for predictions.

For this investigation we have used a recent computer program for lunar ephemerides valid at ancient times (Moshier 1992). Starting at the full moon (lunation 1) that occurred on 20 July 233 B.C., we have calculated[5] the Lunar Four for a series of lunations $1, 2, \ldots, i, \ldots$. At the top of

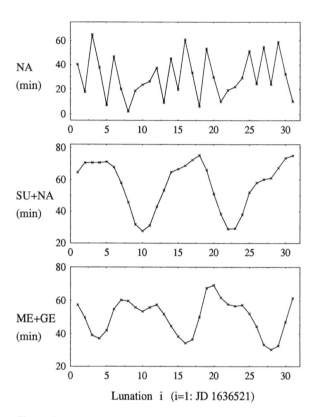

NA
(min)

SU+NA
(min)

ME+GE
(min)

Lunation i (i=1: JD 1636521)

Figure 4
Calculated horizontal phenomena (from top to bottom) *NA*, *ŠÚ + NA* and
ME + GE, as seen from Babylon, over a period of 30 months starting (for *i* = 1) from
JD 1636521 = July 20, 233 B.C.).

figure 4, the calculated values of *NA* for thirty consecutive full moons are
marked by crosses (×), connected by straight lines, and shown as functions
of the lunation number *i*. Below *NA*, the sums *ŠÚ + NA* and *ME + GE*
are shown in the same way. They exhibit a much more regular behavior
than *NA*, demonstrating that some of the dependencies on the variables
Δt, $v_{\mathbb{C}}$, $\lambda_{\mathbb{C}}$, and $\beta_{\mathbb{C}}$ have been partially eliminated by simple addition of the
Lunar Four values.

We now investigate if *NA* or *ŠÚ + NA* and *ME + GE* repeat
themselves after one Saros. Figure 5 compares corresponding values of
these phenomena for lunations 223 synodic months (= 1 Saros) apart.
The crosses mark the same quantities as in figure 4; their values measured
223 synodic months earlier are marked by small circles (o). The curves

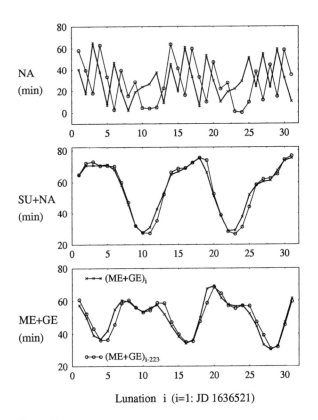

Lunation i (i=1: JD 1636521)

Figure 5
Comparison of the horizontal phenomena from figure 4 (crosses ×, connected by
heavy lines), with those that could be observed 223 months (= 1 Saros) earlier (circles
0, connected by thin lines).

$\v{S}\acute{U} + NA$ and $ME + GE$ obviously repeat themselves after one saros, but
NA does not. Denoting the values of $\v{S}\acute{U}$, NA, and $\v{S}\acute{U} + NA$ established
for the ith lunation as $\v{S}\acute{U}_i$, NA_i, and $(\v{S}\acute{U} + NA)_i$, respectively, we can
express our knowledge in the following way:

$$(\v{S}\acute{U} + NA)_i = (\v{S}\acute{U} + NA)_{i-223},$$

$$(ME + GE)_i = (ME + GE)_{i-223}.$$

In other words, the sums $\v{S}\acute{U} + NA$ and $ME + GE$ are known for each of
the months of year Y (plus for the preceding six months), if a Goal-Year
tablet for this year exists. This can be used in the following way: In a case
where the Babylonians were able to observe, say $\v{S}\acute{U}$, but clouds made the

moon invisible the next morning, the missing NA could of course be established by means of a Goal-Year table composed for the actual year. Let the missing NA be NA_i. In the corresponding Goal-Year tablet, the values for $\check{S}\acute{U}_{i-223}$ and NA_{i-223} can be found and their sum calculated, and hence NA_i is found:

$$NA_i = (\check{S}\acute{U} + NA)_i - \check{S}\acute{U}_i = (\check{S}\acute{U} + NA)_{i-223} - \check{S}\acute{U}_i.$$

Or, inversely: If NA is known, $\check{S}\acute{U}$ can be found. Correspondingly, one can deal with ME and GE: If one of them is measured, the other can be found using the data on an appropriate Goal-Year tablet.

This way of using the sums $\check{S}\acute{U} + NA$ and $ME + GE$ for finding one missing observation was also proposed by van der Waerden (1974, 110). The difference to our interpretation is that he suggested the use of $\check{S}\acute{U} + NA$ from the lunation 19 years earlier instead of using $(\check{S}\acute{U} + NA)_{i-223}$, that is, the value 18 years earlier. The sum $\check{S}\acute{U} + NA$ repeats itself very accurately after 1 Saros (as we have seen in figure 5), but less exactly after 19 years (Brack-Bernsen 1994, figure 2). The sums $\check{S}\acute{U} + NA$ and $ME + GE$ were only recorded during 6 months of the year $Y - 19$, and not during the whole of the year. We are therefore convinced that the Babylonians used the Lunar Six observations 1 Saros before the month in question, and not 19 years before. For the purpose of the data recorded from the year $Y - 19$, we have another proposal (see section 11 below).

In the Diaries, however, there are lunations where both values ($\check{S}\acute{U}$ and NA or ME and GE) are recorded, in spite of remarks saying that in none of the cases had the moon been visible. In the next section, we will investigate how such Lunar Four eventually can be calculated by means of the lunar data observed one Saros earlier.

9 COMPARING $\check{S}\acute{U}_i$ AND $\check{S}\acute{U}_{i-223}$

In sections 4 and 5 we saw that each of the Lunar Four is a function of four variables. Taking $\check{S}\acute{U}$ as an example, we write:

$$\check{S}\acute{U} = \check{S}\acute{U}(\Delta t, v_{\mathbb{C}}, \lambda_{\mathbb{C}}, \beta_{\mathbb{C}}).$$

We concentrate on $\check{S}\acute{U}_i$ and $\check{S}\acute{U}_{i-223}$. The variable Δt is the time difference between the last sunrise before opposition and the opposition itself in the months i and $i - 223$, respectively. The other variables, $v_{\mathbb{C}}$, $\lambda_{\mathbb{C}}$, and $\beta_{\mathbb{C}}$, are the velocity of the moon, its longitude, and its latitude at opposition O_i or O_{i-223}, respectively.

One Saros is defined as 223 synodic months; but in a good approximation it also equals an integer number of anomalistic, sidereal or draconitic months:

223 syn.m. $\approx$ 239 anom.m. $\approx$ 241 sid.m. $\approx$ 242 drac.m. $\approx$ 18 years.

Therefore the three variables $\nu_{\mathbb{C}}$, $\lambda_{\mathbb{C}}$, and $\beta_{\mathbb{C}}$ will have approximately the same magnitudes at oppositions O_i and O_{i-223} situated one Saros apart. The only variable determining $\check{S}\acute{U}$ that might have changed is Δt. We therefore try to find the difference between Δt_{i-223} and Δt_i. Now:

1 Saros = 223 syn.m. = $6585 + \frac{1}{3}$ day.

Therefore, in comparison to sunrise, the opposition O_i will take place $\frac{1}{3}$ day later than was the case at opposition O_{i-223}. Our variable Δt_i, the time difference between the last sunrise before opposition i and the opposition itself, is hence equal to $\Delta t_{i-223} + \frac{1}{3}$ day:

$\Delta t_i = \Delta t_{i-223} + \frac{1}{3}$ days   for $\check{S}\acute{U}$.

We remember that $\check{S}\acute{U}$ is the setting time of the ecliptic arc between $\mathbb{C}_{SU}$ and $\overline{\odot}$, the length of which equals $\Delta t(\nu_{\mathbb{C}} - \nu_{\odot})$. The velocities of the sun and moon are the same at the oppositions O_{i-223} and O_i; only the factor Δt has changed by $\frac{1}{3}$ day. Therefore, at lunation i, arc($\mathbb{C}_{SU}, \overline{\odot}$) will be $\frac{1}{3}$ day $\times (\nu_{\mathbb{C}} - \nu_{\odot})$ larger than at lunation $i - 223$. The time it takes this little arc of length $\frac{1}{3}$ day $\times (\nu_{\mathbb{C}} - \nu_{\odot})$ to set is the difference between $\check{S}\acute{U}_i$ and $\check{S}\acute{U}_{i-223}$. But this setting time is just one third of $\check{S}\acute{U} + NA$. As we saw, $\check{S}\acute{U} + NA$ is the setting time of arc($\mathbb{C}_{SU}, \mathbb{C}_{NA}$) of length 1 day $\times (\nu_{\mathbb{C}} - \nu_{\odot})$. We therefore now have:

$$\check{S}\acute{U}_i - \check{S}\acute{U}_{i-223} = \tfrac{1}{3}(\check{S}\acute{U} + NA)_i = \tfrac{1}{3}(\check{S}\acute{U} + NA)_{i-223}. \tag{1}$$

This can also be demonstrated in a figure. In figure 3, the position of the moon relative to the anti-sun in the days around opposition had been visualized: On a celestial sphere we had drawn the anti-sun at the horizon and for consecutive mornings the position the moon at the moment of anti-sun set (= sunrise). In figure 6, we have twice drawn an enlarged part of the celestial sphere, namely, the situation on the western horizon at the moment of anti-sun set, for the two oppositions O_{i-223} and O_i taking place 1 Saros apart. The movement of the moon relatively to the anti-sun is illustrated. On the morning just before opposition, the moon is in the position $\mathbb{C}_{SU}$, moving slowly along the ecliptic in the direction indicated by the arrow. At opposition, the moon passes the anti-sun and on the next morning, it has reached the position $\mathbb{C}_{NA}$.

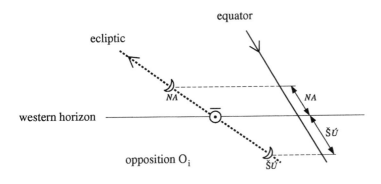

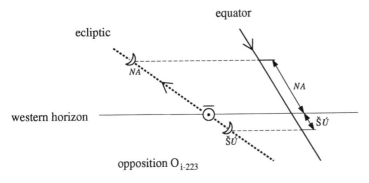

Figure 6

Lower part: Position of the moon and "anti-sun" at the western horizon on two mornings around opposition O_{i-223} at which $\check{S}\acute{U}$ and NA are measured. *Upper part*: The situation at the western horizon one Saros later, that is, at opposition O_i.

The two situations in figure 6 are almost identical: The angles between the ecliptic and the horizon are the same and so are the lengths of arc($\mathbb{C}_{SU}$, $\mathbb{C}_{NA}$), (i.e., the relative displacement of the moon with respect to the anti-Sun during the 24 hours considered). Its rising time in both situations is the same: $(\check{S}\acute{U} + NA)_i = (\check{S}\acute{U} + NA)_{i-223}$. The only difference is the position of the anti-sun $\overline{\odot}$ with respect to $\mathbb{C}_{SU}$ and $\mathbb{C}_{NA}$: At lunation $i - 223$, the moon passes the anti-sun shortly after measuring $\check{S}\acute{U}$ (Δt_{i-223} is small), whereas at lunation i, the opposition takes place 8 hours later ($\Delta t_i = \Delta t_{i-223} + \frac{1}{3}$ day).

In case of NA, the situation is reversed: The distance between $\overline{\odot}$ and $\mathbb{C}_{NA}$ is large at lunation $i - 223$ and smaller at lunation i, by an amount equal to one third of the arc between $\mathbb{C}_{SU}$ and $\mathbb{C}_{NA}$. Therefore, in case of NA the parameter Δt will satisfy the following relation:

$\Delta t_i = \Delta t_{i-223} - \frac{1}{3}$ day   for NA.

We thus conclude: NA_i must be smaller than NA_{i-223} by one third of the amount $(\check{S}\acute{U} + NA)_{i-223}$. (The sum $\check{S}\acute{U} + NA$, however, will remain the same, as we already know.)

We can also explain $\check{S}\acute{U} + NA$ in another way. It is the daily change (i.e., decrease) of $\check{S}\acute{U}$:

$$\Delta \check{S}\acute{U} = -(\check{S}\acute{U} + NA).$$

The time difference between moonset and sunrise measured one day too early, i.e. one day before $\check{S}\acute{U}$, would amount to (approximately) $\check{S}\acute{U} + (\check{S}\acute{U} + NA)$. The same time difference measured one day too late is of course $-NA$ (NA being the time difference between sunrise and moonset): $-NA = \check{S}\acute{U} - (\check{S}\acute{U} + NA)$. The observable NA, measured one day too late, would be roughly $NA + (\check{S}\acute{U} + NA)$. For that very reason, $\check{S}\acute{U} + NA$ is the setting time of the moon's daily elongation arc($\mathbb{C}_{SU}, \mathbb{C}_{NA}$). This elongation arc represents the displacement of the moon relative to the sun on the day of opposition. Without committing grave errors, we can use the same amount for the moon's displacement in the days just before and after opposition.

We summarize: The sum $\check{S}\acute{U} + NA$ can be interpreted as the daily decrease of $\check{S}\acute{U}$ or as the daily increase of NA. Van der Waerden (1974), 110 has the same understanding of $\check{S}\acute{U} + NA$ as the daily retardation of the moon's rising. Concerning the Lunar Six, he proposes to add $\frac{1}{3}$ day to every rising or setting of the moon. As we have shown, however, it is rather the moment of opposition in comparison to sunrise (and to sunset) that is shifted by $\frac{1}{3}$ day.

10   PREDICTION RULES FOR CALCULATING THE LUNAR FOUR

We have seen—equation (1)—how it is possible to calculate $\check{S}\acute{U}$ for an arbitrary month i by means of $\check{S}\acute{U}_{i-223}$ and $(\check{S}\acute{U} + NA)_{i-223}$, quantities written on the relevant Goal-Year tablet. Arguments completely analogous to the ones above will in case of the observables NA, ME, and GE lead to formulae similar to equation (1). We therefore conclude: Whenever we have a Goal-Year tablet for a year Y, we are able to calculate and thus predict all Lunar Four occurring in year Y. The following four equations state this mathematically:

$$\check{S}\acute{U}_i = \check{S}\acute{U}_{i-223} + \tfrac{1}{3}(\check{S}\acute{U} + NA)_{i-223}, \tag{2}$$

$$NA_i = NA_{i-223} - \tfrac{1}{3}(\check{S}\acute{U} + NA)_{i-223}, \tag{3}$$

$$ME_i = ME_{i-223} + \tfrac{1}{3}(ME + GE)_{i-223}, \tag{4}$$

$$GE_i = GE_{i-223} - \tfrac{1}{3}(ME + GE)_{i-223}. \tag{5}$$

These equations are actually correct only if their right-hand sides are positive for equations (3) and (5), not larger than $\check{S}\acute{U} + NA$ for equation (2), and not larger than $ME + GE$ for equation (4), respectively. Otherwise they would correspond to the phenomena observed one day too early and therefore have to be corrected. For example, if $\check{S}\acute{U}_i$ according to equation (2) becomes larger than $(\check{S}\acute{U} + NA)_{i-223} = (\check{S}\acute{U} + NA)_i$, it gives the value of $\check{S}\acute{U}$ observed one morning too early. On the next morning, the Moon would still set before sunrise; the correct time interval $\check{S}\acute{U}_i$ then is obtained by subtracting $(\check{S}\acute{U} + NA)_{i-223}$ from the right-hand side of equation (2). The analogous corrections for the other three quantities are obvious.

In Brack-Bernsen (1994), section 4b, we proposed that it is possible to calculate $\check{S}\acute{U}$ and NA in this way. At the time, we thought that it probably would never be possible to find out if the Babylonians, too, had discovered and used these simple rules. We furthermore had not the slightest idea how the values of KUR and NA_N could be determined.

In the mean time, Hunger has provided us with a translation of the very difficult and important text TU 11. Passages in this text tell us that the Babylonians, indeed, did know and use these rules! But more than that: Other passages contain the procedures for calculating NA_N in an easy and precise way, as we shall see. The procedures are written so briefly that it is doubtful that we would have understood the meaning without knowing the rules expressed in equations (2)–(5).

11   Passages of Text TU 11

The passages relevant for this analysis are written in the lines 29, 30, and 36–38 on the front side of TU 11. In Hunger's preliminary translation we read:[6]

29) *For you to see lack and fullness (of a month). If in an 18(-year period), month I, the 1st day (following a month of 30 days)—there is no addition added to it; month II, which is after it, is full—one-third*

30) *of $\check{S}\acute{U} + NA$ is 6: you subtract this from NA of the 1st day of month II, and if it is less than in month I, which is before it, then month II of your new year is full*

. . .

36) *For you to make the counterpart(?) of the moon in the west. You go back from month I of the 36(-year-period) by 6 months I, and you take(?) 0;40 of $\check{S}\acute{U} + NA$ of month VII and subtract it from NA of the 1st day*

37) of month I of the 36(-year-period). And if it is less than 10 uš, you add ŠÚ + NA completely(?) to it. 0;40 of ŠÚ + NA you subtract from NA of the middle of the month.

38) 0;40 of ME + GE you subtract from GE.

These lines give rules for calculating NA_N (explicitly called "*NA* of the first day"), *NA* (identified as "*NA* in the middle of the month"), and *GE*; the sums *ŠÚ + NA* and *ME + GE* are used, and so is a 36-year period plus the coefficient $0;40 = \frac{2}{3}$. The Saros is, as often in cuneiform texts, just called "18." This justifies the reading: "36 [year period] as 2 Saroi = 446 synodic month."

We first concentrate on the comments on the Lunar Four. Having mentioned 2 Saroi, the text goes on and requests the reader in the last third of line 37 and in line 38 to calculate the following difference:

$$NA - \tfrac{2}{3}(ŠÚ + NA) \quad \text{and} \quad GE - \tfrac{2}{3}(ME + GE).$$

The text does not tell clearly from which month or opposition these values shall be taken. But the passages make sense when we read them as: "In order to find *NA* one has to go 2 Saroi back and then to subtract $\frac{2}{3}$ of *ŠÚ + NA* from *NA*" (both values stemming from the lunation two Saroi earlier than the one, say number *i*, we are concerned with):

$$NA_i = NA_{i-446} - \tfrac{2}{3}(ŠÚ + NA)_{i-446},$$

and analogously for *GE*:

$$GE_i = GE_{i-446} - \tfrac{2}{3}(ME + GE)_{i-446};$$

but these formulae are simply the equations (3) and (5) used twice, namely for two consecutive Saroi.

We consider this as a strong support for our reading. Furthermore, it is a clear confirmation that the Babylonians really did know the formulae (3) and (5), and presumably also formulae (2) and (4).

Lines 36 and 37 deal with *NA* of the first day, which we call NA_N. The text, here more clearly formulated than the passages mentioned above, gives a procedure how to estimate (calculate) NA_N for the first month (*I*) in a year *Y*.

We understand the text, aiming at finding $(NA_N)_I$, as follows: "From month *I* [of your] 36 [year period] (i.e., month *I* − 446); you go 6 months backward (to month *I* − 452) and 0;40 of *ŠÚ + NA* [i.e., $\frac{2}{3}$ of $(ŠÚ + NA)_{I-452}$] you subtract from *NA* of the first day of the 36 [year period] [i.e., from $(NA_N)_{I-446}$].[7]

We write this instruction as an equation:

$$(NA_N)_I = (NA_N)_{I-446} - \tfrac{2}{3}(\check{S}\acute{U} + NA)_{I-452}.$$

This formula calculates NA_N from the NA_N observed 2 Saroi earlier and uses for the daily change of NA_N (which we call ΔNA_N) the quantity $\check{S}\acute{U} + NA$ stemming from a full moon $5\tfrac{1}{2}$ months further back in time. The daily change of NA_N cannot be determined by observation: The setting moon is invisible before conjunction. Using $\check{S}\acute{U} + NA$ for ΔNA_N is a very clever and, as we shall see later, also a very precise method.

The text continues: "... and if it is less than 10 $u\check{s}$, you add the whole $\check{S}\acute{U} + NA$ to it." The Babylonians knew and utilized that $(NA_N)_I$, when not observable on the first evening after conjunction, would be equal to the expected amount plus the sum $\check{S}\acute{U} + NA$ observed 6 months earlier. We know, and they evidently knew too, that the tiny new moon cannot be seen if it is too close to the sun and sets less than, say, 10 $u\check{s} = 40$ minutes after sunset. The new moon will in that case first be visible in the next evening, and its amount will then be enlarged by ΔNA_N. The text really tells us to add the whole $\check{S}\acute{U} + NA$. The Babylonians evidently used $\check{S}\acute{U} + NA$ as the daily change of NA_N.

We derived the above formula for the specific case of the first month (I) of a year, as it occurs in the text TU 11. Obviously, the recipe works equally well for an arbitrary month of the year. We shall therefore continue using the general index i for a month.

If we now instead of 2 Saroi only go back by 1 Saros, the formula would be:

$$(NA_N)_i = (NA_N)_{i-223} - \tfrac{1}{3}(\check{S}\acute{U} + NA)_{i-229}. \tag{6}$$

The correspondingly reconstructed formula for calculating KUR must be:

$$KUR_i = KUR_{i-223} + \tfrac{1}{3}(ME + GE)_{i-229}. \tag{7}$$

The lines 29 and 30 confirm for us that the recipe expressed in equation (6) really was in use by the Babylonians of the Seleucid era. The concern in these lines is to predict whether a month has 29 or 30 days. We are, however, at this place, only interested in the calculation of NA_N. The 18-year period (*one* Saros) is mentioned, and so is NA_N of a month II, and the third of $\check{S}\acute{U} + NA$ is said to be 6. At spring time (month II), $\check{S}\acute{U} + NA$ always is minimum, ranging between 8 and 10 $u\check{s}$; but here is $\check{S}\acute{U} + NA$ indirectly said to be 18. We hence know that this value must stem from fall time, where $\check{S}\acute{U} + NA$ assumes its maximum value. The whole text is consistent, and that brief remark makes sense if we read it as follows:

"Subtract the third of $\check{S}\acute{U} + NA$ (six months earlier than II) from NA_N (of month II)." But this is exactly the rule contained in equation (6).

We saw above that the data necessary for calculating the Lunar Four for the whole year Y are written in the entries for the months of year $Y - 18$ (on the relevant Goal-Year tablet). These data do, however, not suffice for calculating the new moon phenomena NA_N and KUR for the year Y. In order to calculate, for instance, NA_N in month 1 of year Y, one needs the NA_N from the first month's entry of the Goal-Year tablet; but also the sum $\check{S}\acute{U} + NA$ from the full moon six months further back. This sum is, indeed, also recorded in the upper left corner of the corresponding Goal-Year tablet!

We have hereby for the first time understood and explained why at the beginning of each Goal-Year tablet for a year Y, the sums $\check{S}\acute{U} + NA$ and $ME + GE$ valid for the last 6 months of year $Y - 19$ are recorded.

12 The Daily Change of $(NA_N)_i$ and Its Relation to $(\check{S}\acute{U} + NA)_{(i-6)}$

The setting of the moon is not visible in the days before conjunction, and the moonrise is invisible during the first days after conjunction. The daily changes of NA_N and KUR around conjunction therefore cannot be observed. Nevertheless, the new moon data corresponding to $\check{S}\acute{U} + NA$ and $ME + GE$ can be computed. Using the ephemeris code of Moshier (1992), we have computed NA_N, as well as the time differences between sunset and moonset the evening just before, as seen from Babylon in ancient times. The difference of these two gives us ΔNA_N, the daily change of NA_N. In this way it is possible to test the Babylonian approach using $(\check{S}\acute{U} + NA)_{i-6}$ instead of $\Delta(NA_N)_i$.

In figure 7 we have plotted the calculated ΔNA_N as function of the lunation number i (thick line). For comparison, we have in the same figure also plotted the computed values of $\check{S}\acute{U} + NA$ observable $5\frac{1}{2}$ months earlier (dotted line). The two curves are almost identical. This shows that the use of $(\check{S}\acute{U} + NA)_{i-6}$ instead of $\Delta(NA_N)_i$ is, indeed, well justified and very precise.

We can also explain why this Babylonian approach works by comparing figures 1, 2, and 3. In order not to draw more figures we argue with KUR instead of NA_N. Figure 1 shows the situation near the eastern horizon at *sunrise* in the mornings around conjunction. What we would call ΔKUR (corresponding to ΔNA_N) is the rising time of arc($\mathbb{C}_{KUR}$, $\mathbb{C}29$). Half a month later, that is, around opposition, the situation at the eastern

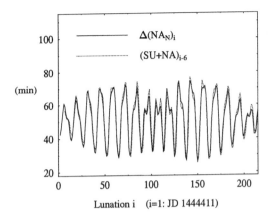

Figure 7

Comparison of calculated values of ΔNA_N (thick line) and of the $\check{S}\acute{U} + NA$ observable 6 months earlier (dotted line), both plotted versus the lunation number i (starting for $i = 1$ on JD 1444411 = August 1, 759 B.C.).

horizon at *sunset* will be as in figure 2. The sun having moved not more than 15° along the ecliptic is about to set; the half of the ecliptic that is above the horizon in figure 1 will now be under the horizon. (For our present concern the small displacement of the sun can be neglected.)

In our example in figure 1 the sun was near the spring equinox, so that the ecliptic at sunrise had a small inclination to the horizon. At sunset half a month later, the fall equinox will be near the eastern horizon and the ecliptic will stand very steep. This illustrates how the parameter $\lambda_{\mathbb{C}}$ has changed from figure 1 to figure 2. Another parameter, the lunar velocity $v_{\mathbb{C}}$, which is crucial for the magnitude of *KUR* and *GE*, may also have changed considerably. In case of *KUR*, $v_{\mathbb{C}}$ is the velocity of the Moon at conjunction; in case of *ME* and *GE* it is the velocity at opposition half a month later. Now, half a synodic month is only about one day longer than one half of the anomalistic month, the period of the variable lunar velocity. If $v_{\mathbb{C}}$ is maximum in figure 1 it will be minimum in figure 2, and vice versa.

We see that the situations in figures 1 and 2 are very different: there is no hope of determining ΔKUR from $ME + GE$. But if we go 6 months backward from this month and imagine what the conditions for $ME + GE$ then will be, we realize that now we will have a situation almost identical with the one in figure 1.

We now are concerned with the opposition taking place 6 months earlier than in figure 2 and consider how the celestial sphere might have looked like at that time. In the course of 6 months, the sun and the anti-sun have moved about 180° and hence have exchanged places. Therefore, at the moment of *anti-Sun rise*, the position of the ecliptic will be as in figure 3, where the anti-sun was setting at the time of figure 2. In order not to draw another figure, we just look at figure 3, now being interested in the front side, imagining the anti-sun placed at the eastern horizon. The elongation arc($\mathbb{C}_{ME}$, $\mathbb{C}_{GE}$) is placed in some way around $\overline{\odot}$. It is the rising time of this elongation arc which the Babylonians used as ΔKUR.

We now understand why that is a good approach: The rising time of arc($\mathbb{C}_{ME}$, $\mathbb{C}_{GE}$) must be approximately the same as the rising time of arc($\mathbb{C}_{29}\mathbb{C}_{30}$) in figure 1: The angles between the horizon and the ecliptic will be the same in the two figures as will the lengths of the two ecliptic arcs. This length is, in fact, determined by the lunar velocity, which is approximately the same in the two cases. The time difference of $5\frac{1}{2}$ months between figure 1 and figure 3 is, indeed, approximately equal to 6 anomalistic months.

13   Tᴇsᴛ ᴏꜰ Bᴀʙʏʟᴏɴɪᴀɴ Pʀᴇᴅɪᴄᴛɪᴏɴ Rᴜʟᴇs ʙʏ Mᴏᴅᴇʀɴ Cᴏᴍᴘᴜᴛᴀᴛɪᴏɴs

Similarly as we have checked the equality of $\Delta(NA_N)_i$ and $(\check{S}\acute{U} + NA)_{i-6}$ in the section above, we now shall test some of the prediction rules discussed in Sects. 10 and 11 by modern computer calculations, in order to control their accuracy.

In figure 8, we show by solid lines the quantities $(NA_N)_i$ (*upper part*) and $\check{S}\acute{U}_i$ (lower part) for 50 successive lunations between 236 and 232 B.C. Their predicted values based on earlier data, according to the right-hand sides of equations (6) and (2), respectively, are shown by the dashed lines. Again, the agreement is excellent, demonstrating that these Babylonian procedures for predicting NA_N and $\check{S}\acute{U}$ were not only clever, but also very precise.

14   Tᴇsᴛ ᴏꜰ Lᴜɴᴀʀ Fᴏᴜʀ Dᴀᴛᴀ ᴏɴ Sᴏᴍᴇ Gᴏᴀʟ-Yᴇᴀʀ Tᴀʙʟᴇᴛs ʙʏ Mᴏᴅᴇʀɴ Cᴏᴍᴘᴜᴛᴀᴛɪᴏɴs

We now know how to predict each single one of the Lunar Six by means of an appropriate Goal-Year table, and we understand the meaning of the lines treated above from TU 11, which we see as a kind of procedure text that confirms our method.

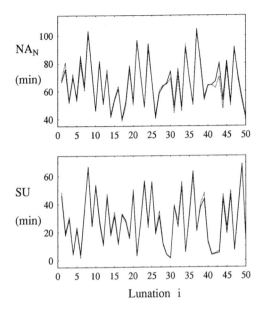

Figure 8
Numerical test of the prediction rules equation (6) for NA_N (upper part) and equation (2) for $\check{S}\acute{U}$ (lower part) for 50 successive lunations between 236 and 232 B.C. The quantities $\check{S}\acute{U}_i$ and $(NA_N)_i$ to be predicted are shown by solid lines; their predictions based on the earlier date, according to the right-hand sides of these equations, are shown by the dashed lines.

If we were to write a "User's Manual" for the Goal-Year tables, we would use more explicit words than the Babylonians, or else we would just give equations (2)–(7) above. This Babylonian method for calculating the Lunar Four must be very precise, provided the data collected on the Goal-Year tablets are sufficiently accurate. In Brack-Bernsen (1994), the data of two texts were analyzed and shown to be surprisingly good: Each of the Lunar Four, drawn as functions of the lunation number, shows the expected irregular behavior, whereas the sums $\check{S}\acute{U} + NA$ and $ME + GE$ form smooth and periodic curves. These curves were tested by modern computations. The sums $\check{S}\acute{U} + NA$ and $ME + GE$ were computed for a long period of time and plotted as functions of the lunation number. Comparison of the Babylonian and the calculated data demonstrated that the sums $\check{S}\acute{U} + NA$ and $ME + GE$ originating from the texts Cambyses 400 (523 B.C.) and the Goal-Year text LBAT 1285 (136 B.C.) were very accurate. This is not always the case. In a forthcoming paper, all the

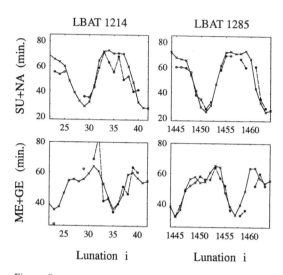

Figure 9

Comparison of the calculated lunar observables $ŠÚ + NA$ and $ME + GE$ (crosses, connected by solid lines, as in figure 4) with the corresponding data recorded on the Babylonian Goal-Year tablets LBAT 1214 and LBAT 1285 (circles, for consecutive months connected by dashed lines).

existing Goal-Year data will be examined in this way. Here we only show the comparison of data from two Goal-Year texts: from LBAT 1214, the one with the worst data, and from LBAT 1285 which contains the best data and which we treated above in section 7.

In figure 9 we have drawn the values of $ŠÚ + NA$ and $ME + GE$, computed with the modern ephemerides code (Moshier 1992), as functions of the lunation number (crosses connected by straight lines), while the values of these sums recorded on the relevant Goal-Year tablets are shown by small circles. The Babylonian data, indeed, form curves with maxima and minima at the same places as the theoretical curves. The data from LBAT 1214 deviate considerably from the computed curves, whereas the data from LBAT 1285 follow the computed curve and even show the same fine structures.

15   Concluding Remarks

We have analyzed the data collected by the Babylonians for predicting lunar phenomena and have seen that it is possible to predict the Lunar Four by means of such data. The text TU 11 proves that the Babylonians did indeed predict the Lunar Four exactly in this way. Furthermore, it also shows us how to predict NA_N (and KUR), the phenomena occurring around conjunction. This Babylonian way of prediction is not only very easy and elegant, but also precise. How far back in time these rules were known, we do not know.

But what we do know is that the Lunar Six are complicated functions depending on four different astronomical variables. It has been pointed out as "one of the most brilliant achievements in the exact sciences of antiquity to have recognized the independence of these influences and to develop a theory which permits the prediction of their combined effects" (Neugebauer 1957, pp. 108, 109).

Our present new findings are almost as impressive. We see them as further evidence that the ancient Babylonians were extremely clever at handling observed data. Their easy and elegant way of predicting the Lunar Six deserves our greatest respect.

ACKNOWLEDGMENT

This research was supported by Deutsche Forschungsgemeinschaft.

NOTES

1. Normally no difference is made in the Babylonian texts between the symbol for NA (new moon) observed after conjunction and that for NA (full moon) observed shortly after opposition. In order to avoid confusion, we shall use the symbol NA_N for the time interval observed after conjunction (new moon).

2. The observed data collected on this tablet, LBAT 1431 (also known as BM 34075), have been evaluated by F. R. Stephenson (1974), who judged them to be rather exact.

3. The position of arc($\mathcal{C}_{KUR}$, $\odot$) is given by $\lambda_{\mathcal{C}}$, the position in the ecliptic at which the conjunction will take place. The length of arc($\mathcal{C}_{KUR}$, $\odot$) is determined by the relative lunar velocity and the time difference between the sunrise we are concerned with and the conjunction.

4. The Lunar Six and all their combinations are measured in $u\check{s}$ = time degrees: 1 $u\check{s}$ = 4 minutes, so that 360 $u\check{s}$ = 1 day (i.e., the time of a whole revolution of the sky, about 360°).

5. I thank my husband, M. Brack, for his help in computing the Lunar Six in the following figures.

6. We warmly thank Prof. H. Hunger for the translation of this very difficult text.

7. The indices i, I, and $I - 452$ refer to the Babylonian months starting on the evening when NA_N is observed. The magnitude of $(NA_N)_{I-446}$ is measured on the first day of month $I - 446$, whereas $(\check{S}\acute{U} + NA)_{I-452}$ is measured $5\frac{1}{2}$ months earlier, namely in the middle of month $I - 452$.

Bibliography

ACT: see Neugebauer (1955).

Brack-Bernsen, L.
1990: "On the Babylonian Lunar Theory: A Construction of Column Φ from Horizontal Observations," *Centaurus*, 33, pp. 39–56.
1993: "Babylonische Mondtexte: Beobachtung und Theorie," in: *Grazer Morgenländische Studien. Band 3: Die Rolle der Astronomie in den Kulturen Mesopotamiens*, edited by H. D. Galter, Graz, rm-Verlag, pp. 331–358.
1994: "Konsistenz zwischen Kolonne Φ und Babylonischen Aufzeichnungen der Lunar Four" in: *AD RADICES, Festband zum fünfzigjährigen Bestehen des Instituts für Geschichte der Naturwissenschaften der Johann Wolfgang Goethe-Universität Frankfurt am Main*, Hrsg. Anton von Gotstedter, Steiner Verlag Stuttgart, pp. 45–64.

Brack-Bernsen, L., and Schmidt, O.
1994: "On the foundations of the Babylonian column Φ: Astronomical significance of partial sums of the Lunar Four," *Centaurus*, 37, pp. 183–209.

Diaries: see Sachs and Hunger (1988, 1989, 1996).

HAMA: see Neugebauer (1975).

Hunger, H., and Pingree, D.
1989: *MUL.APIN, an astronomical compendium in cuneiform*, Archiv für Orientforschung, Beiheft 24.

LBAT: see Pinches, T. G., Strassmaier, J. N., and Sachs, A. J.

Moshier, S. L.
1992: Computer code "AA" (v. 5.1, public domain) for ephemerides of the solar system, using algorithms published in the "Astronomical Almanac" (AA) of the U.S. Government Printing Office. For the moon, special use is made of: M. Chapront-Touzé and J. Chapront, "ELP2000-85: a semi-analytical lunar ephemeris adequate for historical times," *Astronomy and Astrophysics*, 190 (1988), pp. 342–352.

Neugebauer, O.
1947: *Studies in ancient astronomy VIII. The water clock in Babylonian astronomy* ISIS Vol. 37/1-2, nos. 107–108, pp. 37–43.
1955: *Astronomical Cuneiform Texts (ACT)*. Lund Humphries, London, Vols. I–III.
1957: *The Exact Sciences in Antiquity*. 2d ed. Providence, Brown University Press (also Dover 1969).

1975: *A History of Ancient Mathematical Astronomy*. Springer Verlag, New York, Vols. I–III.

Pinches, T. G., Strassmaier, J. N., and Sachs, A. J.
1955: *Late Babylonian Astronomical and Related Texts*. Brown University Providence, Press.

Sachs, A.
1948: "A classification of the Babylonian astronomical tablets of the Seleucid period," *JCS* 2, pp. 271–290.

Sachs, A. J., and Hunger, H.
1988: (1989, 1996:) *Astronomical diaries and related texts from Babylonia. Volume I, II and III: Diaries from 652 B.C. to 61 B.C.* Österreichische Akademie der Wissenschaften, Wien.

Schmidt, O.
1997: "Studies on Ancient Spherical Astronomy," to be published by the Institute for Research in Classical Philosophy and Science, Princeton, N.J.

Stephenson, F. R.
1974: "Late Babylonian observations of 'lunar six'es," in *The Place of Astronomy in the Ancient World, Phil. Trans. R. Soc. Lond.* A 276, pp. 118–121.

Thureau-Dangin, Fr.
1922: "Tablettes d'Uruk," Musée du Louvre, *Textes cunéiformes* 6, Paris.

van der Waerden, B. L.
1949: "Dauer der Nacht und Zeit des Monduntergangs in den Tafeln des Nabû-zuqup-GI.NA.," *ZA* 49, p. 307 ff.
1951: "Babylonian Astronomy. III. The earliest astronomical Computations," *JNES* 10, pp. 20–34.
1974: *Science Awakening II: The Birth of Astronomy*. Oxford University Press, New York, pp. 108–110.

7

A New Mathematical Text from the Astronomical
Archive in Babylon: BM 36849
Asger Aaboe

A cuneiform sexagesimal multiplication table for the principal number p is
a list of pairs of numbers n and $n \times p$, $n = 1, 2, 3, \ldots, 18, 19, 20, 30,$
40, 50. We have quantities of tables of this sort, either with just one value
of p to a tablet—we call these single multiplication tables—or, in the
combined tables, with a selection of two or more such values on one
tablet. In MKT and MCT we find recorded more than 100 single multiplication tables and about 80 of the combined sort. Since a combined
table could contain as many as 37 principal numbers (A 7897, an Old-
Babylonian clay cylinder, now in Chicago), it is surely significant that
there are only 40 different principal numbers in all (thus the Chicago
cylinder is a nearly complete set of multiplication tables).

Table 1 shows the 40 standard principal numbers. With the sole
exception of 7, they are what we call regular sexagesimal numbers; they
contain no prime factors but 2, 3, and 5, so their reciprocals are terminating sexagesimal fractions. Indeed, there is large, though not complete,
agreement between table 1 and the entries in a standard reciprocal table
(the long number 44,26,40 is simply the reciprocal of $1,21 = 81$), and a
reciprocal table is often given at the beginning of a combined multiplication table. The association of reciprocal with multiplication tables reflects,
of course, that the preferred way of dividing by a number was to multiply
by its reciprocal, whenever possible.

The discovery of yet another multiplication table is, then, not likely
to be of great moment; indeed, though MCT roughly doubled the corpus
of multiplication tables already published in MKT, not a single new
principal number was added.

I was then much surprised when in 1968 I found in Istanbul a
combined multiplication table from Uruk (U 91) in which only one
principal number (2,15) out of the ten I could then read was in the standard list, and two of them were irregular to boot ($4,20 = 13 \times 20$ and
$3,30 = 7 \times 30$). I published it with some other tables in JCS in 1969, and
I reproduce my transcription of it as table 2.

Table 1

50	20	8	3,20
48	18	7,30	3
45	16,40	7,12	2,30
44,26,40	16	7	2,24
40	15	6,40	2,15
36	12,30	6	2
30	12	5	1,40
25	10	4,30	1,30
24	9	4	1,20
22,30	8,20	3,45	1,15

Each of the component tables of U 91 ends with the lines:

p a-rá (or gam) p

p^2

ki-2-e p àm

igi p $1/p$

igi $1/p$ p

if the principal number p is regular; if not, the last two lines are replaced by igi p nu tuḫ.

The term ki-2-e is known from another text,[1] but in the context:

n ki-2-e n^2

where it must mean something like "which squared is." In U 91, however, its meaning is rather "whose square root is," a usage without precedent.

The upper right-hand corner of the obverse, now rejoined, was mistakenly included in Neugebauer's *Astronomical Cuneiform Texts* as ACT No. 1017. It contains part of the table for 2,13,20, an important parameter in Babylonian lunar theory. Neugebauer's evidence was a photograph taken during the German excavations at Uruk (Warka) of a tabletop full of fragments of Seleucid clay tablets, some of astronomical content (this photograph is reproduced in Neugebauer's *Exact Sciences in Antiquity* as Plate 6b—our fragment is upside down in the lower left corner). These excavations produced a large number of astronomical texts, now in Istanbul, so our table has some connection with astronomical activities.

In 1992, I was given the opportunity to inspect in the British Museum a selection of texts, mostly small fragments, that all contained numerical material and that all carried accession numbers that began

Table 2

U. 91

Obv.

	I	II	III	IV	V edge	VI
1.		1 28,48	1 11,15	1 6,45	1 3,30	1 [2,13,20]
		2 57,36	2 22,30	2 13,30	2 7	2 [4,26,40]
		3 1,26,24	3 33,45	3 20,15	3 10,30	3 [6,40]
		4 1,55,12	4 45	4 27	4 14	4 [8,53,20]
5.		5 2,24	5 56,15	5 33,45	5 17,30	5 [11,6,40]
		6 2,52,48	6 1,7,30	6 40,30	6 21	6 13,20
		7 3,21,36	7 1,18,45	7 47,15	7 24,30	7 15,33,20]
		8 3,50,24	8 1,30	8 54	8 28	8 17,46,40]
		9 4,19,12	9 1,41,15	9 1,,45	9 31,30	9 20
10.		[10] 4,48	10 1,52,30	10 1,7,30	10 35	10 22,13,20]
		[11] 5,16,48	11 2,3,45	11 1,14,15	11 38,30	11 24,26,40]
		[12] 5,45,36	12 2,15	12 1,21	12 42	12 26,40
		[13] 6,14,24	13 2,26,15	13 1,27,45	13 45,30	13 28,53,20]
		[14] 6,43,12	[14] 2,37,30	14 1,34,30	14 49	14 31,6,40
15.		[15] 7,12	[15] 2,48,45	15 1,41,15	15 52,30	15 33,20
		[16] 7,40,48	[16] 3	16 1,48	16 56	16 35,33,20]
		[17] 8,9,36	[17] 3,11,15	[17] 1,54,45	[17] 59,30	17 37,46,40]
		[18] 8,38,24	[18] 3,22,30	[18] 2,1,30	[18] 1,3	[18 40
		[19] 9,7,12	[19] 3,33,45	[19] 2,8,15	[19] 1,6,30	[19 42,13,20]
20.		[20] 9,36	[20] 3,45	[20] 2,15	[20] 1,10	[20 44,26,40]
		[30] 14,24	[30] 5,37,30	[30] 3,22,30	[30] 1,45	[30 1,6,40]
		[40] 19,12	[40] 7,30	[40] 4,30	[40] 2,20	[40 1,28,53,20]
		[50] 24	[50] 9,22,30	[50] 5,37,30	[50] 2,55	[50 1,41,6,40]
		[28,48 × 28,48]	[11,15 × 11,15]	[6,45 × 6,45]	[3,30 × 3,30]	[2,13,20 × 2,13,20]
25.		[13,49,26,24]	[2,6,33,45]	[45,33,45]	12,15	[4,56,17,46,40]
		[ki-2-e 28,48 àm]	[ki-2-e 11,15 àm]	[ki-2-e 6,45 àm]	[ki-2-e 3,30 àm]	[ki-2-e 2,13,20 àm]
		[igi 28,48 . 2,5]	[igi 11,15 . 5,20]	[igi 6,45 . 8,53,20]	igi 3,30	[igi 2,13,20 . 27]
		[igi 2,5 . 28,48]	[igi 5,20 . 11,15]	[igi 8,53,20 . 6,45]	nu tuk]	[igi 27 . 2,13,20]

(Column I largely destroyed)

Rev.

	I	II	III	IV	V	VI	VII
1.		[1 32]	[1 18,45]	[1 9,22,30]	[1 4,20]	[1 2,15]	
		[2 1,4]	[2 37,30]	[2 18,45]	[2 8,40]	[2 4,30]	
		[3 1,36]	[3 56,15]	[3 28,7,30]	[3 13]	[3 6,45]	
		[4 2,8]	[4 1,15]	[4 37,30]	[4 17,20]	[4 9]	
5.		[5 2,40]	[5 1,33,45]	[5 46,52,30]	[5 21,40]	[5 11,15]	
		[6 3,12]	[6 1,52,30]	[6 56,15]	[6 26]	[6 13,30]	
		[7 3,44]	[7 2,11,15]	[7 1,5,37,30]	[7 30,20]	[7 15,45]	
		[8 4,16]	[8 2,30]	[8 1,15]	[8 34,40]	[8 18]	
		[9 4,48]	[9 2,48,45]	[9 1,24,22,30]	[9 39]	[9 20,15]	
10.		[10 5,20]	[10 3,7,30]	[10 1,33,45]	[10 43,20]	[10 22,30]	
		[11 5,52]	[11 3,26,15]	[11 1,43,7,30]	[11 47,40]	[11 24,45]	
		[12 6,24]	[12 3,45]	[12 1,52,30]	[12 51]	[12 27]	
		[13 6,56]	[13 4,3,45]	[13 2,1,52,30]	[13 55,20]	[13 29,15]	
		[14 7,28]	[14 4,22,30]	[14 2,11,15]	[14 59,40]	[14 31,30]	
15.		[15 8]	[15 4,41,15]	[15] 2,20,37,30	[15] 1,5	[15 33,45]	
		[16 8,32]	[16] 5	16 2,30	16 1,9,20	[16 36]	
		17 [9,4]	[17] 5,18,45	17 2,39,22,30	17 1,13,40	[17 38,15]	[ ]
		18 9,36	18 5,37,30	18 2,48,45	18 1,18	[18 40,30]	1[8]
		19 10,,8	19 5,56,15	19 2,58,7,30	19 1,22,20	[19 42,45]	1[9]
20.		20 10,40	20 6,15	20 3,7,30	20 1,26,40	[20] 45	20[ ]
		30 16	30 9,22,30	30 4,41,15	30 2,10	[30 1,7,30]	30[ ]
		[φ] 21,20	40 12,30	40 6,15	40 2,53,20	[40] 1,30	40[ ]
		[50] 26,40	50 15,37,30	50 7,48,30	50 3,36,40]	[φ] 1,52,30	50[ ]
		[32] gam 32	18,45 gam 18,45	9,22,30 gam 9,22,30	4,20 a-rá 4,20	2,15 a-rá 2,15	
25.		[7,4]	[5,51,33,45]	[1,27]53,24,15	18,46,40	5,3,45	
		[ki-2-e 32 àm]	[ki-2-e 18,45 àm]	[ki-2-e 9,22,30 àm]	ki-2-e 4,20 àm	ki-2-e 2,15 àm	ki-2-e]
		[igi 32 . 1,52,30]	[igi 18,45 . 3,12]	[igi 9,22,30 . 6,24]	igi 4,20	igi 2,15 . 26,40	igi[ ]
		[igi 1,52,30 . 32]	[igi 3,12 . 18,45]	[igi 6,24 . 9,22,30]	nu tuk	igi 26,40 . 2,15	igi[ ]

(Column I destroyed; Column VII largely destroyed)

edge

U. 91

80-6-17. Experience had taught us that on that date, 17 June 1880, the museum's collection had received a group of texts, all unscientifically excavated,[2] that contained an unusually large proportion of astronomical texts from the astronomical archive in Babylon, as we now call it. Two of these fragments, now securely but not trivially joined, turned out to come from a combined multiplication table very like the anomalous U 91.

BM 36849 + 37362 (80-6-17, 589 + 1119)[3]

Transcription: Table 3

Description of Text and Commentary

The two rejoined fragments form the upper third of the right half of the obverse (the lower third of the right half of the reverse) of a tablet that must have been about 7″ (18 cm) wide and 9″ (23 cm) high, as is suggested by curvature and confirmed by content. The writing is neat, and horizontal alignment is obeyed throughout. Though the text is clearly Seleucid, the nine-wedge form of 9 is used (as in U 91). Its four partially preserved columns, separated by vertical rulings, must have continued from the obverse across the lower edge to the reverse and contain parts of a table of squares, probably of all one-digit sexagesimal numbers, 59 in all, as well as of seven multiplication tables. None of the recoverable principal numbers is in the standard list in table 1, and two are irregular (55 = 5 × 11 and 35 = 5 × 7).

The similarity between this text and U 91 is great even, if my memory serves, in handwriting. We find the same arrangement of multiplication tables and the same unprecedented five lines at their ends, though the present text is broken where we expect

igi p nu tuḫ

and with the same peculiar use of ki-2-e. Though their contents overlap, the two texts were clearly not duplicates; indeed, since the British Museum's text's right and bottom edges are preserved, we are sure that 18,45 is the least, and so the last, of its principal numbers in the sense of descending lexicographical ordering that both texts' arrangement of principal numbers obeys, and that is not so in U 91.

I have displayed the disposition of the two overlapping texts in schematic form in table 4. The preserved endings in Column I of the obverse of U 91 agree perfectly with the assumption that we here had a table for 35, as table 4 would have us expect; and further, all that remains

Table 3
BM 36849 + 37362

		I'		II'		III'		IV'
Obv. 1.			1	54	1	35	1	28,48
			2	1,48	2	1,10	2	57,36
			3	2,42	3	1,45	3	1,26,24
			4	3,36	4	2,20	4	1,55,12
5.			5	4,30	5	2,55	5	2,24
		•	6	5,24	6	3,30	6	2,52,48
		•	7	6,18	7	4,5	7	3,21,36
		•	8	7,12	8	4,40	8	3,50,24
		•	9	8,6	9	5,15	9	4,19,12
10.	[10 a-rá 10	1,40]	10	9	10	5,50	10	4[,48
	[11 a-rá 11]	2,1	11	9,54	11	6,25	11	5,16,48⌉
	[12 a-rá]12	2,24	12	10,48	12	7	⌈12⌉	5,45,36
	[13 a-rá 13	2,49	13	11,42	13	7,35	13	6,14,24
	[14 a-rá 14	3,16	14	12,36	14	8,10	14	6,43,12
15.	[15 a-rá 15	3,45	15	13,30⌉	15	[8,45]	15	7,12
	[16 a-rá 16]	4,16	16	14,24⌉	[16	9,20]	16	7,40,48
	[17 a-rá 17	4,49]	17	15,18]	•		17	8,9,36
		•	18	16,12]	•		[18]	8,38,24
		•	[19	17,6]	•		[19]	9,7,12
							[20	9,36
							[30	14,24]]
Rev. 1'.		•		•			[9	2,48,45
		•	[10	6,15]	•		10	3,7,30
		•	[11	6,52,30]	•		11	3,26,15
	[12	11]	[12	7,30⌉	•		12	3,45
5'.	[13	11,55]	13	8,[7,30]	•		13	4,3,45
	[14	12,50]	14	8,45]	[14	7,28]	14	4,22,30
	[15	13,45]	15	9,22,30]	[15	8]	15	4,41,15
	[16	14,40]	16	10	[16	8,32]	16	5
	[17	15,35]	17	10,37,30]	⌈17	⌈9⌉[4]	17	5,18,45
10'.	[18	16,30]	18	11,15	[1]8	9,36	18	5,37,30
	[19	17,25]	19	11,52,30	19	10,8	[19]	5,56,15
	[20	18,20]	20	12,30	20	10,40	20	6,15
	[30	27,30]	30	18,45	30	16	30	9,22,30
	[40	36,40]	40	25	40	21,20	40	12,30
		•	50	31,15	50	26,40	50	15,37,30
15'.		•	[37,30 a-rá 37,30		32 a-rá 32		18,45 a-rá 18,45	
		•	3,26,15		17,4		5,51,33,45	
		•	[ki]2-e 37,30 àm		ki-2-e 32		ki-2-e 18,45	
		•	[igi 37,30 1,36		igi 32 1,52,30		igi 18,45 3,12	
		•	[igi 1,36 37,30		igi 1,52,30 32		igi 3,12 18,45	

BM 36849 + 37362

Table 4

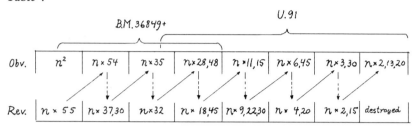

of the same column's reverse is a single terminal 30 where we ought to have 11,52,30 (= 19 × 37,30).

It seems reasonable to assume that BM 36849 had two multiplication tables or 58 lines (57 if a principal number is irregular) per column side. This would agree very well with the tablet's curvature and would allow room for 59 or 60 squares in Column I′ of the obverse—incidentally, horizontal alignment of the columns extends even to the table of squares. It seems equally reasonable to make the same assumption about U 91 and, indeed, two more principal numbers would fit very nicely in each gap of the decreasing sequence, indicated by arrows in table 4, beginning with 55 and ending with 2,13,20 which is as far as U 91 is preserved.

I have little doubt that our text comes from the astronomical archive in the temple *Esagila* in Babylon, and it adds, however slightly, to our knowledge of the activities of those employed in the archive and their commerce with colleagues in Uruk's *Reš* sanctuary.

We have several texts that tell us about the pay, rank (*kalû*-priest), and duties of our astronomers whose professional title was *tupšar Enūma Anu Enlil* (scribe of Enūma Anu Enlil, the astrological series so named after its incipit). The most informative of them is the tablet BM 35559 (published in copy in CT 49 as no. 144),[4] a record of a hearing held in S.E. 193 (119 B.C.) before the administrative assembly of *Esagila*. We learn from it, among other matters, that the same person was expected to carry out celestial observations (i.e., produce the astronomical diaries), and provide *tersētu*-tables (the computed texts of the sort we find in ACT) as well as almanacs.

The latest text we have from the archive is from A.D. 75. It is also the latest dateable cuneiform text to reach us so far, and it was written at a time when technical Akkadian or, for that matter, any kind of Akkadian had long since fallen into disuse everywhere else. It seems to me an unavoidable inference that teaching must have been among the duties of the scholars of the temple, and that mathematics, particularly arithmetic, must have been among the subjects taught. There are, indeed, two astronomical texts, ACT 813 and 817, that as one section—the others concern Jupiter—have a mathematical problem involving a trapezoid, and several mathematical tables surely come from the archive, most notably BM 55557, a list of fourth powers of closely spaced regular numbers published recently by Britton [1993]. The text is heroic in its execution, for the fourth powers run to as many as 25 sexagesimal places, and even more heroic in its decipherment, for we have only the right half of the text with the endings. The regular numbers whose fourth powers are given cannot have been included in the text for want of space, so we are at a loss to discover what purpose could have warranted so large an expenditure of work. I cannot help suspecting, though, that the text may give the answer to a problem set by a lazy or punishing teacher—easy to formulate, but terribly laborious to solve.

Our present text might, of course, be a welcome addition to an arithmetical astronomer's set of tools, though there seems to be no particular agreement between the new principal numbers and significant astronomical parameters, and it might also have had a role to play in the activities of the schoolroom. Whatever the case, it is one more minuscule tile in the incomplete mosaic that is still far from forming a coherent picture of Babylonian astronomy.

NOTES

1. BM 34592, published as LBAT 1637 and as text A in Vaiman (1961); it is also included in Aaboe (1965) with photograph as text IV.

2. For information about the Museum's acquisition of these and other texts from Babylon, but included in the "Sippar" Collections, see Reade (1986).

3. The text is published through the courtesy of the Trustees of the British Museum. I am, once again, indebted to the Museum's Department of Western Asiatic Antiquities, particularly to Mr. C. B. F. Walker, whose active interest has done much to further and facilitate the study of Babylonian astronomical tablets.

4. The text is edited by McEwan (1981); see also the review of van der Spek (1985). It is being re-edited with a new translation by F. Rochberg, who commands the astronomical technical terminology.

BIBLIOGRAPHY

Aaboe (1965): A. Aaboe, *Some Seleucid Mathematical Tables.* . . . JCS XIX, 1965, pp. 79–86.

Aaboe (1969): A. Aaboe, *Two Atypical Multiplication Tables from Uruk*. JCS XII, 1969, pp. 88–91.

Britton (1993): J. P. Britton, *A Table of 4th Powers from Seleucid Babylon*. JCS 43–45, 1991–93, pp. 71–87.

McEwan (1981): G. J. P. McEwan, *Priest and Temple in Hellenistic Babylonia*. FAOS 4, 1981.

Reade (1986): Julian E. Reade, *Introduction* to Erle Leichty, *Catalogue of the Babylonian Tablets in the British Museum, Volume VI: Tablets from Sippar I*, British Museum Publications, 1986.

Vaiman (1961): A. A. Vaiman, *Shumero-Babilonskaya Matematika*, Moskva, 1961, pp. 215 ff.

van der Spek (1985): R. J. van der Spek, review in BiOr 42, 1985, pp. 541–562.

8

Lunar Anomaly in Babylonian Astronomy
John P. Britton

ana tar-sa 1,58,33,42,13,20 tab 4,56 gar-*an*
en 1,58,37,2,13,20 tab tab *u* lal nu tuk[1]

1 Introduction

Babylonian lunar theory comes in two distinct forms, known as System A and System B. Each consists of a set of arithmetic functions, tabulated in columns in ephemerides and auxiliary tables, by which are calculated the times and dates of the syzygies and the magnitudes of eclipses. Both also treat in parallel fashion the same three independent variables, namely:

(1) lunar anomaly, or the effects of the variable lunar velocity on the times and dates of the syzygies;[2]

(2) zodiacal[3] anomaly, or the effects of the longitude of syzygy and its monthly changes on all three objectives; and

(3) nodal motion, which together with the zodiacal anomaly determines the Moon's latitude at syzygy and thus affects both the dates (but not the times) of syzygies and the magnitudes of eclipses.

Otherwise, however, the two theories differ in virtually every respect[4] and nowhere more distinctively than in their treatments of lunar anomaly.

While I shall discuss the treatment of lunar anomaly in System B, my principal subject is the powerful theory found in System A. This theory, whose combination of originality, rigor, and elegance is arguably exceeded only in Ptolemy's *Almagest* 600 years later, appears to have been created in the second half of the 5th century B.C., and in any event before the problem of zodiacal anomaly was solved and a comprehensive theory completed. This makes it, as far as we can tell, the earliest of the theories comprising Babylonian mathematical astronomy and invites the question of whether its invention was not a transforming event in the creation of science as we know it. In any case, it was a remarkable achievement, and what follows seeks to illuminate its structural elements, the quality of its results, and the intellectual power of its unknown author.

Table 1
Distribution of lunar theory texts in ACT

| | Source | | | Date[5] | |
	Babylon	Uruk	Total	Earliest[6]	Latest
System A	94	2	96	−262	+42
System B	27	53	80	−257	−68
Other	11	2	13		
Total	132	57	189		
Earliest	−262	−207			
Latest	+42	−150			

Sources

The prime textual source for Babylonian mathematical astronomy is O. Neugebauer's *Astronomical Cuneiform Texts* (ACT), which was published in 1955 and includes roughly 300 texts and fragments, two-thirds of which came from Babylon and the rest from Uruk. Of these 189 concern the Moon and Sun, distributed by type and origin as shown in table 1. The System A texts come almost exclusively from Babylon, while 29 auxiliary texts swell the number of System B texts from Uruk. For both systems the earliest (−262) and latest (+42) texts are from Babylon, whose archive thus extends over more than three centuries. In contrast the entire Uruk corpus (including planetary texts) covers less than a century from −225 to −150. Interestingly, both the earliest and latest texts are of System A.

Since ACT appeared, roughly 50 additional texts relating to mathematical astronomy have been published, nearly all from collections at the British Museum[7] and originally from Babylon. Of these 42 are from a single archive bearing the receipt date "80-6-17," which was excavated in late 1879 from the vicinity of the Amran mound at Babylon, apparently from a house abutting the inner wall a short distance from the temple of Marduk.[8] This archive is distinctive in both content and scope and is the source of most of the known texts from pre-Seleucid Babylon. These include a solstice and equinox text with entries from the 7th and 6th centuries (B.C.), Jupiter observations from the 6th century, a 5th-century solar eclipse text with System A lunar functions, and a number of proto- and auxiliary System A lunar texts of uncertain date. The latest securely dated text is for S.E. 181 to 199 (−130 to −112), so that the content of this archive extends over the better part of six centuries. In contrast to other archives from the British Museum's collections, System B material is conspicuously absent.

For lunar theory the relevant texts are published in Neugebauer (1957b), Neugebauer and Sachs (1956, 1967, 1969), Aaboe (1968, 1969, 1971), Aaboe and Henderson (1975), Aaboe and Sachs (1969), Aaboe and Hamilton (1979) and Aaboe, Britton, Henderson, Neugebauer and Sachs (1991). Commentaries on various aspects of the theory and related matters are included in these works and also in van der Waerden (1965), HAMA (1975, 474–513), Brack-Bernsen (1980, 1990, 1993), Moesgaard (1980) and Britton (1987, 1989, 1990).

The new textual material yielded some hard-won insights into the workings of the lunar theory, and especially System A. One of the most important was the astronomical significance of the function called Column Φ, unique to System A, which plays a central role in computing the variation due to lunar anomaly in the length of the month, but whose astronomical significance and even units, were completely obscure when ACT was published. Subsequently, Neugebauer (1957b) published a difficult text which linked this function with the so-called "saros" eclipse interval of 223 months,[9] and eventually Aaboe (1968) was able to confirm a conjecture by van der Waerden (1965) and show conclusively that Φ expressed the variable length of this interval in excess of 6585 days. Further work and new texts published by Aaboe (1969, 1971) and Aaboe and Hamilton (1979) finally brought under firm control the interrelations among Φ and its sister functions, G, W and Λ, representing the provisional lengths of 1, 6, and 12 months respectively. Even so, many details remained obscure, including most conspicuously the motivation for the seemingly central role of Φ and all circumstances of its construction.

The new source material also significantly extended the period for which System A was attested. Previously, the earliest evidence for System A had been calculations for S.E. 49 to 60 (-262 to -251). Subsequently, new evidence, and especially the text BM. 40094, showed that the complete System A lunar theory was attested for at least S.E. -8 to S.E. -5 (-318 to -315). More dramatically, two texts published by Aaboe and Sachs (1969) showed evidence of System A functions for dates as early as -474 to -456 and -397, thus pushing back the evidence for System A by more than 200 years. We shall return to the implications of this at a later point.

2   LINEAR ZIGZAG FUNCTIONS

Zigzag functions comprise one of the two principal methods used by the Babylonians to describe variations from uniform behavior and the only one employed in the theories of lunar anomaly. Since they are essential to

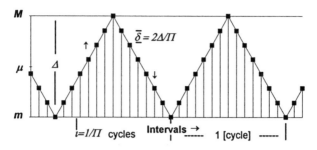

Figure 1
Linear zigzag function $\Pi = 14$

these theories, it will be useful to review their elements and properties before turning to the details of the theories themselves. Simple in concept, such functions consist of four elements:

(1) a *period relation*, which states that the function repeats after Π phenomena, corresponding to Z complete cycles of variation. Π is called the *number period* of the function and is in some respects its most fundamental parameter. Obviously Π and Z must be relatively prime, since otherwise the function would repeat sooner. This, of course, will always be true if Π itself is prime.

(2) an *amplitude*, customarily denoted by Δ, which defines the magnitude of the variation, equal to the difference between the *maximum (M)* and *minimum (m)* of the function;

(3) one *extreme*—i.e., maximum (M) or minimum (m)—or the mean value, $\mu = \frac{1}{2}(M + m)$, of the function, which with the amplitude determines the other two; and

(4) a single *tabular value*, which determines all other values of the function.

The variation described by the function takes place evenly on ascending ($\uparrow$) and descending ($\downarrow$) branches (figure 1), so that over one complete cycle it amounts to twice the amplitude (2Δ). Since the function repeats after Π phenomena, it takes on just Π different values (treating values on different branches as distinct). Distributing these evenly over a single cycle yields a minimum variation,

$$\delta = \pm\frac{2\Delta}{\Pi},\tag{2.1}$$

whose sign is positive on the ascending branch and negative on the descending one, and which corresponds to a minimum interval, $\iota = 1/\Pi$ cycles. In intervals spanning the extremes the sign changes when the extreme is reached, so that around the maximum

$$f(n + \iota)(\downarrow) = M - \{\delta - [M - f(n)(\uparrow)]\}$$
$$= 2M - \delta - f(n)(\uparrow), \tag{2.2a}$$

while around the minimum

$$f(n + \iota)(\downarrow) = m + \delta - [f(n)(\downarrow) - m]$$
$$= 2m + \delta - f(n)(\downarrow). \tag{2.2b}$$

If the function is to be precisely defined at all points, δ must be a terminating sexagesimal number and thus a common factor of Δ and Π. This limits the choices of Δ to multiples of Π whose precision (i.e., number of sexagesimal places) is consistent with computational convenience.

To complete the function, we need a mean value, μ (or either extreme), and one specific value. The mean value is constrained only by the fact that tabular values on at least one branch of the function will usually have as many sexagesimal places as μ. The specific tabular value can be any terminating number within the amplitude, which in turn will lead to tabular *maxima* and *minima* within $\pm \iota/2$ of each extreme. Unlike the function portrayed in figure 1, the extremes need not be taken on and often are not.

We have then a function with Π defined values, which describes a variation with amplitude Δ, which takes place evenly over a cycle divided into Π intervals of length $\iota = 1/\Pi$, with a mean value, $\mu = \frac{1}{2}(M + m)$, and tabular extrema, M' and m', which fall within $\pm \iota/2$ of M and m. This function can be tabulated at any number of intervals, with the increment for q intervals being

$$d_{q\iota} = q\delta \ (\text{modulo } 2\Delta) = q(\text{modulo } \Pi)\,\delta, \tag{2.3}$$

where around the extremes $d_{q\iota}$ must be substituted for δ in expressions (2.2a) and (2.2b). If $d_{q\iota}$ is greater than Δ, reflections will occur at every step, and there may be no evident progression of steady increments, although the function will return to its initial value after Π steps of $q\iota$ intervals. If $q\iota \ll \Pi/2$, however, a derivative linear zigzag function will result with increment $d_{q\iota}$, where the quantity

$$P(q\iota) = \frac{2\Delta}{d_{q\iota}} = \frac{\Pi}{q\,(\text{modulo } \Pi)}, \tag{2.4}$$

is called the *tabular period* of the function and measures the number of intervals, $q\iota$, required for the function to repeat. Such *tabular periods* are characteristic, not of the underlying function, but of the intervals at which it is tabulated and can range from $P(\Pi\iota) = 1$ to $P(\iota) = \Pi$.

If the function is to be tabulated at monthly intervals, one needs to determine the number of minimum intervals (ι) that correspond to one month. This can be found from the period relation, which for lunar theory is typically of the form

$$\Pi \text{ months} = Z \text{ cycles of variation,} \tag{2.5a}$$

whence it follows that

$$1 \text{ month} = \frac{Z}{\Pi} \text{ cycles} = Z \times \iota = Z \text{ minimum intervals } (\iota). \tag{2.5b}$$

Thus the increment corresponding to one month will be

$$d_1 = Z \text{ (modulo } \Pi) \, \delta, \tag{2.6a}$$

and more generally for n months

$$d_n = nZ \text{ (modulo } \Pi) \, \delta, \tag{2.6b}$$

with reflections occurring around the extremes in accordance with equations (2.2a) and (2.2b) with now d_n substituted for δ.

One special case of particular interest is the increment for half a month,

$$d_{\frac{1}{2}} = \frac{Z}{2} \text{ (modulo } \Pi) \, \delta. \tag{2.7}$$

If Z is odd, $d_{\frac{1}{2}}$ will be an odd integer multiple of $\delta/2$, and the value, $f(n + \frac{1}{2}) = f(n) + d_{\frac{1}{2}}$, will not be among the points comprising the original function. Thus if Z is odd, a function tabulated for new and full moons will have distinct values at each type of syzygy, separated by a minimum composite increment of $\delta/2$. This is illustrated in figure 2, which shows a simple function for lunar velocity ($\Pi = 14$ and $Z = 15$) tabulated at intervals first of 1 tithi ($= \frac{1}{30}$ of a mean synodic month, which here is equal to $\frac{30}{28}$ths of an anomalistic month) and then of 1 month. Since in this example $d_1 = \delta$, the tabular values and increments are identical, but the difference in scale shows the difference in phase more clearly.

Example: a Rudimentary Function for Lunar Velocity, F(0)
To see how this works in detail, consider the following hypothetical construction of a simple function depicting the variation in lunar velocity. The process will also illustrate how little is required to build such functions. We begin with the fact that the daily change in the Moon's position repeats after 27 or 28 days, whereas a month takes 29 or 30 days, so that for our period relation we use

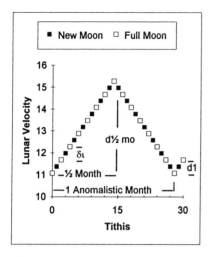
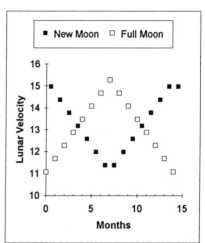

Figure 2
Lunar velocity function, $F(0)$

1 anomalistic month (m_a) $\cong \dfrac{28}{30}$ synodic months (m), or   (2.8a)

$\Pi = 14$ synodic months $= Z = 15$ anomalistic months,   (2.8b)

whence Z (modulo Π) = 1.   (2.8c)

Next, we observe that the moon's velocity varies by roughly 4 degrees per day over the course of an anomalistic month. Since $\Delta = \Pi\delta/2$, the nearest choice for which δ is a *regular* one-place sexagesimal[10] is $\delta = 0;36°/d$, which makes

$\Delta = 4;12°/d$.   (2.9a)

$d_1 = Z$ (modulo Π) $\delta = \delta = 0;36°/d$,   (2.9b)

and, since 223, modulo 14 = -1, also

$d_{223} = -\delta = -d_1 = -0;36°/d$,   (2.9c)

where the negative sign implies a step backward in time (i.e., a variation on either branch corresponding to the time interval $-\iota$).

To complete the function, we need only a mean value (or one extreme) and a single tabular value. To obtain a mean value we might "observe" (*a*) that the Moon returns to its same place after 235 months and 254 rotations, as may be seen from lunar eclipses separated by such intervals, and (*b*) that such eclipses are always separated by at least $6939\frac{1}{3}$

days and not more than 6940 days.[11] Thus we have

6939;20 days < 254 sidereal months < 6940 days, or

254 sidereal months $\cong$ 6939;40 days,   (2.10)

whence

$\mu = 254 \times 360° \div 6939;40$ days

$= 13;10,35,7,..°/\mathrm{d} \cong 13;10,35°/\mathrm{d}$,   (2.11)

which leads to

$$M = \mu + \frac{\Delta}{2} = 15;16,35°/\mathrm{d} \text{ and } m = \mu - \frac{\Delta}{2} = 11;4,35°/\mathrm{d}.$$ (2.12)

For our single tabular value, if we choose either extreme, the other extreme will also be a value (since $\Delta = 7\delta$), and the function will be symmetrical and identical on both branches as shown in figure 2 and section (A) of table 2. This symmetry will also characterize the function for half a month later (A-2), where, however, neither extreme will occur.

Table 2

Two versions of a simple function, F(0), for the variation of lunar velocity. $\Pi = 14$, $Z = 15$, $\Delta = 4;12°/\mathrm{d}$, $\delta = d_1 = 0;36°/\mathrm{d} = -d_{223}$, and $\mu = 13;10,35°/\mathrm{d}$

	A						B					
mo	F(0)			F(0) $+ \frac{1}{2}$ mo			F(0)′			F(0)′ $+ \frac{1}{2}$ mo		
1	11	4	35	14	58	35	11	6	0	15	0	0
2	11	40	35	14	22	35	11	39	10	14	24	0
3	12	16	35	13	46	35	12	15	10	13	48	0
4	12	52	35	13	10	35	12	51	10	13	12	0
5	13	28	35	12	34	35	13	27	10	12	36	0
6	14	4	35	11	58	35	14	3	10	12	0	0
7	14	40	35	11	22	35	14	39	10	11	24	0
8	15	16	35	11	22	35	15	15	10	11	21	10
9	14	40	35	11	58	35	14	42	0	11	57	10
10	14	4	35	12	34	35	14	6	0	12	33	10
11	13	28	35	13	10	35	13	30	0	13	9	10
12	12	52	35	13	46	35	12	54	0	13	45	10
13	12	16	35	14	22	35	12	18	0	14	21	10
14	11	40	35	14	58	35	11	42	0	14	57	10
15	11	4	35	14	58	35	11	6	0	15	0	0

Thus the functions for full and new moons will appear somewhat different and have the practical drawbacks that values on ascending and descending branches will be indistinguishable, and all values end with space-consuming 0;0,35. If, however, we choose values which fall slightly earlier in the cycle, we obtain the "nicer" functions shown in section (B). Here the functions for full and new moons are more nearly similar; while values on the ascending branch are readily distinguished by the endings 0;0,10 from those on the descending branch which contain at most two sexagesimal places. Finally, the function now economizes on space by removing the need for a units column in the third sexagesimal place.

This simple function (which is unattested to my knowledge, but was almost certainly known to the Babylonians[12]) actually gives a decent representation of the variation in lunar velocity, since the true amplitude of this variation (for all elongations) is $3;59^{o/d}$, while the ratio of anomalistic to synodic month lengths is $0;55,59,6,..$ instead of $0;56$ as implied here. We can improve it, moreover, by assuming that $d_{14} =$ some new minimum increment, δ, instead of 0 as in F(0), whose direction, if not magnitude, can be ascertained from $d_{224} = 16\delta$. If now we limit δ to one sexagesimal place and observe that the increment for 223 months, $d_{223} = q\delta$, is sensible but small and in the opposite direction to δ, we obtain the potential period relations and related parameters shown in table 3.

If we allow δ an additional place of sexagesimal fractions, there are of course more alternatives, but otherwise these (and higher values of $-q$ and thus Δ) are the only choices. Of these the second ($q = -2$) yields by far the nicest parameters, first because $Z - \Pi$ and thus d_1 are regular numbers, secondly because Z is odd (indeed both Π and Z are prime), thereby allowing distinct functions for new and full moons, and finally (and rather remarkably) because d_1 is identical with the corresponding increment in our primitive function, F(0), while Δ is virtually so. As it happens, this also defines the function with the most accurate period

Table 3
Potential parameters for F

q	Π	Z	$Z - \Pi$	Δ	δ	d_1	d_{223}	$m_a/\overline{m}$
-1	3,57	4,14	17	3;57	0;2	0;34	$-0;2$	$0;55,59,3,18,...$
-2	4,11	4,29	18	4;11	0;2	0;36	$-0;4$	$0;55,59,6,28,...$
-3	4,25	4,44	19	4;25	0;2	0;38	$-0;6$	$0;55,59,9,17,...$
$\vdots$								

relation (Nature also seeming to prefer nice numbers), whose ratio of anomalistic to mean synodic month length differs from its modern equivalent by less 0;0,0,0,7.

In any event, it was the one adopted by the Babylonians, which we shall encounter explicitly in the System B theory of lunar anomaly, but which appears to have been a starting point for the System A theory as well. What is especially noteworthy is how little observational input is required for its construction. Thus to arrive at the alternatives shown above requires only the qualitative observations that the lunar velocity varies by roughly 4 degrees per day and exhibits some sensible increment after 14 months (readily observable over 224 months) and an increment with opposite sign after 223 months. Whether the choice of $\Pi = 251$ was influenced by any quantitative considerations beyond these is hard to judge from our present knowledge. What is all but certain is that it reflected no precise measurements. Nevertheless, the resulting function turns out to have a period relation with no sensible error and an amplitude with no significant one, and thus is as good as could be found without a quite different level of observational input.

3 SYSTEM B

Before turning to the complexities of System A it will be useful to look first at the theory of lunar anomaly found in System B, which is the simpler, although probably later, of the two. This consists of essentially two functions: Column F, which describes the variable lunar velocity, expressed in degrees per day;[13] and Column G, which describes the variable length of the month in excess of 29 days, expressed in $u\breve{s}$ or time degrees,[14] equivalent to 1/360th of a day or 4 minutes of our time. Both are simple linear zigzag functions, tabulated month by month in ephemerides separately for new and full moons, and both are based on the quite accurate[15] (and now familiar) period relation

$\Pi = 4,11$ (251) synodic months

$= Z = 4,29$ (269) anomalistic months,   (3.1a)

whence

Z (modulo Π) $= 18$.   (3.1b)

In Column F, which appears to be the more fundamental of the two functions, we have

$\Delta = 0;1^{o/d} \times \Pi = 4;11^{o/d}$ (3.2a)

Table 4
Tabular extremes of Column F (System B)

	Tabular maxima (M′)		Tabular minima (m′)	
F_1 (New)	15;14,10 ↑	15;16 ↓	11;6,10 ↑	11;6 ↓
F_2 (Full)	15;15,10 ↑	15;15 ↓	11;5,10 ↑	11;7 ↓

which leads to the increments

$$\delta = 0;2^{o/d} = d_{14} \quad d_1 = 18\delta = 0;36^{o/d} \quad d_{223} = -2\delta = -0;4^{o/d}. \quad (3.2b)$$

For a mean value F uses the value we have already encountered,

$$\mu = 13;10,35^{o/d}, \quad (3.3)$$

an accurate parameter,[16] which also appears in Geminus,[17] and leads to the extremes

$$M = \mu + \frac{\Delta}{2} = 15;16,5^{o/d} \text{ and } m = \mu - \frac{\Delta}{2} = 11;4,5^{o/d}. \quad (3.4)$$

Finally, the initial value was chosen to yield the tabular extremes shown in table 4, which makes the effective amplitude for both new and full moons $4;10^{o/d}$.

The result is an extremely nice and also practical function, which distinguishes values on ascending and descending branches (the former end in 0;0,10 the latter in 0;0,0) and also values for new and full moons (the former have even numbers in the second place, the latter odd numbers). Despite this, however, we still encounter instances where the scribe has mixed up the functions for new and full moons.[18]

With F so defined, and a common period relation, what is left to define G is

(1) a proportional constant, k, such that $\delta_G = k\delta_F$ and, consequently,

$d_nG = kd_nF$ and $\Delta G = k\Delta F$;

(2) its mean value, μ; and

(3) a single tabular extreme, to establish both the remaining tabular values and their phase in relation to F.

The proportional constant, which reflects the variation in G resulting from a given variation in F, is independent of μ and in System B equal to the simple fraction

$$k = \frac{5}{8}(\times 60) = 37;30^{u\check{s}/o/d}. \quad (3.5a)$$

This makes

$$\Delta = 2,36;52,30^{u\tilde{s}} \tag{3.5b}$$

and results in the increments

$$\delta = 1;15^{u\tilde{s}} \quad d_1 = 18\delta = 22;30^{u\tilde{s}} \quad d_{223} = -2\delta = -2;30^{u\tilde{s}}. \tag{3.5c}$$

It seems most likely to have been derived from assuming that the amplitude (i.e., maximum variation due to lunar anomaly in the length) of 223 months is

$$\Delta 223G = 15;0^{u\tilde{s}}, \tag{3.6}$$

and calculating that the amplitude of the corresponding summation of F over 223 months is

$$\Delta 223F = 0;24^{o/d}, \tag{3.7}$$

which leads to the attested value for k. Interestingly, $\Delta 223F$ is independent of both the tabular extrema and mean value of F. Also interesting is the fact that while the actual variation due to lunar anomaly in the length of 223 months is approximately $14;15^{u\tilde{s}}$, making equation (6) a very good approximation, the observable amplitude of 223 months is more than twice as great, the difference being the variation due to solar anomaly.

G's mean value,

$$\mu = 3,11;0,50^{u\tilde{s}}, \tag{3.8}$$

leads to extremes of

$$M = \mu + \frac{\Delta}{2} = 4,29;27,5^{u\tilde{s}} \text{ and } m = \mu - \frac{\Delta}{2} = 1,52;34,35^{u\tilde{s}} \tag{3.9}$$

and corresponds to the more familiar mean synodic month length of $29;31,50,8,20$ days—a value which Ptolemy retained[19] and which is very nearly as accurate as we can compute it today.[20] If, however, one wishes tabular values of G to end in different multiples of $0;0,10^{u\tilde{s}}$ for new and full moons on each branch (as they do), the choices for μ are limited to values that end in certain multiples of $0;0,5^{u\tilde{s}}$, corresponding to $0;0,0,0,50$ days. The nearest alternative values for $\mu(G)$ which satisfy this condition are $3,11;0,40^{u\tilde{s}}$ and $3,11;0,55^{u\tilde{s}}$ corresponding to mean synodic month lengths of $29;31,50,6,40$ days and $29;31,50,9,10$ days respectively. Since the modern value for the mean synodic month at -500 is $29;31,50,8,39$ days, corresponding to $\mu(G) = 3,11;0,51.9^{u\tilde{s}}$, the System B choice is the best of those available, although its high accuracy is to some extent fortuitous. Nevertheless, it is a superb parameter, whose origin is not immediately evident.[21]

Table 5
Tabular extremes of Column G (System B)

	Tabular maxima (M')		Tabular minima (m')	
G_1 (New)	4,28;37,30 ↑	4,29;1,40 ↓	1,53;37,30 ↑	1,52;46,40 ↓
G_2 (Full)	4,29;15,0 ↑	4,28;24,10 ↓	1,53;0 ↑	1,53;24,10 ↓

The tabular extremes of G are shown in table 5, and result in effective amplitudes for both G_1 and G_2 of 2,36;15^{uš}. Furthermore, each function has distinctive endings on each branch, G_1's being 30(↑) and 40(↓) and G_2's being 0(↑) and 10(↓), thereby making the branch and function of any value immediately recognizable to the Babylonian scribe. This, as noted, is a consequence of the initial value chosen [presumably $M'(G_2)↑ = 4,29;15$] and the endings of M and m, which in turn are a function of the endings of $\varDelta$ and μ, and constrain the admissible values of μ.

There remains the difference in phase between F and G. Since F indicates the lunar velocity at a given syzygy and G the length of the preceding month, G should be at its maximum when the average value of F over the preceding 18 intervals is at a minimum, and thus at a point 9 intervals after the minimum of F, that is, when $F(↑) - m(F) = 9\delta$. Instead we find that $M(G)$ follows $m(F)$ by 11;12,10 intervals. Of this amount, the fractional part (0;12,10ι), arises from the effects of choosing tabular extrema such that all values of both functions end in multiples of 0;0,10. This leaves, however, a discrepancy of 2 intervals, which makes the maximum of G (except for the rounding effects) correspond with a value of F 223 months earlier (or 28 months later) than expected. That this should be a mistake seems inconsistent with the competence and consistency of the rest of the theory, but I cannot otherwise explain it.

Daily Motion
System B also includes a function, F*, which describes the *daily* variation in lunar velocity. Since the period relation underlying columns F and G permits their tabulation only at intervals which are multiples (including regular fractions) of $\frac{1}{251}$ of an anomalistic month, F cannot be used for this purpose and a new period relation must be found. From the old period relation (1a) and the value of the mean synodic month implicit in G (3.8) we obtain for the length of the anomalistic month, m_a

$$1m_a = \frac{251}{269} \times 29;31,50,8,20 \text{ days} = 27;33,16,26,57 \dots \text{ days,} \qquad (3.10)$$

which yields as potential period relations:

$$1m_a = \quad 28 \text{ days} \quad -0;26,44, \text{ days} \tag{3.11a}$$

$$2 \cdot 1 = 2m_a = \quad 55 \text{ days} \quad +0; \ 6,32, \text{ days} \ (= 26,44/\mathbf{4};5,) \tag{3.11b}$$

$$4 \cdot 2 + 1 = 9m_a = 248 \text{ days} \quad -0; \ 0,31,57, \text{ days} \ (= 6,32/\mathbf{12};16,). \tag{3.11c}$$

The process can be continued to whatever accuracy is desired—the next in the series being $110m_a = 50,31$ (3031) days—but already in (11c) we find a period relation that is not only reasonably accurate, but has a period number, $\Pi = 4,8$ (248), which is conveniently similar to F's, thereby permitting F* to retain many of F's essential features. Thus letting $\Delta = \Pi/60$ as in F, we obtain the same minimum increment, $\delta_{F^*} = 0;2^{o/d} = \delta_F$, which with μ_F leads to the parameters for F*

$\Pi = 4,8$	$\mu = 13;10,35$	$m'\!\uparrow = 11;7,10$
$Z = 9$	$M = 15;14,35$	$m'\!\downarrow = 11;8$
$\Delta = 4;8$	$m = 11;6,35$	$M'\!\uparrow = 15;13;10$
$\delta = 0;2$	$d_1 = 0;18$	$M'\!\downarrow = 15;14.$

The effective extrema are the same as are found in texts from Uruk (ACT 190–194) and apart from three extreme values (11;5,10↑ and 15;15,10↑ and 15;16↓) are identical with values of F_2 on the ascending branch and F_1 on the descending branch. The same extrema (and thus intermediate values) are also reflected in two late texts from Babylon (ACT 194a and 194b), but with the branches reversed, thereby distinguishing the values of F* from those of F. In either form, however, F* reflects a careful (and successful) effort to achieve as close a correspondence with F as their different period relations will permit.

In sum, the System B theory of lunar anomaly is based on simple linear zigzag functions with accurate mean values for both the synodic month length and lunar velocity, which eventually found their way into Greek astronomy. These parameters and other critical elements of the System B functions are summarized in exhibit 1, together with the maximum amplitudes of important intervals derived from Column G. Except for the amplitude of 223 months, the latter are not very good and reflect the structural limitations of the theory. Thus despite the quality of its parameters and the intuitive familiarity of its structure, System B depicts the variations that actually occur in the lengths of the relevant intervals rather poorly.

Exhibit 1

	System B		
Name	*Column F*	*Column F**	*Column G*
Description	Daily Motion	Daily Motion	1 Month-29 days
Tabulated	Monthly	Daily	Monthly
Units	°/day	°/day	$u\check{s}$ (time degrees)
Number Period, Π	4,11 (251)	4,8 (248)	4,11 (251)
Cycles (m_a), Z	4,29 (269)	9	4,29 (269)
Z, modulo Π	18	9	18
Tabular Period, Π/Z, mod.Π	13;56,50^m	27;33,20^d	13;56,50^m
Amplitude, Δ	4;11	4; 8	2,36;52,30
$\delta = 2\Delta/\Pi$	0; 2	0; 2	1;15
$d_1 = Z(\text{mod}.\Pi)\delta$	0;36	0;18	22;30
d_{14}	0; 2		1;15
d_{223}	−0; 4		−2;30
Mean value, μ	13;10,35	13;10,35	3,11; 0,50
Maximum, M	15;16, 5	15;14,35	4,29;27, 5
Minimum, m	11; 5, 5	11; 6,35	1,52;34,35
Modern μ (−500)	13;10,34,52,41.1$^{°/d}$		3,11; 0,51.9$^{u\check{s}}$
Error $(\mu_{mod} - \mu_{bab})$	−0; 0, 0, 7,19$^{°/d}$		0; 0, 2$^{u\check{s}}$
Extreme tabular values			
$m'\uparrow$	11; 5,10 (F$_2$)	11; 7,10	1,53; 0 (G$_2$)
$M'\uparrow$	15;15,10 (F$_2$)	15;13,10	4,29;15 (G$_2$)
$m'\downarrow$	11; 6 (F$_1$)	11; 8	1,52;46,40 (G$_1$)
$M'\downarrow$	15;16 (F$_1$)	15;14	4,29; 1,40 (G$_1$)
Δ'	4;10	4; 6	2,36;15

Maximum variation (Δn) in n months

	System B	Modern[22]	Error
Δ1m	2,36;15$^{u\check{s}}$	2,15;0$^{u\check{s}}$	−21;15$^{u\check{s}}$
Δ6m	8,56;15	9,27;45	+31;30
Δ12m	4,22;30	4, 7;45	−14;45
Δ223m	15;0	14;15	−0;45
Δ235m	4,33;45	4,23;..	−10;45

4 SYSTEM A—DESCRIPTION

System A is a different matter altogether. Here we find three functions dependent on lunar anomaly—Columns F and G again, describing, as in System B, the variation in the daily lunar motion and length of the month, but also a function named Column Φ, which is precisely in phase with F and serves essentially as an argument of anomaly in the determination of G. In ephemerides both Φ and F are pure zigzag functions, although elsewhere truncated versions appear, which in fact have greater astronomical significance. For most of its extent G is identical with a linear zigzag function, $\hat{G}$, from which it departs by being flattened near its extremes, where it is derived from Φ.

The parameters of Φ, F, and $\hat{G}$ are shown in exhibit 2 and differ dramatically from those of System B. All are based on the period relation

$$6{,}247 \text{ synodic months (m)} = 6{,}695 \text{ anomalistic months (m}_a), \qquad (4.1)$$

which can otherwise be expressed as

$$(223 + \varepsilon)m = (239 + \varepsilon)m_a, \qquad (4.2a)$$

where

$$\varepsilon = \frac{3}{28}. \qquad (4.2b)$$

A striking feature of all three functions is their great number of sexagesimal places, Φ and F having four places of sexagesimal fractions, while G has five. These can be reduced and the resulting numbers made (somewhat) more manageable by stripping the factor $0;17,46,40^{us}$ from Φ and G, which results in the parameters shown under S and M in exhibit 2, where the increment in S for 223 months is now 1 and the minimum increment, $\delta = 0;1,15 = \frac{1}{48}$, corresponds to an interval, $\iota = m_a/6247 = 1;35, ..^{us}$ In contrast with this seeming precision, the mean values of F and $\hat{G}$ are both strangely high, being respectively $0;20°$ per day and 24us higher than System B's essentially accurate values. These differences are far too great to be errors (unless the theory is quite incompetent), and comprise one of our challenges to explain. Other differences include the larger amplitude and monthly difference, $d_1 = 0;42^{o/d}$, of F and the fact that the ratios of the increments for 223 and 14 months are as $-48 : 25$ instead of $-2 : 1$ as in System B.

The rules for calculating G from Φ are shown in exhibit 3 and spelled out in almost identical fashion in several procedure texts. Given

Exhibit 2

System A

Period Relation: 6,247 synodic months (m) = 6,695 anomalistic months (mₐ)

Number period, Π	1,44, 7 = 6,247	
Cycles (mₐ), Z	1,51,35 = 6,695	
Z, modulo Π	7,28 = 448	
Tabular period, Π/Z, mod. Π	$P = 6247/448 = 13;56,39,6, \ldots$ m	

Name	Column Φ	Column $\hat{G}$	Column F
Description	223 Months	1 Month	Daily
	−6585 days	−29 days	Motion
Units	uš (time degrees)		°/day
Amplitude, Δ	19;16,51, 6,40	2,59;57,17, 2,13,20	4;52,49,41,15
$\delta = 2\Delta/\Pi$	0; 0,22,13,20	0; 3,27,24,26,40	0; 0, 5,37,30
$d_1 = 448\delta$	2;45,55,33,20	25;48,38,31, 6,40	0;42
$d_{14} = 25\delta$	0; 9,15,33,20	1;26,25,11, 6,40	0; 2,20,37,50
$d_{223} = -48\delta$	0;17,46,40	2;45,55,33,20	0; 4,30
Mean value, μ	2, 7;26,23,20, 0	3,34;58,23,42,13,20	13;30,29,31,52,30
Maximum, M	2,17; 4,48,53,20	5, 4;57, 2,13,20	15;56,54,22,30
Minimum, m	1,57;47,57,46,40	2, 4;59,45,11, 6,40	11; 4, 4,41,15
Truncated max., M'	2,13;20	[4,56;35,33,20]	15; 0
Truncated min., m'	1,58;31, 6,40	[2,40;0]	11;15
Truncated ampl., Δ'	14;48,53,20	[2,16;35,33,20]	3;45
Modern μ (−500)	1,56;13,2,37..	3,11;0,51.9	13;10,34,52,41.1
$\mu_{Mod} - \mu_{Bab}$	−11;13,..uš	−23;57,..uš	−0;19,54,..°/d

In Units of $\phi = 0;17,46,40$ uš

	$S = \Phi/\phi$	$M = \hat{G}/\phi$
Amplitude, Δ	1,5; 4,22,30	10, 7;20,50
$\delta = 2\Delta/\Pi$	0; 1,15	0;11,40
$d_1 = 448\delta$	9;20	1,27; 6,40
$d_{14} = 25\delta$	0;31,15	4;51,40
$d_{223} = -48\delta$	1	9;20
Mean value, μ	7,10; 6,33,45	12,5;32, 5
Maximum, M (full)	7,42;38,45	17,9;12,30
Minimum, m (new)	6,37;34,22,30	7,1;51,40
Truncated max., M'	7,30	[16,41]
Truncated min., m'	6,40	[9, 0]
Truncated ampl., Δ'	50	[7,41]

Exhibit 3

To Calculate G from Φ

Opposite $\Phi_n(\uparrow\downarrow)$ you put G_n; whatever exceeds ($<>$) $\Phi_n(\downarrow\uparrow)$, multiply by C_n and add to G_n.

n	Φ_n	Branch	Gn	$<>$	C_n	dC_n
1	2 13 20	↓	2 40	<	$\{A*(\Delta/\phi)\}\phi$	+1
2	2 10 40	↓	2 53 20	<	9 20	0/1
3	1 58 31 6 40	↓	4 46 42 57 46 40	<	8 20	−1
4	1 58 13 20	↓	4 49 11 6 40	<	7 20	−1
5	1 57 55 33 20	↓	4 51 21 28 53 20	≦	6 20	−1
6	1 57 58 8 53 20	↑	4 53 14 4 26 40	>	5 20	−1
7	1 58 15 55 33 20	↑	4 54 48 53 20	>	4	−1 20
8	1 58 33 42 13 20	↑	4 56	>	0	−4
9	1 58 37 2 13 20	↑	4 56	>	2	+2
10	1 58 54 48 53 20	↑	4 56 35 33 20	>	0	−2
11	1 59 12 35 33 20	↑	4 56 35 33 20	>	−2	−2
12	1 59 30 22 13 20	↑	4 56	>	−4	−2
13	1 59 48 8 53 20	↑	4 54 48 53 20	>	−5 20	−1 20
14	2 0 5 55 33 20	↑	4 53 14 4 26 40	>	−6 20	−1
15	2 0 23 42 13 20	↑	4 51 21 28 53 20	>	−7 20	−1
16	2 0 41 28 53 20	↑	4 49 11 6 40	>	−8 20	−1
17	2 0 59 15 33 20	↑	4 46 42 57 46 40	>	−9 20	−1/0
18	2 13 8 8 53 20	↑	2 53 20	>	$-\{A*(\Delta/\phi)\}\phi$	+1
19	2 15 48 8 53 20	↑	2 40	>	0	0

$A*(n) \equiv n + \sum_1^{n-1} I(n)$, where $I(n) \equiv$ all positive integers $\leq n$; $\Delta = \Phi - \Phi_n$.

Line 1 (ACT 200:14)

Opposite 2,13,20 decreasing you put 2,40. From 2,13,20 decreasing to 2,10,40 decreasing whatever is less than 2,13,20 decreasing multiply by 3,22,30. All parts of (the result multiply by) 17,46,40 (and) add them together and add (the result) to 2,40.

Line 18 (ACT 206)

Opposite 2,13,8,8,53,20 increasing you put 2,53,20. Anything beyond 2,13,8,8,53,20 increasing to 2,15,48,8,53,20 increasing, multiply by 3,22,30; all parts of (the result multiply by) 17,46,40 (and) add them together and subtract (the result) from 2,53,20.

Line 8 (ACT 200:14)

Opposite 1,58,33,42,13,20 [increasing] put 4,56; until 1,58,37,2,13,20 increasing everywhere 4,56.

some value of Φ, and whether it is on the descending (↓) or ascending (↑) branch, one first calculates the difference between Φ and the value of Φ next preceding it among the 19 listed in column (1) of exhibit 3. Then, except in lines 1 and 18, this difference is multiplied by the coefficient given in column (3) and the result added algebraically to the value for G listed in column (2).

For line 1 the procedure is expressed in ACT 200[23] as follows, and a precisely analogous instruction for line 18 is found in ACT 206.

> Opposite 2,13,20(↓) you put 2,40. From 2,13,20(↓) to 2,10,40(↓), whatever is less than 2,13,20(↓) multiply by 3,22,30. All parts of (the result)—the akkadian word here is *ma-lu-uš-šú*: its fullness[24]—(multiply by) 17,46,40 and add them together and add (the result) to 2,40.

What this highly compressed instruction effectively says is: divide the difference in Φ by 17,46,40 (the reciprocal of 3,22,30), then multiply each of the component integers of the result plus any fraction by the same 17,46,40, and add them together to get the difference to be added to (or subtracted from) the previous G. Essentially, this amounts to computing the additive analog of the factorial of a number—that is, the sum of its integers plus any fraction, a concept that was evidently familiar to the Babylonian authors of these procedure texts.

Between lines 2–3 and 17–18 in exhibit 3 the function G is linear and identical with Ĝ. Near its extremes it is flattened, with second differences in the interpolation coefficients of +1 in the *maluššu* region near its minimum, and (mostly) −1 or −2 near its maximum. With one exception, G is also perfectly symmetrical about its maximum. The exception is the anomalous interval between lines 8 and 9, where G remains constant at 4,56 before rising to its maximum in line 10. This results in irregular second differences and has no counterpart after the maximum of G. Also, outside the *maluššu* region, Φ changes by ±17,46,40,0 between lines except here, where the difference is only 3,20,0 or the fraction 0;11,15 of the normal difference. We shall return later to this curious interval.

So far there is nothing to hint at the astronomical significance of Φ, whose role in the calculation of G could as readily be played by F. Indeed, even the discovery that Φ was the length of 223 months in excess of 6585 days, expressed, like G, in time degrees, did little to clarify the motivation for the Φ : G scheme and contributed the further puzzle that the mean value of Φ is 11[uš] higher than the average length of 223 months. Nevertheless, as we shall see, Φ plays a central role in the construction of the entire theory, which is built up with singular resourcefulness from a few simple, but critical elements.

5   SYSTEM A—STRUCTURE AND MOTIVATION

To illuminate the structure and motivation of this theory, I shall endeavor to trace the process, or more accurately, the progression of logical elements that appear to have gone into its construction. The task will be to construct a scheme that describes the variations in the lengths of the month and principal eclipse intervals resulting solely from the variation in the moon's daily motion. The theory should have an accuracy consistent with that of plausible observations and, finally, make no assumption about the effects of solar anomaly or any other variable that is a function of position.

Before beginning, some remarks on notation may be helpful. M and m have been used to denote the maxima and minima of linear zigzag functions, although for clarity these are sometimes denoted f_{max} or f_{min}. Primes (M', m', Δ') denote effective extremes and corresponding amplitudes, resulting from truncation of functions in System A and from tabular extremes in System B. In general m denotes synodic months, qm the length of q synodic months, and $\mathbf{m}_n$ or $\mathbf{s}_n$ the specific synodic month or "saros" (223 m) associated with syzygy n. M and qM denote *functions* which describe the length of one month or q months in *units of* ϕ, equal to the sarosly change in the length of a saros, while M_n denotes the month length according to M associated with syzygy n. Thus, 12M and 235M refer to functions describing the variable lengths of 12 and 235 months in units of ϕ, while $12M_n$ and $12M_{n-12}$ denote the values of 12M associated with syzygies n and $n - 12$. An exception is made for the function describing 223 months, which is often called a "saros"[9] and denoted by S instead of 223M. At times it will also be desirable to distinguish functional values associated with different syzygies *within* a saros (e.g., S_n and S_{n-12}) and those of the corresponding syzygies some number of saroi distant. In such cases s is used to denote the number of a particular saros within a string of consecutive saroi, so that the notation $qM_{s,n}$ denotes the functional length of q months associated with syzygy n of saros s and similarly for $S_{s,n}$.

5.1   Building Blocks

Apart from a facility with linear zigzag functions, six elements are required for the assemblage of this theory. Three are period relations, and the other three are critical observations (or equivalent empirical data). The period relations include: the so-called saros cycle of 223 months, in which there is near—but not quite complete—return in anomaly; the anomalistic cycle of 251 months, already encountered in System B; and finally, the 19-year cycle of 235 months in which the Sun and Moon return in longitude. In brief, these period relations state that

Saros cycle:

223 synodic months $\cong$ 239 anomalistic months,   (5.1.1)

Anomalistic cycle:

251 synodic months = 269 anomalistic months,   (5.1.2)

19-year cycle:

235 synodic months = 254 sidereal months = 19 years.   (5.1.3)

The Saros cycle also reflects near returns in nodal elongation and longitude and was probably first identified and certainly more familiar as an eclipse cycle. Significantly, however, 235-month intervals are also frequently bounded by lunar eclipses, a circumstance which permits the return in (sidereal) longitude of the Moon and Sun to be observed directly.

The critical observations, all of extrema, are that: the amplitude—that is, maximum variation—of 235 months does not exceed $4,30^{u\tilde{s}}$; the maximum length of 235 months does not exceed 6940 days $+ 10^{u\tilde{s}}$; and the maximum duration of 223 months does not exceed 6585 days $+ 2,15^{u\tilde{s}}$, that is,

$\Delta(235\text{ m})$ $\qquad \le 4,30^{u\tilde{s}}$,   (5.1.4)

$\text{Max}(235\text{ m}) - 6940^{d} \le 10^{u\tilde{s}}$,   (5.1.5)

$\text{Max}(223\text{ m}) - 6585^{d} \le 2,15^{u\tilde{s}}$.   (5.1.6)

In place of (5.1.5) the mean lunar velocity, $\mu = 13;10,35^{o/d}$, could be substituted, but otherwise these are the necessary and sufficient elements for the construction of this theory.

5.2 Fundamental Relationships

Two relationships are fundamental to the entire theory. The first, which I shall call the "Interval Rule," is general and concerns the differences in the length between overlapping strings of (A) months separated by some interval of (B) months. The second, which I shall call the "Saros Rule," concerns the saros cycle specifically and is derived from its period relation.

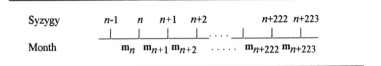

Figure 3a

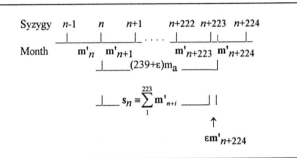

Figure 3b

Interval Rule   Consider the series of months, $\mathbf{m}_n$, $\mathbf{m}_{n+1}$, ... $\mathbf{m}_{n+223}$ (figure 3a), and let s_n equal the saros comprising months $\mathbf{m}_{n+1}$ to $\mathbf{m}_{n+223}$ so that

$$s_n = \sum_{i=n+1}^{n+223} \mathbf{m}_i. \tag{5.2.1}$$

Then, since the intervening months are contained in both S_n and S_{n-1}, it follows that

$$d_1 s_n \equiv s_n - s_{n-1} = \sum_{i=n+1}^{n+223} \mathbf{m}_i - \sum_{i=n}^{n+222} \mathbf{m}_i$$

$$= \mathbf{m}_{n+223} - \mathbf{m}_n \equiv d_{223}\mathbf{m}_n \equiv d_s\mathbf{m}_n, \tag{5.2.2}$$

which is to say that the monthly change in the length of the saros associated with a given month ($d_1 s_n$) is equal to the sarosly change in the length of that month ($d_s\mathbf{m}_n$).

A similar argument addressed to the series of months, $\mathbf{m}_n$, $\mathbf{m}_{n+1}$, ... $\mathbf{m}_{n+a}$, ... $\mathbf{m}_{n+b}$, ... $\mathbf{m}_{n+b+a}$, leads to the more general result that for two intervals,

$$A_n = \sum_{i=n+1}^{n+a} \mathbf{m}_i \quad \text{and} \quad B_n = \sum_{i=n+1}^{n+b} \mathbf{m}_i, \quad \text{where } b > a,$$

$$d_a B_n = d_b A_n \quad \text{[Interval Rule]}, \tag{5.2.3}$$

or the change in B over a months equals the change in A over b months. We will encounter this rule principally in the cases for which $b = 223$ months and $a = 1$ or 12 months. Nevertheless, I have no doubt that it was understood in its full generality by the Babylonian author of our theory.

Saros Rule   The Interval Rule applies to any sequence of months, regardless of their variability. The Saros Rule is less general and concerns changes in a sequence of abstract months, whose length is assumed to vary only with lunar anomaly. Such months are here denoted by $\mathbf{m}'$ and effectively assume uniform motion of the position of syzygy.

Definitions: $\displaystyle s_n \equiv \sum_{1}^{223} \mathrm{m}'_{n+i}, \mathrm{d_q s}_n \equiv s_{n+q} - s_n$, and $\mathrm{d_s} \equiv \mathrm{d}_{223}$.

The saros cycle (5.1.1) says that 223 synodic months very nearly equal 239 anomalistic months. This can be expressed more precisely as the period relation,

$(223 + \varepsilon)$ synodic months $(\mathbf{m}')$

$\qquad = (239 + \varepsilon)$ anomalistic months $(\mathrm{m_a})$,   (5.2.4a)

where ε is a rational fraction which will be small if the approximation is a good one, as indeed it is. Since the anomalistic month does not vary, expression (5.2.4a) can be restated as

$$223\mathbf{m}' = (239 + \varepsilon)\mathrm{m_a} - \varepsilon\mathbf{m}' = \mathbf{C} - \varepsilon\mathbf{m}' \qquad (5.2.4b)$$

and understood to mean that the length of any 223 months $(\mathbf{m}') + \varepsilon$ times the length of the 224th such month is constant. Thus, as illustrated in figure 3b,

$$(223 + \varepsilon)\mathbf{m}' \equiv \mathbf{s}_n + \varepsilon\mathbf{m}'_{n+224} = \mathbf{C},$$

$$\mathbf{s}_n = \mathbf{C} - \varepsilon\mathbf{m}'_{n+224}, \qquad (5.2.5a)$$

and consequently,

$$
\begin{aligned}
d_1 \mathbf{s}_n &= -\varepsilon(\mathbf{m}'_{n+225} - \mathbf{m}'_{n+224}) \\
&= -\varepsilon(\mathbf{m}'_{n+2} + d_s\mathbf{m}'_{n+2} - [\mathbf{m}'_{n+1} + d_s\mathbf{m}'_{n+1}]) \\
&= -\varepsilon([\mathbf{m}'_{n+2} - \mathbf{m}'_{n+1}] + [d_s\mathbf{m}'_{n+2} - d_s\mathbf{m}'_{n+1}]) \\
&= -\varepsilon(d_1\mathbf{m}'_{n+1} + d_1[d_s\mathbf{m}'_{n+1}]) \\
&= -\varepsilon(d_1\mathbf{m}'_n + d_1[d_1\mathbf{m}'_{n+1}] + d_1[d_s\mathbf{m}'_{n+1}]). \qquad (5.2.5b)
\end{aligned}
$$

If now the month length $\mathbf{m}'_n$ is replaced by the corresponding value, M_n, of a linear zigzag function, $d_1[d_1M_{n+1}]$ and $d_1[d_sM_{n+1}]$ will equal 0 whenever M_{n+223} and M_{n+2} are on the same branch.[25] Replacing $\mathbf{s}_n$ with $S_n = \sum_1^{223} M_{n+i}$, we find that

$$d_1S_n = -\varepsilon d_1M_n, \quad \text{if } d_1M_{n+223} = d_1M_{n+2}, \tag{5.2.6a}$$

and more generally

$$d_qS_n = -\varepsilon d_qM_n, \quad \text{if } d_1M_{n+223} = d_1M_{n+1+q}. \tag{5.2.6b}$$

This is the Saros Rule and second fundamental relationship, which states that for any syzygy n the change over q months in the length of the following sequence of 223 such monthly values is equal to $-\varepsilon$ times the change over the same interval in the length of q such months (provided that the relevant values are on the same branch).

Combined relationships   Combining the Saros and Interval Rules we obtain

$$\begin{aligned}
d_1S_n &= -\varepsilon d_1M_n \quad (< \text{Saros Rule}) \\
&= d_sM_n \quad (< \text{Interval Rule});
\end{aligned} \tag{5.2.7}$$

$$\begin{aligned}
d_sS_n &= -\varepsilon d_sM_n \quad (< \text{Saros Rule}) \\
&= -\varepsilon d_1S_n \quad (< \text{Interval Rule}) \\
&= \varepsilon^2 d_1M_n. \quad (< \text{Saros Rule})
\end{aligned} \tag{5.2.8}$$

These are precisely the relationships that we find between the monthly and sarosly differences in columns Φ and $\hat{G}$ in System A. Moreover, they also imply that *the variation in the month can be derived from a function describing the variation in the Saros in terms of the sarosly change in its length.* This is the central premise of the System A theory.

5.3   A Function for s
To construct a zigzag function, S, which will describe the variation in the length of the saros in terms of its sarosly difference, we begin by defining

$$d_sS \equiv d_{223}S \equiv \pm 1. \tag{5.3.1}$$

To obtain the amplitude of S we turn to the Saros period relation (5.2.4a) where

$$\Pi = 223 + \varepsilon \quad Z = 239 + \varepsilon \quad Z, \text{mod}.\Pi = 16. \tag{5.3.2}$$

This leads to

$$P_1 = \frac{\Pi}{Z, \mathrm{mod}.\Pi} = \frac{223 + \varepsilon}{16} = \frac{2\Delta S}{d_1 S}, \tag{5.3.3a}$$

$$\Delta S = d_1 S \frac{223 + \varepsilon}{32\varepsilon}, \tag{5.3.3b}$$

where d_1 is the monthly difference in S and P_1 the corresponding period. Since we have defined the sarosly difference in S as equal to 1, it follows from (5.2.8) that

$$|d_1 S| = \frac{|d_s S|}{\varepsilon} = \frac{1}{\varepsilon}, \tag{5.3.4}$$

and thus from (5.3.3b) that

$$\Delta S = \frac{223 + \varepsilon}{32\varepsilon}, \tag{5.3.5}$$

expressed in units of the sarosly change in S, hereafter called "units of ϕ" to distinguish them from the time degrees ($u\check{s}$) in which the tabulated functions are expressed.

5.4 Establishing ε

We establish ε by equating the saros and anomalistic period relations, that is, setting

$$\frac{223 + \varepsilon}{239 + \varepsilon} = \frac{251}{269}, \tag{5.4.1a}$$

from which

$$\varepsilon = \frac{1}{9} \text{ (precisely).} \tag{5.4.1b}$$

This implies (from 5.3.5) an amplitude, $\Delta S = 62;45$ units of ϕ, and thus a period of 125;30 sarosly steps. This would be fine but for the fact that the resulting function contains only 251 discrete values, far too few to describe adequately changes that will, in instances, be magnified by at least a factor of 81 (i.e., $1/\varepsilon^2$). Consequently we shall need a value for ε that approximates $\frac{1}{9}$, but yields a larger number period.

Our choices are:

$$\frac{2}{17} \text{ and } \frac{2}{19}; \frac{3}{26} \text{ and } \frac{3}{28}; \frac{4}{35} \text{ and } \frac{4}{37}; \ldots \text{ and so forth}, \tag{5.4.2}$$

the accurate value being between $\frac{3}{26}$ and $\frac{1}{9}$. In fact, $\frac{3}{28}$ was chosen, perhaps because of all the alternatives, it contains the smallest irregular number as a factor, or possibly because 14 turns up elsewhere in the theory of lunar anomaly. Whatever the reason, the choice of an approximate ε introduces a deliberate distortion into the resulting period relation, which consequently should not be viewed as a meaningful empirical relationship.

With $\varepsilon = \frac{3}{28}$, the Saros relationship becomes the System A period relation,

$$\Pi = 1,44,7(6247)\text{m} = Z = 1,15,35(6695)\text{m}_a,$$

$$Z, \bmod. \Pi = \frac{48}{\varepsilon} = 7,28(448) \tag{5.4.3}$$

whose number period is now conveniently large ($\Pi \cong 505$ years) and which partially defines the function, S, whose established parameters in units of ϕ now include:

$$\Delta S \text{ (from 5.3.5)} = \frac{223 + \varepsilon}{32\varepsilon} = 1,5;4,22,30^{\phi}$$

$$\delta = \frac{2\Delta}{\Pi} = \frac{1}{48} = 0;1,15^{\phi}. \quad d_1 = 7,28\delta = \frac{1}{\varepsilon} = 9;20^{\phi} \tag{5.4.4}$$

$$d_{223} = d_s = -\varepsilon d_1 = -1^{\phi},$$

where the minus sign indicates that d_s is negative when d_1 and δ are positive, that is, for S values on the ascending branch.

5.5   Avoiding Anomalies of Position

It remains to establish the units, one extreme, and some initial value of this function. To do so, however, we must first find a way to circumvent the effects of all variations (e.g., length of daylight and horizon angle as well as solar anomaly) that depend on the position of syzygy (for which we may not yet even have a uniform scale). We do this by using the 19-year cycle, the last of our assumed period relations, which states that the Sun and Moon return to their initial positions after 235 months and thus implies that the length of 235 months is unaffected by any variation depending on position. Thus we shall seek to construct a function, which describes the variation in the length of 235 months, from the function S that we have just established to describe the Saros, using the Interval Rule to accomplish this.

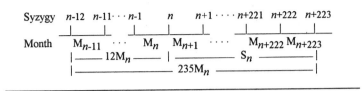

Figure 4

We begin with the fact that 235 months is 12 months more than a saros. From the Interval Rule the sarosly change in 12 months is equal to the 12-month change in 223 months, and thus

$$d_s 12M = d_1 2S = 12 d_1 S (\mathrm{mod}.2\Delta)$$

$$= \frac{12}{\varepsilon} - 2\Delta = -18;8,45^\phi. \tag{5.5.1a}$$

Here the negative sign is associated with S values on the ascending branch so that

$$S_{n+12}(\uparrow) - S_n(\uparrow) = S_n(\uparrow) - S_{n-12}(\uparrow) = -18;8,45^\phi, \text{ and} \tag{5.5.1b}$$

$$S_{n-12}(\uparrow) - S_n(\uparrow) = d_{-12}S = +18;8,45^\phi. \tag{5.5.1c}$$

Next we observe that if we associate with each syzygy, n, the 12 months preceding it and the 223 months following it (figure 4), the resulting 235-month interval will always equal the sum of the associated 12- and 223-month intervals, and similarly for the sarosly changes of these intervals, that is,

$$235 M_n = 12 M_n + 223 M_n = 12 M_n + S_n, \tag{5.5.2a}$$

and thus

$$d_s 235M = d_s 12M + d_s S_n. \tag{5.5.2b}$$

Therefore, we now construct two saros functions, S_n and S_{n-12}, separated by 12 months as shown in columns (3) and (2) of exhibit 4, and let each progress by sarosly steps, denoted by s. For simplicity and symmetry we begin with values of S_n and S_{n-12} which are equidistant from S_{max}, but on different branches, and let S_n be on the ascending branch so that $d_s S(\uparrow) = -1$. Consequently, $S_{1,n}$ and $S_{1,n-12}$, will be identical except for branch and equal to

$$S_{1,n}(\uparrow) = S_{1,n-12}(\downarrow) = S_{max} - \tfrac{1}{2} d_1 2S = S_{max} - 9;4,22,30, \tag{5.5.3a}$$

and

$$d_{12}S_{1,n} = S_{1,n}(\uparrow) - S_{1,n-12}(\downarrow) = 0. \tag{5.5.3b}$$

Proceeding in this way we calculate in columns (4) and (7) the sarosly differences in 12M and 235M step by step over a full period from

$$d_s 12M_{s,n} = d_{12}S = S_{s,n} - S_{s,n-12}, \tag{5.5.4a}$$

and

$$d_s 235M_{s,n} = d_s 12M_{s,n} + d_s S_{s,n}. \tag{5.5.4b}$$

Finally, in column (9) we compute the cumulative sum of all the sarosly differences, $d_s 235M_{s,n}$, to obtain the function $235M - \mathrm{Max}(235M)$, whose amplitude thus corresponds to the maximum variation in 235 months.

The approach is promising, but unsatisfactory for several reasons that are evident in exhibit 4. The most important is that the sarosly differences in 12M are asymmetrical, owing to the different numerical effects of reflecting S_n and S_{n-12} at their extremes. Consequently both 12M and 235M are asymmetrical about their extremes, which are also not clearly defined. Because of this, a second problem is that 235M does not return precisely to its starting point in step 130 before turning down again. Finally, the second differences of 235M are abruptly discontinuous, jumping from -2 near its maximum to 0 on its linear stretch to $+2$ near its minimum.

5.6 Fixing the Problems

The manner in which the author of System A solved these problems is one of the most remarkable features of the entire theory. The solution begins with an observation that all asymmetries disappear if, and only if, the values of S_n are so chosen that S_n and S_{n-12} are mirror images of each other—that is, they contain the same numbers but in reverse order and on opposite branches so that $S_{s,n}(\uparrow\downarrow) = S_{131-s,n-12}(\downarrow\uparrow)$. This in turn requires that

$$S_{s,n}(\uparrow) = S_{131-s,n}(\downarrow)$$
$$= 2S_{max} - d_{-12}S - S_{1,n} - 2\Delta S + 131 - 2s, \tag{5.6.1a}$$

or

$$S_{max} - S_{s,n} = \Delta S + \tfrac{1}{2}d_{-12}S - 65;30 + s = 8;38,45 + s, \tag{5.6.1b}$$

and thus that

$$S_{1,n} = S_{max} - 9;38,45. \tag{5.6.1c}$$

(1) s	(2) S(s, n − 12)		(3) S(s, n)		(4) d12S = ds[12M]	(5) ΔΔ ds[12M]	(6) dsS(n)	(7) ds[235M] = ds[12M] + dsS	(8) ΔΔ ds[235M]	(9) Σ(7) [235M] − Max
1	Smax-9;4,22,30	→	Smax-9;4,22,30	↑	0	−2	−1	−1	−2	−1
2	S(1, n − 12) + 1	→	S(1, n) − 1	↑	−2	−2	−1	−3	−2	−4
3	+2	→	−2	↑	−4	−2	−1	−5	−2	−9
...	...		...		...	...	...	...	...	...
9	+8	→	−8	↑	−16	−2	−1	−17	−2	−81
10	+9	→	−9	↑	−18	−0;8,45	−1	−19	−0;8,45	−100
11	+ 8; 8,45	←	−10	↑	−18; 8,45	0	−1	−19; 8,45	0	−119; 8,45
12	+ 7, 8,45	←	−11	↑	−18; 8,45	0	−1	−19; 8,45	0	
...	...		...		...	...	...	...	...	...
19	− 0; 8,45	←	−18	↑	−18; 8,45	0	−1	−19; 8,45	0	
20	− 0;51,45	←	−19	↑	−18; 8,45	0	−1	−19; 8,45	0	
...	...		...		...	...	...	...	...	...
56	−36;51,15	←	−55		−18; 8,45	0	−1	−19; 8,45	+2	−1063; 1,15
57	−37;51,15	←	−56	min	−18; 8,45	+2	+1	−17; 8,45	+2	−1062;10
58	−38;51,15	←	−55	→	−16; 8,45	+2	+1	−15; 8,45	+2	−1059;18,45
59	−39;51,15	←	−54	→	−14; 8,45	+2	+1	−13; 8,45	+2	
...	...		...		...	...	...	...	...	...
65	−45;51,15	←	−48	→	−2; 8,45	+2	+1	−1; 8,45	+2	
66	−46;51,15	←	−47	→	−0; 8,45	+2	+1	+0;51,15	+2	
67	−47;51,15	←	−46	→	+1;51,15	+2	+1	+2;51,15	+2	
68	−48;51,15	←	−45	→	+3;51,15	+2	+1	+4;51,15	+2	
...	...		...		...	...	...	...	...	...
75	−55;51,15	←	−38	→	+17;51,15	+0;17,30	+1	+18;51,15	+0;17,30	
76	−55; 8,45	→	−37	→	+18; 8,45	0	−0;51,15	+19; 8,45	0	
77	−54; 8,45	→	−36	→	+18; 8,45	0	−1	+19; 8,45	0	
...	...		...		...	...	...	...	...	...
121	−10; 8,45	→	+8	↑	+18; 8,45	0	+1	+19; 8,45	−1;51,15	−66;37,50
122	−9; 8,45	→	+9	↑	+18; 8,45	−1;51,15	−1	+17;17,30	0	−51;20,10
123	−8; 8,45	→	+8; 8,45	←	+16;17,30	−2	−1	+15;17,30	−2	−38; 2,40
124	−7; 8,45	→	+7; 8,45	←	+14;17,30	−2	−1	+13;17,30	−2	
...	...		...		...	...	...	...	...	...
130	−1; 8,45	→	+1; 8,45	↑	+2;17,30	−2	−1	+1;17,30	−2	−0;17,30
131	−0; 8,45	→	+0; 8,45	↑	+0;17,30	−2	−1	−0;42,30		−1;0
132	+0;51,15	←	−0;51,15	↑	−1;42,30					

This solves the first problem and also the second, since with the restoration of symmetry, 235M now returns to its starting point in step 130. It also means that $S_{\max}$ will be among the tabular values of S.

The third problem can be solved, and the second differences made much tidier and more agreeable, by truncating S near its extremes. If the superior truncation is greater than 9;38,45 this will have the effect of introducing second differences of -1 between the maximum and linear stretch of 235M and of $+1$ between the linear stretch and some point near the minimum. More importantly, however, the amount of the superior truncation will also establish the number of steps which will be required for 235M to return to its maximum, and thereby the effective period over which the variation in that function takes place.

This means that by truncating S at a distance $9;38,45 + q$ steps from $S_{\max}$, that is, at

$$S'_{\max} = S_{\max} - 9;38,45 - q, \tag{5.6.2}$$

where q is an integer ≥ 0, the variation in 235M will take place over $130 - 2q$ steps, which becomes the effective anomalistic period of the function. Thus the superior truncation permits the effective anomalistic period to be established at any value consistent with $130 - 2q$ steps and thus both to remove the effects of introducing an artificial value of ε to lengthen the number period and to improve the initial period, if we wish. For $\varepsilon = \frac{1}{9}$, characteristic of the original anomalistic period relation, the effective period is

$$P_s = \frac{223 + \varepsilon}{16\varepsilon} = 125;30 \text{ steps}, \tag{5.6.3}$$

which compares with the period $P_s = 130;8,45$ steps for our function S based on $\varepsilon = \frac{3}{28}$. Thus from the original period relation we would expect a truncation at $q = 2$, resulting in an effective period of 126 steps. Instead, however, the author of the System A theory chose an effective period of 124 steps. What is interesting about this choice (which is not a particularly "nice" number,[26] and indeed less so than 126) is that the modern ratio of anomalistic to mean synodic month for -500 yields an effective period of 123;26 steps. Put another way, since

$$P'_s = 130 - 2q = \frac{223 + \varepsilon'}{16\varepsilon'}, \quad \varepsilon' = \frac{223}{16P' - 1}, \tag{5.6.4a}$$

the choice of $P'_s = 124$ steps corresponds to the period relation,

Table 6
Ratios of anomalistic and mean synodic month lengths

Modern	$m_a/\overline{m} =$	$= 0;55,59,6,35^-,$
System A (implicit)	$m_a/\overline{m} = \dfrac{27652}{29635}$	$= 0;55,59,6,33^-$
System B	$m_a/\overline{m} = \dfrac{251}{269}$	$= 0;55,59,6,28..$
System A (explicit)	$m_a/\overline{m} = \dfrac{6247}{6695}$	$= 0;55,59,6,13..$

$$223 \tfrac{223}{1983}\, m = 239 \tfrac{223}{1983}\, m_a, \tag{5.6.4b}$$

or

$$27652(7,40,52)m = 29635(8,13,55)m_a,$$

whence

$$m_a/\overline{m} = 0;55,59,6,33^-. \tag{5.6.4c}$$

The excellence of this relationship can be seen from the following comparison. Thus the truncation chosen not only improves on the initial anomalistic period relation, but comes as close as the structure of the theory permits to the accurate value of this parameter.

The inferior truncation has no such implication as long as it is less than $\tfrac{1}{2}d_{12}S$, merely serving to smooth the second differences about the minima of 235M and 12M. Since $q = 3$, we find from (5.6.2) that the superior truncation amounts to 12;38,45 and thus that the distance from the truncated maximum to S_{min} is

$$S'_{max} - S_{min} = \mathit{\Delta} - 12;38,45 = 52;25,37,30. \tag{5.6.5}$$

Consequently, the smallest truncation which leads to a regular number of steps between the truncated maximum and minimum is 2;25,37,30 which results in a truncated amplitude of 50 steps and a nice distribution of second differences around the minimum.

In this fashion we achieve the function 235M shown in exhibit 5, which is symmetrical about its extremes, has graceful second differences, and reflects an effective anomalistic period, which not only removes the effect of the adjustment introduced to lengthen the period number, but actually improves upon the original period relation. Its amplitude, furthermore, is the nicely symmetrical number 15,6;6,15, and it wants only the definition of its units and a maximum to be complete.

Exhibit 5

(1) s	s'	(2) S'(s, n − 12)		(3) S'(s, n)		(4) d12S' = ds[12M]	(5) ΔΔ ds[12M]	(6) dsS'(n)	(7) ds[235M] = ds[12M] + dsS'	(8) ΔΔ ds[235M]	(9) Σ(7) [235M] − Max
4	1	Smax-12;38,45		Smax-12;38,45		0	−1	−1	−1	−1	0
5	2	S'(1,n − 12) − 0		S'(1,n) − 1		−1	−1	−1	−2	−1	−1
6	3	−0		−2		−2	−1	−1	−3	−1	−3
⋮	⋮	⋮									
21	18	−0		−17		−17		−1	−18		−2,33
22	19	−0		−19		−18	−1	−1	−19	−1	−2,51
23	20	− 0;51,45	←	−19	←	−18; 8,45	−0; 8,45	−1	−19; 8,45	−0; 8,45	−3,10
24	21	− 1;51,45	←	−20	←	−18; 8,45	0	−1	−19; 8,45	0	−3,29; 8,45
⋮	⋮	⋮									
34	31	−11;51,45	←	−30	←	−18; 8,45	0	−1	−19; 8,45	0	−6,40;36,15
35	32	−12;51,45	←	−31	←	−18; 8,45	0	−1	−19; 8,45	0	−6,59,45
⋮	⋮	⋮									
53	50	−30;51,15	←	−49	←	−18; 8,45		−1	−19; 8,45		
54	51	−31;51,15	←	−50		−18; 8,45	+1	0	−18; 8,45	+1	
55	52	−32;51,15	←	−50		−17; 8,45	+1	0	−17; 8,45	+1	
56	53	−33;51,15	←	−50		−16; 8,45	+1	0	−16; 8,45	+1	
57	54	−34;51,15	←	−50		−15; 8,45	+1	0	−15; 8,45	+1; 8,45	
58	55	−35;51,15	←	−50	→	−14; 8,45	+1; 8,45	0; 8,45	−14	+2	
59	56	−36;51,15	←	−49;51,15	→	−13	+2	+1	−12	+2	
60	57	−37;51,15	←	−48;51,15	→	−11	+2	+1	−10	+2	
61	58	−38;51,15	←	−47;51,15	→	−9	+2	+1	−8	+2	
62	59	−39;51,15	←	−46;51,15	→	−7	+2	+1	−6	+2	
63	60	−40;51,15	←	−45;51,15	→	−5	+2	+1	−4	+2	−15, 0; 6,15
64	61	−41;51,15	←	−44;51,15	→	−3	+2	+1	−2	+2	−15, 4; 6,15
65	62	−42;51,15	←	−43;51,15	→	−1	+2	+1	−0	+2	−15, 6; 6,15
66	63	−43;51,15	→	−42;51,15	→	+1	+2	+1	+2	+2	−15, 6; 6,15
67	64	−44;51,15	→	−41;51,15	→	+3	+2	+1	+4	+2	−15, 4; 6,15
68	65	−45;51,15	→	−40;51,15	→	+5	+2	+1	+6	+2	−15, 0; 6,15
⋮	⋮	⋮									
107	104	−20	→	− 1;51,45	→	+18; 8,45	0; 8,45	+1	+19; 8,45	−0; 8,45	−3,29; 8,45
108	105	−19	→	− 0;51,45	→	+18; 8,45	−1	+0;51,15	+19	−1	−3,10
109	106	−19		− 0		+18	−1	0	+18	−1	−2,51
⋮	⋮	⋮									
125	122	− 2	→	− 0	→	+2	−1	0	+2		−3
126	123	− 1		− 0		+1	−1	0	+1		−1

5.7   Establishing the Units

To find the magnitude of our units we use the observation (5.1.4) that the maximum variation of 235 months does not exceed $4{,}30^{u\check{s}}$. We divide this by $15{,}6{;}6{,}15$ units—the corresponding amplitude of 235M—obtaining

$$1^{\phi} \le 4{,}30^{u\check{s}} \div 15{,}6{;}6{,}15 \cong 0{;}17{,}52{,}..^{u\check{s}}. \tag{5.7.1}$$

However, if we wish the resulting number to be regular, we have only three choices in the vicinity with less than four sexagesimal places, namely those shown below with their reciprocals ($\bar{n}$) and corresponding amplitudes:

$$1^{\phi} = 0{;}18^{u\check{s}}\overline{(3{,}20)}: \qquad \Delta 235M = 4{,}31{;}49{,}52{,}30^{u\check{s}}$$

$$1^{\phi} = 0{;}17{,}46{,}40^{u\check{s}}\overline{(3{,}22{,}30)}: \qquad \Delta 235M = 4{,}28{;}28{,}31{,}6{,}40^{u\check{s}} \tag{5.7.2}$$

$$1^{\phi} = 0{;}17{,}21{,}40^{u\check{s}}\overline{(3{,}27{,}21{,}36)}: \quad \Delta 235M = 4{,}22{;}10{,}58{,}30^{u\check{s}}$$

Of these $0{;}18^{u\check{s}}$ is obviously the most convenient, while $0{;}17{,}21{,}40^{u\check{s}}$ is the most accurate. However, $0{;}17{,}46{,}40^{u\check{s}}$ was the one chosen, being the closest to the value found in (5.7.1) and consistent with the assumed inequality. This defines what amounts to a new natural unit of time, the sarosly change in a saros, which corresponds to $1/(20{,}15)$th of a day or slightly more than a minute of our time ($1{;}11{,}6{,}40^{\min}$). It also adds 2 sexagesimal places in transforming S into Φ.

5.8   Establishing the Maxima

With the units determined we can now establish the maxima of 235M and S and thus also of 12M. We transform our second observation—that the maximum *length* of $235M \le 6{,}10^{u\check{s}} + 6939^{d}$—into units of φ, finding that

$$235M_{max} - 6939^{d} \le 6{,}10^{u\check{s}} \cong 1248{;}45^{\phi} = 20{,}48{;}45^{\phi} \tag{5.8.1}$$

which we round, observing the inequality, to

$$235M_{max} - 6939^{d} = 1245^{\phi} = 20{,}45^{\phi}, \tag{5.8.2}$$

which corresponds to $6{,}8{;}53{,}20^{u\check{s}}$.

Since the median value of 235M corresponds to step $31\frac{1}{2}$, we can compute the average month length and daily lunar progress corresponding to this value as follows. From the sum of the sarosly differences, we find that the difference between the maximum of 235M and its value at step $31\frac{1}{2}$ is

$$235M_{max} - 235M_{31\frac{1}{2}} = 6{,}50{;}10{,}37{,}30^{\phi},$$

and thus that

$$235M_{31\frac{1}{2}} = 20,45 - 6,50;10,37,50 = 13,54;49,22,30^{\phi}$$
$$= 6939^{d} + 4,7;21,17,46,40^{u\check{s}}, \tag{5.8.3}$$

which corresponds to an average month length of

$$M_{avg} = 29;31,50,6,0,45.. \text{ days.} \tag{5.8.4}$$

From the 19-year cycle the lunar progress in this interval is

$$\Delta\lambda_{moon} = 254 \cdot 360° \tag{5.8.5}$$

which yields almost precisely the familiar mean value for the daily lunar motion,

$$\mu\left(\frac{\Delta\lambda}{\Delta t}\right) = 13;10,34,59,30^{o/d} \cong 13;10,35^{o/d}, \tag{5.8.6}$$

which we have previously encountered in System B. Whether this was the origin of this parameter, or whether the derivation started with it and introduced a small rounding of $235M_{max}$ is difficult to judge.[27]

Finally, from our last observation—that is, that the maximum duration of 223 months is less than $2,15^{u\check{s}}$ in excess of whole days—we compute that

$$S'_{max} \leq 7,35^{\phi}, \tag{5.8.7a}$$

which—again observing the inequality—we round to

$$S'_{max} = 7,30^{\phi} = 2,13;20^{u\check{s}}. \tag{5.8.7b}$$

This leaves only $12M_{max}$ to be determined, which follows readily from

$$12M_{max} = 235M_{max} - S'_{max} = 13,15^{\phi} = 3,55;33,20^{u\check{s}}. \tag{5.8.8}$$

The resulting functions: S_{n-12}, S_n, $12M$ and $235M$ expressed in units of ϕ and the corresponding functions: Φ, Λ, and Θ expressed in time degrees are shown in exhibits 6A and 6B over the 124 steps used in this process. In Exhibit 6B the columns corresponding to Φ and Λ are identical with the first three columns of Text E [BM. 36311 (= 80-6-17, 37 + 321)], published in Aaboe (1968), which in fact unlocked the structure of the entire scheme. It is also noteworthy that the syzygies associated with each of the Φ (or S) values in these exhibits, are not only for full moon (hence Φ_2), but also syzygies at which lunar eclipses occur.[28] Given the symmetry considerations discussed above (section 5.6), this strongly suggests that the entire theory was built around lunar eclipse observations, and that the corresponding functions for new moon were derived from their full moon counterparts. Finally the relative magnitudes of S, 12M, and 235M are shown graphically in figure 5.

5.9   Constructing M and G from S and Φ

It remains to construct the function for the variation in the length of one month from the variation in the length of a saros, which was the original objective of our theory. Having established the units, we could do this directly from Φ, but the structure will appear more clearly if we stay with units of ϕ and derive M from S.

We begin as before, invoking first the Interval Rule (5.2.3), which says that

$$d_s M = d_1 S, \tag{5.9.1}$$

and then the Saros Rule (5.2.6), which says that, absent truncation

$$|d_1 S| = \frac{|d_s S|}{\varepsilon} = \frac{1}{\varepsilon} = 9;20. \tag{5.9.2}$$

If we start at S'_{max} on the ascending branch, however, $d_1 S$ will be unaffected by the truncation of S and simply equal to $+1/\varepsilon$, and thus the corresponding value of M will be on its linear stretch, instead of at its minimum.

We finesse this problem by simply reversing the direction of time and proceeding in steps of $-s$ or by -223 months. This implies the following equalities,

$$d_{-s} S = -d_s S = +\varepsilon d_1 S = -1, \quad \text{whence}$$

$$d_1 S = -\frac{1}{\varepsilon} \text{ and thus } S = S(\downarrow); \tag{5.9.3a}$$

$$d_{-s} S = -\varepsilon d_{-1} S = -\varepsilon d_{-s} M, \text{ so that}$$

$$d_{-s} M = +\frac{1}{\varepsilon}, \text{ and} \tag{5.9.3b}$$

$$d_{-s} M = d_{-1} S = -d_1 S = S_{n-1} - S_n. \tag{5.9.3c}$$

What these say in effect is that all relationships are preserved if we begin at S'_{max} but on the descending branch, and reverse the sign of the monthly difference in S, that is, make

$$d_{-s} M = d_{-1} S = S_{n-1} - S_n, \quad \text{in place of } d_s M = d_1 S = S_n - S_{n-1}. \tag{5.9.4}$$

This allows us to use the same procedure and to begin with the same values for S as we used to construct 235M, but leaves us with new

Exhibit 6a

(1) s′	(2) S′(s′, n − 12)	(3) S′(s′, n)	(4) +12M	(5) = 235M	(6) d12[S′] = ds[12M]	(7) ds[S′]	(8) ds[235M]
1	7 30	7 30 ↑	13 15	20 45	0	− 1	− 1
2	7 30	2 29 ↑	13 15	20 44	− 1	− 1	− 2
3	7 30	7 28 ↑	13 14	20 42	− 2	− 1	− 3
4	7 30	7 27 ↑	13 12	20 39	− 3	− 1	− 4
5	7 30	7 26 ↑	13 9	20 35	− 4	− 1	− 5
6	7 30	7 25 ↑	13 5	20 30	− 5	− 1	− 6
7	7 30	7 24 ↑	13 0	20 24	− 6	− 1	− 7
8	7 30	7 23 ↑	12 54	20 17	− 7	− 1	− 8
9	7 30	7 22 ↑	12 47	20 9	− 8	− 1	− 9
10	7 30	7 21 ↑	12 39	20 0	− 9	− 1	− 10
11	7 30	7 20 ↑	12 30	19 50	− 10	− 1	− 11
12	7 30	7 19 ↑	12 20	19 39	− 11	− 1	− 12
13	7 30	7 18 ↑	12 9	19 27	− 12	− 1	− 13
14	7 30	7 17 ↑	11 57	19 14	− 13	− 1	− 14
15	7 30	7 16 ↑	11 44	19 0	− 14	− 1	− 15
16	7 30	7 15 ↑	11 30	18 45	− 15	− 1	− 16
17	7 30	7 14 ↑	11 15	18 29	− 16	− 1	− 17
18	7 30	7 13 ↑	10 59	18 12	− 17	− 1	− 18
19	7 30	7 12 ↑	10 42	17 54	− 18	− 1	− 19
20	7 29 8 45 ↑	7 11 ↑	10 24	17 35	− 18 8 45	− 1	− 19 8 45
21	7 28 8 45 ↑	7 10 ↑	10 5 51 15	17 15 51 15	− 18 8 45	− 1	− 19 8 45
⋮	⋮	⋮	⋮	⋮	⋮	⋮	⋮
50	6 59 8 45 ↑	6 41 ↑	1 19 37 30	8 0 37 30	− 18 8 45	− 1	− 19 8 45
51	6 58 8 45 ↑	6 40 ↑	1 1 28 45	7 41 28 45	− 18 8 45	0	− 18 8 45
52	6 57 8 45 ↑	6 40	0 43 20 0	7 23 20 0	− 17 8 45	0	− 17 8 45
53	6 56 8 45 ↑	6 40	0 26 11 15	7 6 11 15	− 16 8 45	0	− 16 8 45
54	6 55 8 45 ↑	6 40	0 10 2 30	6 50 2 30	− 15 8 45	0	− 15 8 45
55	6 54 8 45 ↑	6 40	−0 5 6 15	6 34 53 45	− 14 8 45	+ 0 8 45	− 14
56	6 53 8 45 ↑	6 40 8 45 ↓	−0 19 15 0	6 20 53 45	− 13	+ 1	− 12
57	6 52 8 45 ↑	6 41 8 45 ↓	−0 32 15 0	6 8 53 45	− 11	+ 1	− 10
58	6 51 8 45 ↑	6 42 8 45 ↓	−0 43 15 0	5 58 53 45	− 9	+ 1	− 8
59	6 50 8 45 ↑	6 43 8 45 ↓	−0 52 15 0	5 50 53 45	− 7	+ 1	− 6
60	6 49 8 45 ↑	6 44 8 45 ↓	−0 59 15 0	5 44 53 45	− 5	+ 1	− 4
61	6 48 8 45 ↑	6 45 8 45 ↓	−1 4 15 0	5 40 53 45	− 3	+ 1	− 2
62	6 47 8 45 ↑	6 46 8 45 ↓	−1 7 15 0	5 38 53 45	− 1	+ 1	− 0

(1) s'	(2) S'(s', n − 12)	(3) S'(s', n)	(4) +12M	(5) = 235M	(6) d12[S'] = ds[12M]	(7) ds[S']	(8) ds[235M]
63	6 46 8 45 ↑	6 47 8 45 ↓	−1 8 15 0	5 38 53 45	+ 1	+ 1	2
64	6 45 8 45 ↑	6 48 8 45 ↓	−1 7 15 0	5 40 53 45	+ 3	+ 1	4
65	6 44 8 45 ↑	6 49 8 45 ↓	−1 4 15 0	5 44 53 45	+ 5	+ 1	6
66	6 43 8 45 ↑	6 50 8 45 ↓	−0 59 15 0	5 50 53 45	+ 7	+ 1	8
67	6 42 8 45 ↑	6 51 8 45 ↓	−0 52 15 0	5 58 53 45	+ 9	+ 1	10
68	6 41 8 45 ↑	6 52 8 45 ↓	−0 43 15 0	6 8 53 45	+ 11	+ 1	12
69	6 40 8 45 ↑	6 53 8 45 ↓	−0 32 15 0	6 20 53 45	+ 13	+ 1	14
70	6 40	6 54 8 45 ↓	−0 19 15 0	6 34 53 45	+ 14 8 45	+ 1	15 8 45
71	6 40	6 55 8 45 ↓	−0 5 6 15	6 50 2 30	+ 15 8 45	+ 1	16 8 45
72	6 40	6 56 8 45 ↓	0 10 2 30	7 6 11 15	+ 16 8 45	+ 1	17 8 45
73	6 40	6 57 8 45 ↓	0 26 11 15	7 23 20 0	+ 17 8 45	+ 1	18 8 45
74	6 40 ↓	6 58 8 45 ↓	0 43 20 0	7 41 28 45	+ 18 8 45	+ 1	19 8 45
75	6 41 ↓	6 59 8 45 ↓	1 1 28 45	8 0 37 30	+ 18 8 45	+ 1	19 8 45
:	:	:	:	:	:	:	:
104	7 10 ↓	7 28 8 45 ↓	9 47 42 30	17 15 51 15	+ 18 8 45	+ 1	19 8 45
105	7 11 ↓	7 29 8 45 ↓	10 5 51 15	17 35	+ 18 8 45	+ 0 51 15	19
106	7 12 ↓	7 30	10 24	17 54	+ 18	0	18
107	7 13 ↓	7 30	10 42	18 12	+ 17	0	17
108	7 14 ↓	7 30	10 59	18 29	+ 16	0	16
109	7 15 ↓	7 30	11 15	18 45	+ 15	0	15
110	7 16 ↓	7 30	11 30	19 0	+ 14	0	14
111	7 17 ↓	7 30	11 44	19 14	+ 13	0	13
112	7 18 ↓	7 30	11 57	19 27	+ 12	0	12
113	7 19 ↓	7 30	12 9	19 39	+ 11	0	11
114	7 20 ↓	7 30	12 20	19 50	+ 10	0	10
115	7 21 ↓	7 30	12 30	20 0	+ 9	0	9
116	7 22 ↓	7 30	12 39	20 9	+ 8	0	8
117	7 23 ↓	7 30	12 47	20 17	+ 7	0	7
118	7 24 ↓	7 30	12 54	20 24	+ 6	0	6
119	7 25 ↓	7 30	13 0	20 30	+ 5	0	5
120	7 26 ↓	7 30	13 5	20 35	+ 4	0	4
121	7 27 ↓	7 30	13 9	20 39	+ 3	0	3
122	7 28 ↓	7 30	13 12	20 42	+ 2	0	2
123	7 29 ↓	7 30	13 14	20 44	+ 1	0	1
124	7 30 ↓	7 30	13 15	20 45	0	0	0

Exhibit 6b

$s' =$ Text E: line	I $\Phi'(s', n-12)$ 223 m > (n − 12)	II $\Phi'(s', n)$ 223 m > (n)	III $\Lambda(s', n)$ 12 m < (n)	− $\Theta(s', n)$ 235 m <> (n)
1	2 13 20	2 13 20 0 0 ↑	3 55 33 20 0	6 8 53 20 0
2	2 13 20	2 13 2 13 20 ↑	3 55 33 20 0	6 8 35 33 20
3	2 13 20	2 12 44 26 40 ↑	3 55 15 33 20	6 8 0 0 0
4	2 13 20	2 12 26 40 0 ↑	3 54 40 0 0	6 7 6 40 0
5	2 13 20	2 12 8 53 20 ↑	3 53 46 40 0	6 5 55 33 20
6	2 13 20	2 11 51 6 40 ↑	3 52 35 33 20	6 4 26 40 0
7	2 13 20	2 11 33 20 0 ↑	3 51 6 40 0	6 2 40 0 0
8	2 13 20	2 11 15 33 20 ↑	3 49 20 0 0	6 0 35 33 20
9	2 13 20	2 10 57 46 40 ↑	3 47 15 33 20	5 58 13 20 0
10	2 13 20	2 10 40 0 0 ↑	3 44 53 20 0	5 55 33 20 0
11	2 13 20	2 10 22 13 20 ↑	3 42 13 20 0	5 52 35 33 20
12	2 13 20	2 10 4 26 40 ↑	3 39 15 33 20	5 49 20 0 0
13	2 13 20	2 9 46 40 0 ↑	3 36 0 0 0	5 45 46 40 0
14	2 13 20	2 9 28 53 20 ↑	3 32 26 40 0	5 41 55 33 20
15	2 13 20	2 9 11 6 40 ↑	3 28 35 33 20	5 37 46 40 0
16	2 13 20	2 8 53 20 0 ↑	3 24 26 40 0	5 33 20 0 0
17	2 13 20	2 8 35 33 20 ↑	3 20 0 0 0	5 28 35 33 20
18	2 13 20	2 8 17 46 40 ↑	3 15 15 33 20	5 23 33 20 0
19	2 13 20	2 8 0 0 0 ↑	3 10 13 20 0	5 18 13 20 0
20	2 13 4 48 53 20 ↑	2 7 42 13 20 ↑	3 4 53 20 0	5 12 35 33 20
21	2 12 47 2 13 20 ↑	2 7 24 26 40 ↑	2 59 30 44 26 40	5 6 55 11 6 40
:	:		:	
50	2 4 11 28 53 20 ↑	1 58 48 53 20 ↑	23 35 33 20 0	2 22 24 26 40 0
51	2 3 53 42 13 20 ↑	1 58 31 6 40 ↑	18 12 57 46 40	2 16 44 4 26 40
52	2 3 35 55 33 20 ↑	1 58 31 6 40 ·	12 50 22 13 20	2 11 21 28 53 20
53	2 3 18 8 53 20 ↑	1 58 31 6 40	7 45 33 20 0	2 6 16 40 0 0
54	2 3 0 22 13 20 ↑	1 58 31 6 40	2 58 31 6 40	2 1 29 37 46 40
55	2 2 42 35 33 20 ↑	1 58 31 6 40	− 1 30 44 26 40	1 57 0 22 13 20
56	2 2 24 48 53 20 ↑	1 58 33 42 13 20 ↓	− 5 42 13 20	1 52 51 28 53 20
57	2 2 7 2 13 20 ↑	1 58 51 28 53 20 ↓	− 9 33 20 0	1 49 18 8 53 20
58	2 1 49 15 33 20 ↑	1 59 9 15 33 20 ↓	− 12 48 53 20	1 46 20 22 13 20
59	2 1 31 28 53 20 ↑	1 59 27 2 13 20 ↓	− 15 28 53 20	1 43 58 8 53 20
60	2 1 13 42 13 20 ↑	1 59 44 48 53 20 ↓	− 17 33 20 0	1 42 11 28 53 20
61	2 0 55 55 33 20 ↑	2 0 2 35 33 20 ↓	− 19 2 13 20	1 41 0 22 13 20
62	2 0 38 8 53 20 ↑	2 0 20 22 13 20 ↓	− 19 55 33 20	1 40 24 48 53 20

$s' =$ Text E: line	I $\Phi'(s', n - 12)$ $223\ m > (n - 12)$	Text E-Column: II $\Phi'(s', n)$ $223\ m > (n)$	III $\Lambda(s', n)$ $12\ m < (n)$	– $\Theta(s', n)$ $235\ m <> (n)$
63	2 0 20 22 13 20 ↑	2 0 38 8 53 20 ↓	− 20 13 20 0	1 40 24 48 53 20
64	2 0 2 35 33 20 ↑	2 0 55 55 33 20 ↓	− 19 55 33 20	1 41 0 22 13 20
65	1 59 44 48 53 20 ↑	2 1 13 42 13 20 ↓	− 19 2 13 20	1 42 11 28 53 20
66	1 59 27 2 13 20 ↑	2 1 31 28 53 20 ↓	− 17 33 20 0	1 43 58 8 53 20
67	1 59 9 15 33 20 ↑	2 1 49 15 33 20 ↓	− 15 28 53 20	1 46 20 22 13 20
68	1 58 51 28 53 20 ↑	2 2 7 2 13 20 ↓	− 12 48 53 20	1 49 18 8 53 20
69	1 58 33 42 13 20 ↑	2 2 24 48 53 20 ↓	− 9 33 20 0	1 52 51 28 53 20
70	1 58 31 6 40	2 2 42 35 33 20 ↓	− 5 42 13 20	1 57 0 22 13 20
71	1 58 31 6 40	2 3 0 22 13 20 ↓	− 1 30 44 26 40	2 1 29 37 46 40
72	1 58 31 6 40	2 3 18 8 53 20 ↓	0 2 58 31 6 40	2 6 16 40 0 0
73	1 58 31 6 40	2 3 35 55 33 20 ↓	0 7 45 33 20 0	2 11 21 28 53 20
74	1 58 31 6 40	2 3 53 42 13 20 ↓	0 12 50 22 13 20	2 16 44 4 26 40
75	1 58 48 53 20 ↓	2 4 11 28 53 20 ↓	0 18 12 57 46 40	2 22 24 26 40 0
:	:	:	:	:
104	2 7 24 26 40 ↓	2 12 47 2 13 20 ↓	2 54 8 8 53 20	5 6 55 11 6 40
105	2 7 42 13 20 ↓	2 13 4 48 53 20 ↓	2 59 30 44 26 40	5 12 35 33 20
106	2 8 0 0 0 ↓	2 13 20	3 4 53 20 0	5 18 13 20 0
107	2 8 17 46 40 ↓	2 13 20	3 10 13 20 0	5 23 33 20 0
108	2 8 35 33 20 ↓	2 13 20	3 15 15 33 20	5 28 35 33 20
109	2 8 53 20 0 ↓	2 13 20	3 20 0 0 0	5 33 20 0 0
110	2 9 11 6 40 ↓	2 13 20	3 24 26 40 0	5 37 46 40 0
111	2 9 28 53 20 ↓	2 13 20	3 28 35 33 20	5 41 55 33 20
112	2 9 46 40 0 ↓	2 13 20	3 32 26 40 0	5 45 46 40 0
113	2 10 4 26 40 ↓	2 13 20	3 36 0 0 0	5 49 20 0 0
114	2 10 22 13 20 ↓	2 13 20	3 39 15 33 20	5 52 35 33 20
115	2 10 40 0 0 ↓	2 13 20	3 42 13 20 0	5 55 33 20 0
116	2 10 57 46 40 ↓	2 13 20	3 44 53 20 0	5 58 13 20 0
117	2 11 15 33 20 ↓	2 13 20	3 47 15 33 20	6 0 35 33 20
118	2 11 33 20 0 ↓	2 13 20	3 49 20 0 0	6 2 40 0 0
119	2 11 51 6 40 ↓	2 13 20	3 51 6 40 0	6 4 26 40 0
120	2 12 8 53 20 ↓	2 13 20	3 52 35 33 20	6 5 55 33 20
121	2 12 26 40 0 ↓	2 13 20	3 53 46 40 0	6 7 6 40 0
122	2 12 44 26 40 ↓	2 13 20	3 54 40 0 0	6 8 0 0 0
123	2 13 2 13 20 ↓	2 13 20	3 55 15 33 20	6 8 35 33 20
124	2 13 20	2 13 20	3 55 33 30 0	6 8 53 20 0

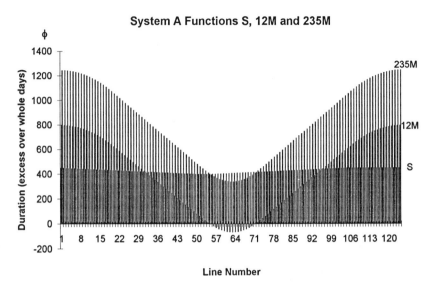

Figure 5
Relative magnitudes of S, 12M and 235M

asymmetries. These were eliminated in the construction of 235M by a procedure which made S_n complementary to S_{n-12} and resulted in fractional endings of S that were either 0 or 0;8,45. Now, however, since $d_1 S = 9;20$, different fractional endings will result from reflecting at the extremes. Consequently, S_n will not everywhere be complementary to S_{n-1}, with resulting asymmetries similar to those first encountered in constructing S.

Since we have already established S'_{max}, we are left with no alternative but to force the best solution we can manage. This is to gather all the asymmetries together into one adjustment, leaving the rest of M otherwise symmetrical about its extremes. We accomplish this by shifting S as it emerges from its inferior truncation by a fraction of a step equal to

$$\gamma S \approx 0;20 - 0;8,45 = 0;11,15 \text{ step}, \tag{5.9.5}$$

which, as we saw in Exhibit 3, is precisely the shift encountered in the $\Phi : G$ scheme.

5.10   Establishing the Minimum of M

We now need to establish the minimum of M, which we do as follows. First we calculate the average of $M - M_{min}$ step by step over its full period of 130 steps, from

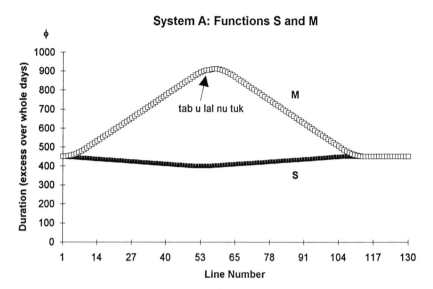

Figure 6
Relative magnitudes and variations of S and M

$$Avg(M - M_{min}) \equiv M_{avg} - M_{min} = \frac{\sum (M_s - M_{min})}{130},$$

which yields the surprisingly nice ratio

$$\frac{\sum (M_s - M_{min})}{130} = \frac{7,5,0;0^\phi}{2,10} = 3,16;9,13,..^\phi. \tag{5.10.1}$$

Next we equate M_{avg} with the average month length, derived in (5.8.4) from the 19-year cycle and the median value of 235M, and thus set

$$M_{avg} = 29;31,50,6,0 \text{ days},$$

$$= 10,44;39,31,30^\phi + 29 \text{ days}. \tag{5.10.2}$$

Combining this with (5.10.1) gives us the minimum of M, namely

$$M_{min} = 10,44;39,31,30 - 3,16;9,13,50.. = 7,28;30,..^\phi, \tag{5.10.3}$$

which rounds—stunningly—to

$$M_{min} = 7,30 \text{ units of } \phi, \tag{5.10.4}$$

the very same value as the effective maximum of S! (It is easy to imagine the author of the theory contemplating this extraordinary result with satisfaction, if not wonderment.)

This completes the definition of M in terms of S, which is shown graphically in figure 6. The resulting relationship is summarized in exhibit 7 in the same format as the $\Phi : G$ scheme (exhibit 3), with S, however, replaced by the untruncated zigzag function, $\hat{S}$ (which is the strict analog of Φ). Compared with the $\Phi : G$ scheme the numbers here are obviously both shorter and nicer, and the additive interpolation $A^*(n)$) is simpler to calculate. Otherwise, the two schemes are identical, but for units and a constant term.

To obtain Φ and G we would expect simply to multiply $\hat{S}$ and M by ϕ; however, G corresponds to a function which everywhere reflects the addition of $26;40^{u\check{s}}(= 1,30^{\phi})$. This is the adjustment (rounded from $1,29;..^{\phi}$) required for G to equal the length of the month assuming solar progress between syzygies of $30°$ per month, as found in the System A scheme for zodiacal anomaly. The calculation of this adjustment[29] involves nothing beyond the present theory, but it is difficult to judge whether it was part of the original theory or a later addition. It clearly reflects both an awareness of zodiacal anomaly and the assumption that $30°$ is the maximum monthly solar progress, which is consistent with establishing S at its maximum. It does not, however, imply or require a specific solution or solar model, such as that found in System A.

In sum, we find that the functions Φ and G as tabulated in System A derive from their underlying functions, $\hat{S}$ and M by

$$\Phi = \phi \cdot \hat{S} \tag{5.10.5}$$

and

$$G = \phi \cdot (M + 1,30). \tag{5.10.6}$$

The resulting functions for both $M + 1,30^{\phi}$ and for G are shown in exhibit 8, where the shifts in S and Φ are clearly shown.

5.11   Six Months

We have now only the variation in six months left to be established. This we do using S and M (or Φ and G) as follows.[30] Beginning, as in the construction of M with the initial value

$$S'_{\max}(\downarrow) = 7,30(\downarrow) \quad \text{or} \quad \Phi_{1,n} = \Phi'_{\max} = 2,13;20^{u\check{s}}(\downarrow) \tag{5.11.1}$$

we proceed as before in steps of $-s$ and calculate for each syzygy corresponding to S_n the sum of M_n plus the lengths of the 5 preceding months as determined from the function M. In this way we define the function 6M as

$$6M_n \equiv \sum_{n-5}^{m} M_n \tag{5.11.2}$$

Exhibit 7

To Compute M from S

Opposite $S_n(\uparrow\downarrow)$ you put M_n; whatever exceeds $(<>)S_n(\downarrow\uparrow)$, multiply by C_n and add to M_n.

---

n	$\hat{S}_n$	Branch	M_n	$<>$	C_n	dC_n
1	7 30	↓	7 30	<	$A^*(\Delta)$	+1
2	7 21	↓	8 15	<	9 20	0/1
3	6 40	↓	14 37 40	<	8 20	−1
4	6 39	↓	14 46	<	7 20	−1
5	6 $\underline{\underline{38}}$	↓	14 53 20	$\underline{\underline{\leq}}$	6 20	−1
6	6 38 8 45	↑	14 59 40	>	5 20	−1
7	6 39 8 45	↑	15 5	>	4	−1 20
8	6 40 8 45	↑	15 9	>	0	−4
9	6 40 20	↑	15 9	>	2	+2
10	6 41 20	↑	$\underline{\underline{15\ 11}}$	>	0	−2
11	6 42 20	↑	$\overline{15\ 11}$	>	−2	−2
12	6 43 20	↑	15 9	>	−4	−2
13	6 44 20	↑	15 5	>	−5 20	−1 20
14	6 45 20	↑	14 59 40	>	−6 20	−1
15	6 46 20	↑	14 53 20	>	−7 20	−1
16	6 47 20	↑	14 46	>	−8 20	−1
17	6 48 20	↑	14 37 40	>	−9 20	−1/0
18	7 29 20	↑	8 15	>	−$A^*(\Delta)$	+1
19	7 38 20	↑	7 30	>	0	0

$A^*(n) \equiv n + \sum_1^{n-1} I(n)$, where $I(n) \equiv$ all positive integers $\leq n$; $\Delta = S - S_n$.

Line 1 (after ACT 200:14)

Opposite 7,30 decreasing you put 7,30. From 7,30 decreasing to 7,21 decreasing whatever is less than 7,21 decreasing add all of the parts together and add (the result) to 7,30.

Line 18 (after ACT 206)

Opposite 7,29,20 increasing you put 8,15. Anything beyond 7,29,20 increasing to 7,38,20 increasing add all the parts together and subtract (the result) from 8,15.

Line 8 (after ACT 200:14)

Opposite 6,40,8,45 [increasing] put 15,9; until 6,40,20 increasing everywhere 15,9.

Exhibit 8a

ln(−s)	S'(n − 1)	S'(n)	S'(n − 1) − S'(n)	M(n) + 1,30
1	7 30 (7 39 20 0 →)	7 30 →	0	9 0 0
2	7 30 (7 38 20 0 →)	7 29 →	1	9 1 0
3	7 30 (7 37 20 0 →)	7 28 →	2	9 3 0
4	7 30 (7 36 20 0 →)	7 27 →	3	9 6 0
5	7 30 (7 35 20 0 →)	7 26 →	4	9 10 0
6	7 30 (7 34 20 0 →)	7 25 →	5	9 15 0
7	7 30 (7 33 20 0 →)	7 24 →	6	9 21 0
8	7 30 (7 32 20 0 →)	7 23 →	7	9 28 0
9	7 30 (7 31 20 0 →)	7 22 →	8	9 36 0
10	7 30 (7 30 20 0 →)	7 21 →	9	9 45 0
11	7 29 20 →	7 20 →	9 20	9 54 20
12	7 28 20 →	7 19 →	9 20	10 3 40
⋮	⋮	⋮	⋮	⋮
49	6 51 20 →	6 42 →	9 20	15 49 0
50	6 50 20 →	6 41 →	9 20	15 58 20
51	6 49 20 →	6 40 →	9 20	16 7 40
52	6 48 20 →	6 40 (6 39 0 0 →)	8 20	16 16 0
53	6 47 20 →	6 40 (6 38 0 0 →)	7 20	16 23 20
54	6 46 20 →	6 40 (6 38 8 45 ↑)	6 20	16 29 40
55	6 45 20 →	6 40 (6 39 8 45 ↑)	5 20	16 35 0
56	6 44 20 →	6 40 20 (6 40 8 45 ↑)*	4 0	16 39 0
57	6 43 20 →	6 41 20 ↑ *{ + 11 15}	2 0	16 41 0
58	6 42 20 →	6 42 20 ↑	0 0	16 41 0
59	6 41 20 →	6 43 20 ↑	−2 0	16 39 0
60	6 40 20 →	6 44 20 ↑	−4 0	16 35 0
61	6 40 (6 39 20 0 →)	6 45 20 ↑	−5 20	16 29 40
62	6 40 (6 38 20 0 →)	6 46 20 ↑	−6 20	16 23 20
63	6 40 (6 37 48 45 ↑)	6 47 20 ↑	−7 20	16 16 0
64	6 40 (6 38 48 45 ↑)	6 48 20 ↑	−8 20	16 7 40

Row	col 1	col 2	col 3	col 4
65	6 40 (6 39 48 45 ↑)	6 49 20 ↑	−9 20	15 58 20
66	6 41 (6 40 48 45 ↑)*	6 50 20 ↑	−9 20	15 49 0
67	6 42 ↑ *{+ 11 15}	6 51 20 ↑	−9 20	15 39 40
⋮	⋮	⋮	⋮	⋮
103	7 18 ↑	7 27 20 ↑	−9 20	10 3 40
104	7 19 ↑	7 28 20 ↑	−9 20	9 54 20
105	7 20 ↑	7 29 20 ↑	−9 20	9 45 0
106	7 21 ↑	7 30 (7 30 20 0 ↑)	−9	9 36 0
107	7 22 ↑	7 30 (7 31 20 0 ↑)	−8	9 28 0
108	7 23 ↑	7 30 (7 32 20 0 ↑)	−7	9 21 0
109	7 24 ↑	7 30 (7 33 20 0 ↑)	−6	9 15 0
110	7 25 ↑	7 30 (7 34 20 0 ↑)	−5	9 10 0
111	7 26 ↑	7 30 (7 35 20 0 ↑)	−4	9 6 0
112	7 27 ↑	7 30 (7 36 20 0 ↑)	−3	9 3 0
113	7 28 ↑	7 30 (7 37 20 0 ↑)	−2	9 1 0
114	7 29 ↑	7 30 (7 38 20 0 ↑)	−1	9 0 0
115	7 30 ↑	7 30 (7 39 20 0 ↑)	0	9 0
116	7 30 (7 31 0 0 ↑)	7 30 (7 40 20 0 ↑)	0	9 0
117	7 30 (7 32 0 0 ↑)	7 30 (7 41 20 0 ↑)	0	9 0
118	7 30 (7 33 0 0 ↑)	7 30 (7 42 20 0 ↑)	0	9 0
119	7 30 (7 34 0 0 ↑)	7 30 (7 41 57 30 →)	0	9 0
120	7 30 (7 35 0 0 ↑)	7 30 (7 40 57 30 →)	0	9 0
121	7 30 (7 36 0 0 ↑)	7 30 (7 39 57 30 →)	0	9 0
122	7 30 (7 37 0 0 ↑)	7 30 (7 38 57 30 →)	0	9 0
123	7 30 (7 38 0 0 ↑)	7 30 (7 37 57 30 →)	0	9 0
124	7 30 (7 39 0 0 ↑)	7 30 (7 36 57 30 →)	0	9 0
125	7 30 (7 40 0 0 ↑)	7 30 (7 35 57 30 →)	0	9 0
126	7 30 (7 41 0 0 ↑)	7 30 (7 34 57 30 →)	0	9 0
127	7 30 (7 42 17 30 ↑)	7 30 (7 33 57 30 →)	0	9 0
128	7 30 (7 42 17 30 →)	7 30 (7 32 57 30 →)	0	9 0
129	7 30 (7 41 17 30 →)	7 30 (7 31 57 30 →)	0	9 0
130	7 30 (7 40 17 30 →)	7 30 (7 30 57 30 →)	0	9 0

Exhibit 8b

ln(−s)	Φ'(n − 1)		Φ'(n)		Φ'(n − 1) − Φ'(n)	G(n)
1	(2 16 5 55 33 20 →)	2 13 20	2 13 20 →		0	2 40
2	(2 15 48 8 53 20 →)	2 13 20	2 13 2 13 20 →		0 17 46 40	2 40 17 46 40
3	(2 15 30 22 13 20 →)	2 13 20	2 12 44 26 40 →		0 35 33 20	2 40 53 20 0
4	(2 15 12 35 33 20 →)	2 13 20	2 12 26 40 0 →		0 53 20 0	2 41 46 40 0
5	(2 14 54 48 53 20 →)	2 13 20	2 12 8 53 20 →		1 11 6 40	2 42 57 46 40
6	(2 14 37 2 13 20 →)	2 13 20	2 11 51 6 40 →		1 28 53 20	2 44 26 40 0
7	(2 14 19 15 33 20 →)	2 13 20	2 11 33 20 0 →		1 46 40 0	2 46 13 20 0
8	(2 14 1 28 53 20 →)	2 13 20	2 11 15 33 20 →		2 4 26 40	2 48 17 46 40
9	(2 13 43 42 13 20 →)	2 13 20	2 10 57 46 40 →		2 22 13 20	2 50 40 0 0
10	(2 13 25 55 33 20 →)	2 13 20	2 10 40 0 0 →		2 40 0 0	2 53 20 0 0
11		2 13 8 8 53 20 →	2 10 22 13 20 →		2 45 55 33 20	2 56 5 55 33 20
12		2 12 50 22 13 20 →	2 10 4 26 40 →		2 45 55 33 20	2 58 51 51 6 40
⋮			⋮		⋮	⋮
49		2 1 52 35 33 20 →	1 59 6 40 0 →		2 45 55 33 20	4 41 11 6 40 0
50		2 1 34 48 53 20 →	1 58 48 53 20 →		2 45 55 33 20	4 43 57 2 13 20
51		2 1 17 2 13 20 →	1 58 31 6 40 →		2 45 55 33 20	4 46 42 57 46 40
52		2 0 59 15 33 20 →	1 58 31 6 40	(1 58 13 20 0 0 →)	2 28 8 53 20	4 49 11 6 40 0
53		2 0 41 28 53 20 →	1 58 31 6 40	(1 57 55 33 20 0 0 →)	2 10 22 13 20	4 51 21 28 53 20
54		2 0 23 42 13 20 →	1 58 31 6 40	(1 57 58 8 53 20 8 53 20 ↑)	1 52 35 33 20	4 53 14 4 26 40
55		2 0 5 55 33 20 →	1 58 31 6 40	(1 58 15 55 33 20 55 33 20 ↑)	1 34 48 53 20	4 54 48 53 20 0
56		1 59 48 8 53 20 →	1 58 37 2 13 20 / 1 58 33 42 13 20 ↑ *		1 11 6 40	4 56 0 0 0
57		1 59 30 22 13 20 →	1 58 54 48 53 20 *{+ 3 20 0 0}		0 35 33 20	4 56 35 33 20 0
58		1 59 12 35 33 20 →	1 59 12 35 33 20 ↑		0 0 0 0	4 56 35 33 20 0
59		1 58 54 48 53 20 →	1 59 30 22 13 20 ↑		0 −35 33 20	4 56 0 0 0
60		1 58 37 2 13 20 →	1 59 48 8 53 20 ↑		−1 11 6 40	4 54 48 53 20 0
61	(1 58 19 15 33 20 →)	1 58 31 6 40	2 0 5 55 33 20 ↑		−1 34 48 53 20	4 53 14 4 26 40
62	(1 58 1 28 53 20 →)	1 58 31 6 40	2 0 23 42 13 20 ↑		−1 52 35 33 20	4 51 21 28 53 20
63	(1 57 52 13 20 0 ↑)	1 58 31 6 40	2 0 41 28 53 20 ↑		−2 10 22 13 20	4 49 11 6 40 0
64	(1 58 10 0 0 0 ↑)	1 58 31 6 40	2 0 59 15 33 20 ↑		−2 28 8 53 20	4 46 42 57 46 40

Line	Φ (attested)	Φ (restored)	(restored, arrows)	col I	fraction	sign	far-right
65	1 58 31 6 40	(1 58 27 46 40 0 ↑) *	(2 1 17 2 13 20 ←	4 43 57	45 55 33 20	−2	2 13 20
66	1 58 48 53 20	(1 58 45 33 20 0 ↑) *	(2 1 34 48 53 20 ←	4 41 11	45 55 33 20	−2	6 40
67	1 59 6 40	*{ + 3 20 0 0 }	(2 1 52 35 33 20 ←	4 38 25	45 55 33 20	−2	6 40
(..)							
103	2 9 46 40	(2 13 37 46 40 0 ↑)	(2 12 32 35 33 20 ↑	2 58 51 51	45 55 33 20	−2	2 40
104	2 10 4 26 40	(2 13 55 33 20 0 ↑)	(2 12 50 22 13 20 ↑	2 56 5 55	45 55 33 20	−2	2 40
105	2 10 22 13 20	(2 14 13 20 0 0 ↑)	(2 13 8 8 53 20 ↑	2 53 20 0	45 55 33 20	−2	2 40
106	2 10 40 0	(2 14 31 6 40 0 ↑)	(2 13 25 55 33 20 ↑	2 50 40 0	40 0 0	−2	2 40
107	2 10 57 46 40	(2 14 48 53 20 0 ↑)	(2 13 43 42 13 20 ↑	2 48 17 46 40	22 13 20	−2	2 40
108	2 11 15 33 20	(2 15 6 40 0 0 ↑)	(2 14 1 28 53 20 ↑	2 46 13 20	4 26 40	−2	2 40
109	2 11 33 20	(2 15 24 26 40 0 ↑)	(2 14 19 15 33 20 ↑	2 44 26 40	46 39 0	−1	2 40
110	2 11 51 6 40	(2 15 42 13 20 0 ↑)	(2 14 37 2 13 20 ↑	2 42 57 46 40	28 53 20	−1	2 40
111	2 12 8 53 20	(2 16 0 0 0 0 ↑)	(2 14 54 48 53 20 ↑	2 41 46 40	11 6 40	−1	2 40
112	2 12 26 40	(2 16 17 46 40 0 ↑)	(2 15 12 35 33 20 ↑	2 40 53 20	53 19 0	−0	2 40
113	2 12 44 26 40	(2 16 35 33 20 0 ↑)	(2 15 30 22 13 20 ↑	2 40 17 46 40	35 33 20	−0	2 40
114	2 13 2 13 20	(2 16 53 20 0 0 ↑)	(2 15 48 8 53 20 ↑	2 40	17 46 40	−0	2 40
115	2 13 20		(2 16 5 55 33 20 ↑	2 40		0	2 40
116	2 13 20		(2 16 23 42 13 20 ↑	2 40		0	2 40
117	2 13 20		(2 16 41 28 53 20 ↑	2 40		0	2 40
118	2 13 20		(2 16 59 15 33 20 ↑	2 40		0	2 40
119	2 13 20		(2 16 52 35 33 20 ↓	2 40		0	2 40
120	2 13 20		(2 16 34 48 53 20 ↓	2 40		0	2 40
121	2 13 20		(2 16 17 2 13 20 ↓	2 40		0	2 40
122	2 13 20		(2 15 59 15 33 20 ↓	2 40		0	2 40
123	2 13 20		(2 15 41 28 53 20 ↓	2 40		0	2 40
124	2 13 20		(2 15 23 42 13 20 ↓	2 40		0	2 40
125	2 13 20		(2 15 5 55 33 20 ↓	2 40		0	2 40
126	2 13 20		(2 14 48 8 53 20 ↓	2 40		0	2 40
127	2 13 20		(2 14 30 22 13 20 ↓	2 40		0	2 40
128	2 13 20	(2 16 58 31 6 40 ↓)	(2 14 12 35 33 20 ↓	2 40		0	2 40
129	2 13 20	(2 16 40 44 26 40 ↓)	(2 13 54 48 53 20 ↓	2 40		0	2 40
130	2 13 20	(2 16 22 57 46 40 ↓)	(2 13 37 2 13 20 ↓	2 40		0	2 40

over a full period of 130 steps, from which we calculate intermediate values simply by interpolation. This formulation incorporates the singular asymmetry introduced in the construction of M, but otherwise introduces no new elements to the theory.

5.12   Lunar Velocity and Column F

This concludes our theory of the variations due to lunar anomaly in the lengths of the month and important eclipse intervals, and it remains only to construct Column F, the daily lunar velocity. This function does not come into play in determining the times of the syzygies, but is used in computing the visibility phenomena around the middle and ends of the months (and thus the dates of the syzygies). Its construction proceeds as follows.

First we assume a maximum velocity of $15^{o/d}$, which we identify with $S'_{max} = 7,30^{\phi}$. Next, we assume an average velocity of $13;30^{o/d}$, which we identify with the mean value, $\mu(S)$, rounding the difference

$$S'_{max} - \mu(S) = 19;53,26,15^{\phi} \cong 20^{\phi}. \tag{5.12.1}$$

Finally, the ratio,

$$\frac{F'_{max} - \mu(F)}{S'_{max} - \mu(S)} \Delta S' = \Delta F', \tag{5.12.2}$$

yields

$$\Delta F' = \frac{1;30 \cdot 5}{20} = 3;45^{o/d} \tag{5.12.3}$$

for the effective amplitude of F'. This corresponds to the truncated minimum for F of

$$F'_{min} = 11;15^{o/d}, \tag{5.12.4}$$

and leads to the characteristic differences shown below and seen earlier in exhibit 2, namely that

$$d_s F = \frac{1;30}{20} = 0;4,30^{o/d}, \tag{5.12.5a}$$

and

$$d_1 F = \frac{d_s F}{\varepsilon} = 0;42^{o/d}. \tag{5.12.5b}$$

In the process F's mean value is slightly shifted, but it remains in phase with S.

Table 7
Lunar velocity at syzygy

	Anomaly	Velocity
Maximum	0°	15;16,8$^{o/d}$
Minimum	180°	11;49,39$^{o/d}$
Mean	86;40°	13;26,59$^{o/d}$
Average (3)		13;30,54$^{o/d}$
Amplitude, Δ		3;26,29$^{o/d}$

This leaves the question: why the mean value 13;30$^{o/d}$—a value far removed from the relatively accurate value 13;10,35$^{o/d}$, which not only was clearly known, but implicit in other parts of the theory? The answer, by now unsurprising, is that 13;30$^{o/d}$ is a very good approximation of the actual mean lunar velocity *at syzygy*, which is what F portrays in System A, and which differs from the mean lunar velocity at other elongations.[31] This is due mainly to the inequality known as the Variation, which at syzygy adds a constant increment of roughly +0;16,21$^{o/d}$ to the lunar velocity, an effect which is slightly magnified at the extremes of velocity by the contributions of other inequalities.[32]

Table 7 shows the lunar velocities at syzygy computed from Brown's theory[33] and a mean motion corresponding to −500 (13;10,34,52,41.1$^{o/d}$) for lunar anomalies of 0°, 180°, and 86;40° (the last corresponding to the minimum contribution of all terms depending on anomaly). These exhibit an average—13;30,54$^{o/d}$—which is almost precisely the mean value of Column F as well as an amplitude of approximately 3;26$^{o/d}$.

It is evident that F is a derivative function in System A, in contrast System B, where it is the fundamental function of the theory. Instead, F_A's critical parameters derive from the truncated function S in yet another example of both the central role of S and of the rounding of computed values of S to nice integral numbers. Since F is attached to S and Φ by procedures which are convenient, but neither logically compelling nor embedded in the structure of the underlying theory, it is not surprising that we find several variants of F, reflecting small differences in extrema, mean values and even phase. In System A, however, all are distinguished by monthly increments (d_1F_A) of approximately 0;42$^{o/d}$, in contrast to System B where $d_1F_B \cong 0;36^{o/d}$, with corresponding differences in the amplitudes and extremes. Interestingly, except for $F(0)_A$, the variants from both systems all reflect increments for 14 months of roughly 0;2$^{o/d}$.

Table 8
Variants of F: critical parameters

	F_B	F'_A	F_A	F_A(abbr.)	$F(0)_A$
Π	4,11	4,11	1,44,7	4,53	14
Z, mod. Π	18	18	7,28	21	1
Δ	4;11	4;54,8,26,15	4;52,49,41,15	4;53	4;54
$\delta = 2\Delta/\Pi$	0;2	0;2,20,37,30	0;0,5,37,30	0;2	0;42
d_1	0;36	0;42,11,15	0;42	0;42	0;42
d_{14}	0;2	0;2,20,37,30	0;2,20,37,30	0;2	0
d_{223}	−0;4	−0;4,41,15	−0;4,30	−0;10	−0;42
μ	13;10,35	13;17,27,39,22,30	13;30,29,31,52,30[34]	13;30,30	13;30

Table 9
Extrema and amplitudes of S, M, and related functions

	S	M	235M	$\hat{S}$	$\hat{M}$
Max	7,30	15,11	20,45	7,42;38,45	15,39;12,30
min	6,40	7,30	5,38;53,45	6,37;34,22,30	5,31;52,40
Δ	50	7,41	15,6;6,15	1,5;4,22,30	10,37;19,50

Table 8 summarizes the significant parameters of the attested variants, which reflect a wide range of approximating techniques. Thus F'_A adopts the same period relation as we find in System B (supporting the assumption of its underlying role in System A), but keeps d_{14} the same as F_A, which results in small changes in d_1, d_{223} and Δ. In contrast F_A(abbr.) uses rounded values for μ and Δ, keeps d_1 unchanged, and—as in System B—lets $\Pi = 1,0\Delta$. This results in a poor period relation, as can be seen in the anomalous value for d_{223}, but keeps d_{14} (= 0;2) the same as in System B.

5.13   Conclusion

This concludes our reconstruction of the central functions in the System A theory of lunar anomaly. It is evident that the entire theory is built up from the functions expressed in units of ϕ—that is, S and its relatives M and 235M—which in the process of its construction repeatedly undergo small roundings to nice numbers that disappear from view when these functions are dressed in units of *uš*. The central function is, of course, S whose untruncated version, $\hat{S} = \Phi/\phi$, could have as easily served the role of Φ in the theory. Even S, however, loses its apparent simplicity in transformation into $\hat{S}$ as may be seen in table 9, which summarizes the extrema and amplitudes of these functions and their untruncated siblings.

6   TEXTUAL EVIDENCE FOR S

The primary evidence for the functions S and M is found in the functions Φ and G, which, as we have seen, are simply $\hat{S}$ and $M + 1,30$ cloaked in units of $u\check{s}$. Furthermore, as also noted, the text BM 36311 (= 80-6-17, 37 + 321), published by Aaboe (1968) as Text E, contains precisely those values of Φ_2 and Λ corresponding to S_{n-12}, S_n, and $12M_n$ in our development of 235M, which is simply the sum of $\Phi_{s,n} + \Lambda_{s,n}$, stripped of its units of $u\check{s}$. Text E also reflects not only the values found for these functions, but also their tabulation over 124 steps, corresponding to a complete anomalistic period of the function 235M.

Further evidence comes from the so-called "Saros Text",[35] which still has many difficulties of interpretation but which contributes two critical pieces of information. The first of these is an explicit value for the length of the anomalistic month, equal to 27;33,16,30 days[36] or, as the text implies,[37] 2,45,19;39$^{u\check{s}}$. This plays no direct role in the theory regarding the variations in the lengths of different intervals. Nevertheless, it is a fundamental parameter that defines the (invariable) unit of time which serves as the theory's independent variable.

The second bit of evidence, which is more directly relevant to the role of function S, is the repeated appearance of its amplitude, $\Delta S = 1,5;4,22,30$. This parameter first appears in the second line of the text, suggesting its fundamental importance. From the damage on either side of it we can only be sure that it was to be multiplied by some number, but from the parallel phrase preserved in rev.13, lines 2 and 3 seem most likely to have read

obv.2   [. . . . . (;)17,46,40 the increment for 18 years ti]mes 1,5;4,22,30 mak[es

obv.3   19;16,51,6,40 ($u\check{s}$) . . . . 2,13,]20 to 1,58;31,6,40 m[u

which would thus reflect a conversion of ΔS to $\Delta \Phi$, or from units of ϕ to units of $u\check{s}$.

The role of ΔS becomes even more explicit in the next section, which begins by transforming $d_1 S$ into $d_1 \Phi$, and then states the half-length of the anomalistic month, first in days then $u\check{s}$. Then, after comparing the approximate length of 223 months with the precise length of 239 anomalistic months, the text proceeds with an illustrative calculation of the length of the mean synodic month from ΔS and the length of the anomalistic month. The calculation is based on recognizing that the variation ΔS takes half an anomalistic month and begins by computing the time required for a variation equal to one unit of ϕ, namely

$$\frac{dt}{dS} = \frac{\frac{1}{2}m_a}{\Delta S} = \frac{1,22,39;49,30}{1,5;4,22,30} = 1,16;13,10,11,24,36^{u\check{s}}/\phi \tag{6.1}$$

a result which is not quite accurate in its last two places,[38] but used consistently thereafter throughout the text. The next step is to multiply this result by the monthly increment of S ($d_1 S$) to obtain the amount of time required to produce this increment, and thereby the difference in length between a mean synodic and anomalistic month. Thus the calculation proceeds as

$$\frac{dt}{dS} d_1 S = 1,16;13,10,11,24,36 \times 9;20 = 11,51;22,55,6,29,36^{u\check{s}}$$

$$\times 0;0,10^{day}/u\check{s} = 1;58,33,49,11,4,56^{d}, \tag{6.2a}$$

whence

$$m = m_a + \frac{dt}{dS} \cdot d_1 S = 27;33,16,30^{d} + 1;58,33,49,11,4,56^{d} \tag{6.2b}$$

$$= 29;31,50,19,11,4,56^{d}, \text{``the result reached for the month at mean velocity month by month,''}$$

$$= 29 \text{ days (and) } 3,11,1,6, 29,36^{u\check{s}}. \tag{6.2c}$$

The procedure clearly reflects the central role of the function S, whose conversion to Φ is a recurring issue in the text. In particular, following this calculation in section 3 we find instructions to multiply $\phi = 0;17,46,40$: (a) by 12;38,45—the amount (in units of ϕ) by which S is truncated near its maximum; (b) by 50—the effective amplitude of the truncated function S'; and (c) by 2;25,37,30—the amount by which S is truncated near its minimum, the sum of the three intervals being $\Delta S = 1,5;4,22,30$.[39] The parameter $dt/dS = 1,16;13,10,11,24,36^{u\check{s}}/\phi$ also appears in at least two other places in the text, once (in section 6) in computing that the time required to cover 50 steps of variation—that is, the effective amplitude $\Delta S'$—is $1,3,30;59,29,30,30^{u\check{s}}$, the other (in section 11) being simply a restatement of its derivation.

We have, therefore, in the Saros Text, clear evidence of a familiarity with the function S underlying Column Φ together with some facility in its use. The same text, however, exhibits no awareness of the subtleties underlying the structure of these functions, such as the distortion introduced in the calculated value for the mean synodic month by the adjustment of ε.

7 Commentary

The 19-Year Cycle

Several aspects of this theory bear further comment. One is the use of the 19-year cycle, which otherwise is nowhere apparent in either lunar theory, to eliminate all effects due to zodiacal anomaly and indeed any dependence on longitude as an argument, such as would be required in comparing observations with different evening epochs. Thus the theory is completely independent of any measurement of position and, consequently, requires neither a zodiac nor any other positional reference system for its construction. To accomplish the same end, Ptolemy had first to establish a solar model, which—based on less secure observations—ultimately proved the weakest element in the *Almagest*.

While not really part of this story, it seems probable that the more accurate parameters found in Systems A and B for the mean motions of the Sun and Moon all derive from a small refinement of the 19-year cycle, namely, an observation that (lunar) eclipses separated by 235 months recur 3 fingers or $0;15°^{40}$ short of their initial position, and thus that

$$235 \text{ months} = 19^{\text{rev}}(\text{Sun}) - 0;15°$$
$$= 254^{\text{rev}}(\text{Moon}) - 0;15°. \qquad (7.1)$$

This adjustment together with the mean synodic month of $29;31,50,6,0$ days found in (5.8.4) leads to the parameters shown in table 10, which subsequently are further modified in Systems A and B by various small roundings. In particular, System A's parameters are governed by internal needs that require that the ratio of months to years be that of two relatively prime integers, Π/Z, where Z is regular and not so large as to make $30/Z$ inconveniently small. This actually limits the choice quite severely, and the pair actually chosen—$\Pi = 46,23$ (2783), $Z = 3,45$ (225), $\Pi/$

Table 10

19-year cycle parameters

Parameter	Unadjusted	Adjusted
1 year	$12;22,6,18,56,..$ months	$12;22,7,56,35,..$ months
$\Delta\lambda_1$	$29;6,22,58,..°/m$	$29;6,19,8,55,..°/m$
$\Delta\lambda_{223}$	$10;43,24,..°$	$10;29,10,10,..°$
$\Delta\lambda_{235}$	$0;0°$	$-0;15°$
μ_m	$13;10,34,59,30,35,..°/d$	$13;10,34,51,43,..°/d$
μ_s	$0;59,8,17,12..°/d$	$0;59,8,9,25,33,..°/d$

$Z = 12;22,8$ months—is the one whose ratio most closely approximates the adjusted parameter (table 10, line 1) from among all those where Z is regular and $< 1,0,0$. In contrast, System B seems to have assumed that $\Delta\lambda_{223} \cong 10;30°$ and rounded the implicit $\Delta\lambda_1(29;6,19,22..°)$ to $29;6,19,20°$. Neither theory, however, shows any evidence of going beyond the accuracy implied by the ostensible observation (7.1).[41]

General Accuracy

A second point to be emphasized is the remarkable accuracy of this theory, a feature hidden by the apparent inaccuracy of its period relation, ostensible mean synodic month length, and various other mean values, which result from the deliberate adjustments to ε and M. The most dramatic example is the accuracy of the computed amplitudes of the principal intervals, which agree exceptionally well with their counterparts computed from modern theory. This may be seen in Exhibit 9, which shows the amplitudes (Δ) of the relevant intervals and their respective errors, expressed in $u\check{s}$ and calculated: first, from System B; then from System A assuming three different values of ϕ, namely:

$\phi = 0;18^{u\check{s}}$,   which corresponds to $\Delta235M$ implicit in System B, thus allowing comparison of the errors arising from the structures of the two theories, as distinct from their parameters;

$\phi = 0;17,46,40^{u\check{s}}$,   which is the value actually adopted in System A;

$\phi = 0;17,21,40^{u\check{s}}$,   which corresponds to a very nearly accurate value of $\Delta235M$, and thus reflects the irreducible errors of the theory from structural sources.

Also shown, for comparison, are the actual amplitudes computed from modern theory, followed by a tabulation of the corresponding errors for each of the relevant intervals and parameters.

Two points stand out from the table of errors, and especially from the sums of the absolute errors. The first is that the errors in the amplitudes from System A are substantially less for every interval than those from System B. This is true even with comparable parameters (i.e., $\phi = 0;18^{u\check{s}}$), where the aggregate error ($\Sigma|\delta|$) for the five intervals computed from System B is nearly 3 times as great that resulting from System A's procedures. This illustrates the severe structural limitations of System B, its accurate mean values notwithstanding. Particularly noteworthy are the relatively gross errors in the amplitudes of 1 and 6 months from System B, compared with corresponding errors from System A less than a

Exhibit 9

Amplitudes and Their Errors

Amplitudes(Δ)	System B	0;18,0	System A : ϕ =			Modern[22]
			0;17,46,40	0;17,21,40		
Δ1M	2,36;15us	2,18;18us	2,16;35,33,20^{us}	2,13;23,28,20^{us}		2,15; 0us
Δ6M	8,56;15	9,33; 0	9,25;55,48, 8, / 52,20	9,12;40		9,27;45
Δ12M	4,22;30	4,18;58,30	4,15;46,40	4, 9;46,58,30		4, 7;45
Δ223M	15; 0	15; 0	14;48,53,20	14;28, 3,20		14;15
Δ235M	4,33;45	4,31;49,52,30	4,28;28,31,6,40	4,22;10,58,30		4,23

Errors (δ)	System B	0;18,0	System A : ϕ =	
			0;17,46,40	0;17,21,40
$\delta\Delta$1M	−21;15us	−3;18us	−1;36us	+1;37us
$\delta\Delta$6M	+31;30	−5;15	+1;50	+13; 5
$\delta\Delta$12M	−14;45	−11;13	−8; 2	−2; 2
$\delta\Delta$223M	−0;45	−0;45	−0;34	−0;13
$\delta\Delta$235M	−10;45	−8;50	−5;28	+0;49
$\Sigma|\delta|$	79; 0us	29;21us	17;30us	17;47us
rel $|\delta|_b$	4.5	1.7	1.00	1.02

Table 11
Accuracy of assumed "observations"

	Calculated (whole days +)			Error	
	Actual	System A	"Observed"	System A	Observed
Max (235m)	$11;30^{uš}$	$8;53..^{uš}$	$\leq\ 40^{uš}$	$+2;37^{uš}$	$1;30^{uš}$
min (235m)	1,48;45	1,40;25	[>1,40]	+8;20	[+ <8;45]
Δ (235m)	4,22;45	4,28;28.	<4,30	−5;43	[− <7;15
Max (223m)	2,14;0	2,13;20	≤2,15	+0;40	−1;0

tenth as large. The comparison not only highlights the power and flexibility of System A's structure and procedures but shows its uniform superiority to System B.

A second point is that while the most accurate of the three values of φ further reduces the errors by roughly 40%, it gives no better—in fact slightly worse—results than the value actually adopted. Thus φ = $0;17,46,40^{uš}$ gives as good results as can be obtained from the theory and leads to errors in these amplitudes which in no case exceed $\pm10^{uš}$ and average roughly $3\frac{1}{2}^{uš}$, corresponding to 12^{ϕ} or 15 minutes of our time. Given the fact that all intervals except 235 months are substantially affected by variations due to solar anomaly, and thus cannot be directly observed, this can only be regarded as an extraordinary result.

Finally, it should be noted that the assumed "observations" reflect a similar level of accuracy, as shown in table 11. All of these data are also essentially accurate to within $10^{uš}$, and the most critical data—the maxima of 235M and 223M—are accurate to within roughly $2;30^{uš}$, or 10 minutes. This is perhaps not surprising for the maximum of 235 months, which is a small interval (in excess of whole days) but is striking in the maximum of 223 months, which is a relatively large interval and all but dead accurate. In any event all of these parameters are more accurate than one would expect from direct measurement of eclipse times relative to sunset, and seem more likely to have been derived from an analysis of the abundant measurements of lunar visibilities near full moon—ŠÚ, NA, ME, and GE$_6$.

Other critical parameters of the theory reflect a similarly high accuracy. We have already noted the essentially accurate mean value of F as the lunar velocity *at syzygy*, which is more difficult to evaluate than the mean sidereal motion, since one cannot simply average over long periods. A more fundamental parameter is the length of the anomalistic month, encountered in the Saros Text, which is the sole month-length in the

Table 12
Summary of System A critical parameters

	System A	Modern (−500)
anomalistic month, m_a	27;33,16,30^d	27;33,16,30,*34*..^{d}
$m_a/\overline{m}$ (from truncation)	0;55,59,6,33$^-$	0;55,59,6,*35*$^-$
mean synodic month, $\overline{m}$		
from $m_a/\overline{m}$	29;31,50,9,1..^{d}	29;31,50,8,*39*..^{d}
from 235M	29;31,50,6,0..^{d}	29;31,50,8,*39*..^{d}
mean lunar velocity	13;10,35°$^{/d}$	13;10,34,52,41..°$^{/d}$
″ at syzygy	13;30°$^{/d}$	≈13;27°$^{/d}$

System A theory that is explicitly defined, and also the only month-length that is constant. This parameter,

$m_a = 27;33,16,30$ days

is essentially identical with its modern equivalent (for −500) of 27;33,16,30,*34* days. It plays no apparent role in the construction of the theory, but combining it with the ratio of mean synodic to anomalistic month implied in the 124 line truncation (5.6.5) yields an implicit value for the mean synodic month, $\overline{m} = 29;31,50,9,1..$days, which is about as accurate as System B's. The other value for the mean synodic month underlying the System A theory is that found from the median value of 235M (5.8.4), namely,[42]

$\overline{m} = 29;131,50,6,0$ days $= 29^d + 3,11;0,36^{u\tilde{s}}$,

which also closely approximates the mean value of G in System B. These parameters and their modern counterparts are summarized in table 12. They show—together with the consistently small errors in the amplitudes—that in all important respects the System A theory is exceptionally well founded, notwithstanding its adjusted period relation, and the evident rounding of key parameters at critical steps in its construction.

Style

Finally, there are the elements of style that pervade this theory. These include an evident mastery of all aspects of the properties and behavior of linear zigzag functions, an affinity for algebraic formulation, and a disciplined sense of rigor governing all aspects of its construction. Examples abound, but my personal favorites include the conditions for symmetry in the construction of S, the truncation of S so as to correct the intentional distortion introduced into the period relationship, and the reversal of time in the construction of M from S. Beyond these, however, is an aesthetic sensibility in the structure of the theory, expressed in an evident preference for symmetry and simplicity. In general, meaningless precision is avoided in favor of nice numbers, especially in S and M, but with apparent care that this not compromise the fundamental accuracy of the theory. Lastly, there is an air of privacy in the theory, where one senses that the subtleties of its structure are not intended to be seen, hidden as they are under the cloak of several additional sexagesimal places. Whatever the case, this powerful theory is also one of great elegance. It is a pity that we lack its author's name.

8 Chronology

We are left with the question of when was this done, a question that at present can be delimited but not definitively answered. The earliest evidence for Columns Φ and F appears in a text known as Text S,[43] which gives calculated values for various functions together with observational remarks for a saros of (38) solar eclipse possibilities from -474 to -456. The functions attested include: Φ and F; a function for the longitude of syzygy, reflecting a scheme for essentially uniform motion with reference to zodiacal signs; and functions for the magnitude and duration of eclipses which are far more primitive than anything found in Systems A or B. No function corresponding to G or 6M is apparent, although the times of the syzygies are probably recorded in the poorly preserved date column. There is no question about the dates of events described, which correspond precisely with the conventional Saros Cycle for lunar eclipses (SC 16), which extends from -472 to -454.[44]

The text was obviously written after the events described—in pre-Seleucid times this is almost invariably the case[45]—and probably after the end or the corresponding lunar saros cycle. At the same time details of the function for eclipse magnitudes make it unlikely that this function was computed much more than one saros after the events described, and more probably during the Saros Cycle which immediately followed.[46] This would place the composition of Text S in the interval -456 to -418 at the widest, and most probably in the years -454 to -436. Obviously the theory was invented earlier, but it seems unlikely to have materially predated the zodiac, which seems to have appeared between -463 and -453. On balance, if we assign its invention to -440 ± 15 years, we should not be too far off.

Two other early texts, one dated (with qualifications) to -397[47] and the second concerning two saroi of lunar eclipses from -416 to -380,[48] also provide evidence that by these dates the problem of zodiacal anomaly had not yet been solved, or at least not fully. The first (Text A) includes computed values of Φ and a variant of F,[49] together with a variant System A scheme for the longitudes of syzygies, a primitive scheme for the length of daylight, and a function perhaps for the month length whose construction is obscure. No functions of latitude appear. All in all, the text seems to reflect a stage of development where the tools were at hand for solving the problem of zodiacal anomaly, but not yet fully applied, let alone integrated with a theory of latitudes and eclipse magnitudes. There is no evidence for its date of composition, but no reason to suppose this much different from the date of the events described.

The second text (Text L) contains eclipse magnitudes and (mean) longitudes of the syzygies, the latter computed by means of the same scheme used in Text S. The scheme for eclipse magnitudes is more sophisticated than that in Text S and approximately reflects the variation in magnitude due to zodiacal anomaly, but linked to the nodes rather than to the zodiac. As with the scheme in Text S, it also appears to have been empirically adjusted and is much more accurately centered relative to the nodes.[50] On balance it seems to reflect better data and techniques, but still an incomplete understanding of the integration of nodal motion and zodiacal anomaly. From the accuracy of its data and a structure that precludes computing back any distance, it seems likely to have been composed not long after the events describe and almost certainly within the following saros.

Texts A and L reflect an evolution of techniques for treating the effects of zodiacal anomaly on both the times of the syzygies and eclipse magnitudes, but also that by as late as -380 the connection between nodal motion and zodiacal anomaly was not yet fully understood. Thus it appears that the completion of a fully integrated lunar theory occurred sometime between this date and -315, when we have secure evidence of the complete System A, making the probable date for the completion of System A circa -350 ± 25 years. This implies that at least half, and possibly three quarters of a century separated the invention of the theory of lunar anomaly and the completion of the whole theory, making it unlikely that this was the work of one person. Given the internal consistency, coherence, and logical rigor that characterize the complete System A lunar theory, this is a surprise, which, for the time being at least, must remain unresolved.

In sum, the System A theory of lunar anomaly seems likely to have been created around or slightly after the middle of the 5th century B.C., shortly after the introduction of the zodiac, and perhaps half a century after intercalations began to reflect official use of the 19-year cycle. We know little of the state of astronomy at this time, but what we do know goes scarcely beyond familiarity with period relations and simple schemes to predict the dates of solstices and equinoxes. In any event, nothing remotely like this theory appears to have existed either then or for a long time after, making it a radical departure from all previous efforts to harness the power of mathematics to the process of prediction.

ABBREVIATIONS

ACT   O. Neugebauer, *Astronomical Cuneiform Texts*, 3 vols. (London, 1955), Lund Humphries. Reprinted (1983), Springer-Verlag.

AD   A. J. Sachs and H. Hunger, *Astronomical Diaries and Related Texts from Babylon* (Vienna, Österreich. Akad. d. Wiss.), Vol. I (−651 to −261), 1988; Vol. II (−260 to −164), 1989: Vol. III (−163 to end), 1995.

AHES   *Archive for History of the Exact Sciences*, Springer-Verlag.

Alm.   *Ptolemy's Almagest*, translated and annotated by G. J. Toomer (New York, Berlin, Heidelberg, Tokyo, 1984), Springer-Verlag.

GMS3   *Die Rolle der Astronomie in den Kulturen Mesopotamiens*, Beiträge zum 3. Grazer Morgenländischen Symposion (23–27 September, 1991), ed. by Hannes D. Galter (Graz, 1993).

FsAaboe   *From Ancient Omens to Statistical Mechanics: Essays on the Exact Sciences presented to Asger Aaboe*, ed. by J. L. Berggren and B. R. Goldstein (Copenhagen, 1987), Univ. Library.

FsSachs   *A Scientific Humanist: Studies in Honor of Abraham Sachs*, ed. by E. Leichty, M. deJ. Ellis, P. Gerardi, (Philadelphia, 1988), Univ. Museum.

HAMA   O. Neugebauer, *A History of Ancient Mathematical Astronomy* (New York, Heidelberg, Berlin, 1975), Springer-Verlag.

JCS   *Journal of Cuneiform Studies* (New Haven, Cambridge, MA, Philadelphia, Ann Arbor).

KDVSMM   *Kongelige Danske Videnskabernes Selskab, Matematisk-fysiske Meddelelser* (Copenhagen).

NOTES

1. ACT 207c$^+$ (Aaboe and Hamilton, 1979, 21).

2. Neither system assumes that lunar anomaly affects the longitudes of the syzygies and thus either latitudes or eclipse magnitudes.

3. The so-called "annual equation" in the Moon's motion contributes a small, but not insensible (0;14°), inequality to that resulting from the solar anomaly. Thus, strictly speaking, the term "zodiacal anomaly" is preferable to "solar anomaly." It also conforms to the thrust of Babylonian theory which treats a single anomaly resulting from the longitude of syzygy.

4. In ACT Neugebauer distinguishes between systems according to whether they use step-functions (System A) or linear zigzag functions (System B) to compute the longitudes of syzygies. This characterization, however, tends to obscure the almost complete lack of common elements in the two systems at every level of detail. Except for the common assumption of a 3 : 2 ratio between longest and shortest daylight and the (covert) appearance of 12;22,8—the System A period for the year—as an implicit parameter in Column H of System B, the two systems are relentlessly different in both the functions they employ and the parameters these are based on.

5. Here "earliest" and "latest" refer to the dates of the preserved contents of the texts in question. Dates are expressed according to the convention that year 1 B.C. = 0, and thus year $-n = n + 1$ B.C.

6. The earliest unambiguously dated text is ACT 70, which gives latitudes at full moon from at least S.E. 49 to S.E. 60 (-262 to -251). Subsequently, Aaboe and Henderson (1975, 210) showed that a lunar latitude given as an example in the procedure text ACT 200 corresponds to that for S.E. -7 month III (-318). Furthermore a new text, BM. 40094, published in Aaboe (1969) reflects elements of the complete System A for the years SE -8 to -5 (-318 to -315), confirming the pre-Seleucid completion of System A. This makes it likely that ACT 18a ($= 80$-6-$17,341 + 571$), should be dated to S.E. 41 (-270) instead of S.E. 266, and at least possible that ACT 51a ($= 80$-6-$17,807$) covered the years SE -19 to SE -5 (-330 to -316) instead of 225 years later as in ACT.

7. The exceptions are two texts from Uruk (W 22925 and W 22801 (+) 22805), published by Hunger (1991), which reflect schemes for the computation of solstices for the periods -360 to -323 and (at least) -611 to -555.

8. According to Reade (1986) tablets bearing the receipt date 80-6-17 were a consignment of tablets found mainly at Babylon (a few are from Borsippa) between October 1879 and January 1880. Rassam had returned to London at that time, and the excavations were carried on under the supervision of Daud Thoma, apparently at two sites in Babylon, one on the Amran mound which covers the temple of Marduk, and the other at the inner wall between Amran and the village of Jumjuma. On October 31 Daud reported, "we now have about 100 [tablets] large and small ... we are still working in the direction of the wall, which will shortly end, when we will shift and work opposite the Imam Amran." Subsequently, Col. Miles, Consul General

in Baghdad, visited the site and reported, "I found Mr. Daood Thoma excavating the mounds opposite the little village of Jumjumeh where he has laid bare a long wall of great thickness which he considered to be the inner face of the inner rampart. *Contiguous to this wall were the debris of houses in which were discovered the tablets and other articles forwarded to you . . .* (emphasis added) I also noticed a large tablet almost perfect about 9 or 10 inches long." While the evidence is inconclusive, this report suggests a private library, which would also be consistent with the distinctive nature of the archive. Tablet finds from the vicinity of the wall (Jumjuma) continued to be reported but in diminishing numbers after January 1880. See Reade (1986, xvii–xx, xxx) for details.

9. See Neugebauer (1957a, 141–43) and HAMA, 497 n. 2, for the history of the modern usage, beginning with Halley in 1691, of the term "saros" in reference to the eclipse cycle of 223 months. Here I use "saros" to mean simply 223 months, :"saros cycle" to refer the eclipse and quasi-anomalistic cycle of 223 months, and "Saros Cycle" to refer to a specific group of 223 months that is consistent with those described in Aaboe, Britton, Henderson, Neugebauer, and Sachs (1991, 3–31 and esp. 21).

10. Nothing requires that δ be regular. If otherwise, $\Delta = 4;5$ ($\delta = 0;35$), $\Delta = 3;58$ ($\delta = 0;34$) and $\Delta = 3;51$ ($\delta = 0;33$) all are closer to our assumed estimate, $4^{o/d}$. For $\delta = 0;32$, the next closest regular value, $\Delta = 3;44$.

11. A more precise estimate based on modern calculations places the minimum interval at approximately 6939 days $+1,49$ $uš$ and the maximum at 6940 $+11$ $uš$, yielding the same mean value of 6939;40 days.

12. An analogous function for System A is preserved in MNB 1856, published as Text H in Neugebauer and Sachs (1969), 92–94. Its parameters are: $\Pi = 14$, $Z = 15$, $\Delta = 4;54$, $\delta = d_1 = 0;42$, $\mu = 13;30$, $M = 15;57$, $m = 11;3$. Thus it has the same Π and Z, and but different μ and Δ, and consequently δ. Interestingly, despite the difference in amplitude, the effective amplitude—that is, difference between maximum and minimum tabular values (15;36 − 11;24 = 4;12)—is the same as Δ in our example.

13. Ephemerides from Uruk invariably use a variant, F', with the same parameters as F but expressed in degrees (or minutes of arc) per time-degree (= 1/360th of a day). Ephemerides from Babylon frequently tabulate both forms, but if only one is given, it is F rather than F'.

14. In ACT Neugebauer unfortunately introduced the artificial unit "large hour" ($1^H = 1,0^{uš}$), for typographical convenience. It should be emphasized that this unit has no counterpart in Babylonian practice, and should be abandoned.

15. From modern theory for (-500) $269m_a - 251m = 1;29^{uš} \cong 6$ minutes $\approx 0;3,15°$ of argument of anomaly. Thus it would take 375 years for the cumulative error in this relationship to amount to the equivalent of $1°$ of argument of anomaly. In contrast the explicit System A period relation results in an error corresponding to slightly more than $4°$ of argument of anomaly in 505 years or roughly $1°$ every 125 years.

16. The mean sidereal lunar velocity calculated from modern theory for -500 is $13;10,34,52,41^{o/d}$.

17. Geminus (*Isagoge*, Manitius, 205) attributes this parameter to the "Chaldeans," whom he claims derived it from the "exeligmos" or triple saros cycle, in which the moon purportedly moves $260,312°$ (723 sidereal rotations $+32°$) in 19756 days. These numbers seem more likely, however, to have resulted from the reverse calculation using $\mu(F)$ and the System B mean synodic month length with roundings to whole degrees and days. Geminus also reports details (maximum, minimum, daily difference) of the System B function for the daily variation in lunar velocity, designated F^* in ACT, which has the same mean value as F but different extrema. See HAMA, 585–6 and 602–3.

18. For example, ACT 102 has F_2 correct, but $F_1 = F_2$; 103 and 104 have F_1, but with the branches reversed. All three texts are from Uruk.

19. Both this value and the period relation (3.1a) are attributed by Ptolemy to Hipparchus, and it was Kugler (1900) who first recognized their Babylonian origin. Since then a considerable literature on the subject of Babylonian parameters in Greek astronomy has appeared, notably, *inter alia*, in Aaboe (1955), HAMA, Toomer (1980 and *Alm.*), and Jones (1983 and 1993). Recently Neugebauer (1989a) discovered a tabulation of Column G from System B in a fragment of Greek papyrus from Roman Egypt, which implies, as Neugebauer (1989b, 399) notes, that most, if not all, of the Babylonian System B procedures as well as parameters were known to Greek speaking scholars.

20. The modern equivalent for -500 is $\mu = 3,11;0,51.9$ uš corresponding to 29;31,50,8,39 days.

21. Combining the 19-year cycle (235 m $= 254$ $m_s = 19$ years) with $\mu(F) = 13;10,35°/d$ yields a value for the mean synodic month, $\bar{m} = 29;31, 50, 4, 54..$days. Combining the System A value for the length of the anomalistic month, 27;33,16,30 days, and the period relation m $= 269/251$ m_a yields the value $\bar{m} = 29;31,50,11,36$ days. Averaging the two yields $\bar{m} = 29;31,50,8,15$, which would round to $\mu(G)$ under the constraints described. Such an origin, however, would make μ's accuracy wholly fortuitous, which is difficult to accept in light of the consistent excellence of System B's parameters generally.

22. Modern values for Δn are from Brack-Bernsen (1980, 46) except $\Delta 235$, which is calculated from syzygy times in Goldstine (1973).

23. Also ACT 207a.

24. I am indebted to Professors Erica Reiner and Hermann Hunger for the explanation of this term. In ACT Neugebauer transcribes this word as *lu-uš-šú*, having attached the preceding "−*ma*" to the "DU" before it as a normal enclitic particle. Intrigued by a word which was not in any of the dictionaries, but whose meaning was clear from its context, I asked Erica Reiner if she could shed any light on this. She passed my question on to Hermann Hunger, who was at the *Dictionary* at the time, and who explained that the "*ma*" should be attached to *luššu* rather than "DU," as confirmed by the use of "DIR," the logogram for "*malu*," in the analogous phrase in ACT 206. Erica Reiner recalled two other instances of the word itself, in different contexts but with congruent meanings, and the whole question was thereby tidied up.

25. If M_{n+223} and M_{n+2} are on the same branch, so also will be M_{n+224}, M_{n+225}, M_n, and M_{n+1}, which lie between them.

26. "Nicer" choices would be either 128 (2^7) or 126 ($2 \cdot 3^2 \cdot 7$).

27. If we reverse the derivation and start with $\mu(F) = 13;10,35^{o/d}$, we get $235M_{31(1/2)} = 13,53;21,5,8^{\phi}$ instead of $13,54;49,22,30^{\phi}$ as here and in System A (Aaboe, 1968, Text E, $[\Phi + \Lambda]/\phi$). This leads to $235M_{max} = 20,43;31..$ which would round to $20,45$ as we find. Thus it is possible that the critical parameter was $\mu(F) = 13;10,35^{o/d}$, effectively modified by the small rounding of $235M_{max}$.

28. The syzygies associated with the Φ values tabulated in Text E extend from -745 Jan (Nabonassar 1**, month VI) through $+248$ Aug and correspond to lunar eclipse possibilities EP- 32 to EP- 36 in the Saros scheme described in Aaboe, Britton, Henderson, Neugebauer and Sachs (1991, 3–31 and esp. 21) for up to 56 saros cycles (EP-s 33–35). Lunar eclipses occurred at all of these syzygies except those at -744 Jul (EP-35, SC 1) and $+248$ Aug (EP-36, SC 56). See Britton (1990, 66) for details.

29. The calculation simply increases M by the difference between 30° and the mean synodic arc calculated from the 19-year cycle, $\Delta\lambda = 29;6,22,58..$ divided by the mean motion in elongation, $\mu(\eta) = 12;11,26,42..$ with appropriate adjustment of units. Thus

$$\delta M = (30° - 29;6,22,..°) \div 12;11,26,42,.. \times 360 \div \phi = 1,29;3,..^{\phi}.$$

Using more accurate System A parameters leads to $\delta M = 1,29;10,..^{\phi}$.

30. The procedure for 6M was discovered by Aaboe and Hamilton, who describe it and the textual evidence for it in Aaboe and Hamilton (1979).

31. The value $13;30^{o/d}$ for $\mu(F)$ appears in MNB 1856 (Neugebauer and Sachs, 1969, Text H, 92-3) as the mean value of an approximate function for lunar velocity with a period of 14 months, but the System A increment, $d_1 = 0;42$. See note 12.

32. This explanation for the high mean value of F was first suggested to A. Aaboe in the 1970s by K. P. Moesgaard, who also discussed it in detail at the same Dibner Institute symposium in May 1994 at which this paper was originally presented. The development described here was undertaken independently of Moesgaard's analysis, but with awareness of his earlier suggestion through conversations with Prof. Aaboe.

33. Brown (1919, 3–28).

34. Variant: ACT 92 has $\mu(F) = 13;26,18,16,52,30^{o/d}$, a shift downward from $\mu(F_A)$ of $-0;4,11,15^{o/d}$, an amount equal to $d_{223}F'_A$. All other parameters are the same as F_A's. The motivation and significance of this adjustment are obscure, but the resulting mean value is essentially accurate.

35. Originally published by Neugebauer (1957) as BM 36705 (= 80-6-17, 437 + 458), to which Aaboe (1968) joined the important fragment BM 37474 (80-6-17, 1241). It was A. Sachs' discovery of the phrase "17,46,40 (is) the increment (tab u lal) for 18 years ..." in this text (rev. 13 and 16) that led to the beginning of an understanding of the astronomical significance of column Φ. The phrase also recurs in the

corner fragment added by Aaboe, where 17,46,40 is described as simply the "difference (*tašpiltu*) for 18 years."

36. Obv. 4–6, 13, 29, 32.

37. The text actually computes the length of half an anomalistic month, equal to 1,22,39;49,30^{uš}.

38. Accurate calculation leads to 1,16;13,10,11,25,*46,9*,.. The author of BM 36705 was evidently aware of this, because he demonstrates the quality of the approximation by computing the product 1,5;4,22,30 × 1,16;13,10,11,24,36 = 1,22,39;49,29,58,43, 55,7,30. The approximation affects the last two places of the resulting month-length, which should be ..,6,45.. instead of ..,4,56.

39. This was first recognized by Aaboe (1968, 37), who identifies and discusses these intervals along with corrections to the commentary in Neugebauer (1957, 18).

40. This of course is equivalent to half the lunar diameter, but it is not certain that the Babylonians regarded the lunar diameter as equal to 0;30° (cf. Britton, 1989, 42). Finger (šu-si), on the other hand, was a conventional and constant unit of measurement for small angular distances, equal to 0;5°, which is attested in the earliest preserved Diary (A.D., −651). The estimated shortfall itself is high by a third, the accurate value being −0;11° ± 0;10°. The error would be virtually insensible over three 19-year cycles, which is the maximum number of consecutive cycles likely to be bounded by eclipses. All in all it is a very decent estimate.

41. An unpublished fragment of a System B procedure text from Birmingham, shown to me by Dr. Wayne Horowitz, contains the following explicit and hitherto unattested mean motions of the Moon and Sun:

$\mu_m = 13;10,34,51^{o/d}$ $\mu_s = 0;59,8,9,48,40^{o/d}$.

The first agrees with the value obtained here from the adjusted 19-year cycle, while the second appears to have derived from the assuming that the sun advances 10;30° in 223 months, combined with the System B value for the month length.

42. This value is not the same as the average value of M, which is effectively raised by roughly 0;27^{uš} as a result of rounding the minimum of M to 7,30$^\phi$ (from 7,28;30..). Obviously this was regarded as of no practical consequence.

43. Published in Aaboe and Sachs (1969) as Texts B and C (+) D, with an additional fragment published in Aaboe, Britton, Henderson, Neugebauer, and Sachs (1991, 69–7) as Text G. Discussed with revised transcription in Britton (1989).

44. See Aaboe, Britton, Henderson, Neugebauer and Sachs (1991, 3–31) for a discussion of the conventional Saros Cycle and related issues.

45. A problem of dating in regnal years is that dates beyond the current regnal year are uncertain, if not inauspicious. The only instance I know of where an apparent date materially exceeds a king's reign is an obscure reference to 36 Kandalanu in Atypical Text C (Neugebauer and Sachs, 1967, 193). Conversely, the only pre-Seleucid text known to me to compute predicted phenomena for future years is a solstice scheme

found in W 22925 (Hunger, 1991), which contains the dates of summer solstices from −360 (45 Artaxerxes II) to −323 (13 Alexander III). The text explicitly cites years 44 and 45 of Artaxerxes II (when it was apparently written), thereafter giving only the months and number of years since 44 Artaxerxes II as part of the calculation. Thus it appears to confirm the assumed tradition of not citing regnal years in advance of the current one.

46. The magnitude function in Column III of Text S reflects a period relation which has a secular error of −1;18 units of magnitude per saros, corresponding to an error in nodal elongation of ca −0;52°. This function, however, which appears to have been subject to empirical adjustment reflects an average error in nodal elongation for the 38 syzygies of only 0;26°. This is consistent with observations in either the saros in question or the following saros, but not later without gross error. See Britton (1989, 33ff.) for details.

47. Published in Aaboe and Sachs (1969) as Text A.

48. Published in Aaboe, Britton, Henderson, Neugebauer, and Sachs (1991, 43–62) as Text L.

49. Not only do the extrema differ modestly, but this variant is also out of phase with Φ.

50. The mean error in nodal elongation from all 76 data is negligible (0;0,33°) and less than the uncertainty from modern theory. Cf. Aaboe, Britton, Henderson, Neugebauer, and Sachs (1991, 59–61) for details.

REFERENCES

Aaboe, A.
1955   "On the Babylonian Origin of Some Hipparchian Parameters," *Centaurus*, 4, 122–5.
1968   "Some Lunar Auxiliary Tables and Related Texts from the Late Babylonian Period," *KDVSMM*, 36:12.
1969   "A Computed List of New Moons for 319 B.C. to 316 B.C. from Babylon: B.M. 40094," *KDVSMM*, 37:3.
1971   "Lunar and Solar Velocities and the Length of Lunation Intervals in Babylonian Astronomy," *KDVSMM*, 38:6.

Aaboe, A., Britton, J. P, Hendersen, J. A., Neugebauer, O., and Sachs, A. J.
1991   "Saros Cycle Dates and Related Babylonian Astronomical Texts," *Transactions* of the American Philosophical Society, 81:6.

Aaboe, A., and Hamilton, N. T.
1979   "Contributions to the Study of Babylonian Lunar Theory," *KDVSMM*, 40:6.

Aaboe, A., and Henderson, J. A.
1975   "The Babylonian Theory of Lunar Latitude and Eclipses According to System A," *Archives Internationales d'Histoire des Sciences*, 25, 181–222.

Aaboe, A., and Sachs, A. J.
1969 "Two Lunar Texts of the Achaemenid Period from Babylon," *Centaurus*, 14, 1–22.

Beaulieu, P.-A., and Britton, J. P.
1994 "Rituals for an Eclipse Possibility in the 8th Year of Cyrus," *JCS*, 46/1, 73–86.

Brack-Bernsen, L.
1980 "Some Investigations on the Ephemerides of the Babylonian Moon Texts, System A," *Centaurus*, 24, 36–50.
1990 "On the Babylonian Lunar Theory: A Construction of Column Φ from Horizontal Observations," *Centaurus*, 33, 39–56.
1993 "Babylonische Mondtexte: Beobachtung und Theorie," *GMS3*, 331–58.

Britton, J. P.
1987 "The Structure and Parameters of Column Φ," *FsAaboe*, 23–36.
1989 "An Early Function for Eclipse Magnitudes in Babylonian Astronomy," *Centaurus*, 32, 1–52.
1990 "A Tale of Two Cycles: Remarks on Column Φ," *Centaurus*, 33, 57–69.

Brown, E. W.
1919 *Tables of the Motion of the Moon*, Sect. I, (New Haven), Yale Univ. Press.

Goldstine, H. H.
1973 "New and Full Moon's from 1001 B.C. to A.D. 1651," *Memoirs* of the American Philosophical Society, 94.

Hunger, H.
1991 "Schematische Berechnungen der Sonnenwenden," *Baghdader Mitteilungen*, 22, 513–19.

Kugler, F. X.
1900 *Die Babylonische Mondrechnung*, (Freiburg im Breisgau).

Jones, A.
1983 "The Development and Transmission of 248-Day Schemes for Lunar Motion in Ancient Astronomy," *AHES*, 29.1, 1–36.
1990 "Babylonian and Greek Astronomy in a Papyrus Concerning Mars," *Centaurus*, 33, 97–114.
1993 "Evidence for Babylonian Arithmetical Schemes in Greek Astronomy," GMS3, 77–94.

Moesgaard, K. P.
1980 "The Full Moon Serpent. A Foundation Stone of Ancient Astronomy," *Centaurus*, 24, 51–96.

Neugebauer, O.
ACT   *Astronomical Cuneiform Texts*, 3 vols. (London, 1955), Lund Humphries. Reprinted (1983), Springer-Verlag.
HAMA *A History of Ancient Mathematical Astronomy*, (New York, Heidelberg, Berlin, 1975), Springer-Verlag.

1957a *The Exact Sciences in Antiquity*, 2nd ed., (Providence), Brown Univ. Press. Reprinted (New York, 1962), Harper & Bros.

1957b "Saros" and Lunar Velocity in Babylonian Astronomy. *KDVSMM*, 31:4.

1989a "A Babylonian Lunar Ephemeris from Roman Egypt," *FsSachs*, 301–304.

1989b "From Assyriology to Renaissance Art," *Proceedings* of the American Philosophical Society, 133, 391–403.

Neugebauer, O., and Sachs, A. J.

1956   "A Procedure Text Concerning Solar and Lunar Motion: B.M. 36712," *JCS*, 10, 131–6.

1967   "Some Atypical Astronomical Cuneiform Texts I," *JCS*, 21, 183–218.

1969   "Some Atypical Astronomical Cuneiform Texts II," *JCS*, 22, 92–113.

Reade, J. E.

1986   "Rassam's Babylonian Collection: the Excavations and the Archives," Introduction to E. Leichty, *Catalogue of the Babylonian Tablets in the British Museum, Volume VI: Tablets from Sippar I*, (London), British Museum Publications.

Sachs, A. J., and Hunger, H.

AD   *Astronomical Diaries and Related Texts from Babylon*, (Vienna, Österreich. Akad. d. Wiss.), Vol. I (−651 to −261), 1988; Vol. II (−260 to −164), 1989: Vol. III (−163 to end), 1995.

Toomer, G. J.

Alm.   *Ptolemy's Almagest*, translated and annotated by G. J. Toomer, (New York, Berlin, Heidelberg, Tokyo, 1984), Springer-Verlag.

1980   "Hipparchus' Empirical Basis for His Lunar Mean Motions," *Centaurus*, 24, 97–109.

van der Waerden, B. L.

1965   *Die Anfänge der Astronomie: Erwachende Wissenschaft II*, (Groningen), Noordhoff.

9

The Derivation of the Parameters of Babylonian Planetary Theory with Time as the Principal Independent Variable
N. M. Swerdlow

This paper is devoted to the following subject: The Astronomical Diaries (ADT) contain records of two classes of observations of planets; the first and more numerous of conjunctions with the moon, with other planets, and with fixed stars; the second of heliacal phenomena, first and last visibility, first and second station, and acronychal rising. All were considered ominous except, it appears, acronychal rising. The only information common to all records is the date, while the specification of location is not consistent. Records of conjunctions usually have measurements of separation in units of cubits and fingers; first visibility, a location by zodiacal sign, occasionally also with a separation from a star or another planet; last visibility, usually a location by zodiacal sign; stations, usually a separation from a star, sometimes only a location by zodiacal sign; acronychal rising has no location at all. Our concern is with heliacal phenomena, for the planetary ephemerides (ACT) are used to compute the dates and locations of the same heliacal phenomena recorded in the Diaries, and thus it is reasonably assumed that they are in some way based upon the kind of information contained in the Diaries. There is no class of texts for the computation of conjunctions, but the dates of conjunctions of planets with stars, although not with the moon or other planets, could be predicted—if that is the right word—as could the separations, by means of goal-year periods after which the conjunctions would recur. The purpose of this paper is to set out a method by which the information recorded in the Diaries for heliacal phenomena, principally dates and secondarily location by zodiacal sign, may be used to derive the parameters of the ephemerides. Since there are no texts explaining how these parameters were derived, the study can only be speculative, showing that such a method is possible, and here the method will be presented theoretically, without considering the empirical foundation in observations recorded in the Diaries and related collections, which are, however, considered in detail in a longer study (Swerdlow, 1998) from which this paper is adapted and in Swerdlow (1999).

The principle underlying the method is that, with some exceptions, the numerical difference between synodic time and synodic arc in the ephemerides is constant. In this way, synodic arc, the arc between phenomena of the same kind, as first visibility, which the observations in the Diaries do not give either directly or with any degree of accuracy—location by zodiacal sign, even specifying beginning or end of the sign, is too imprecise, separations of planets from stars poses considerable if not insuperable difficulties—can be found from synodic time, the time between the same phenomena, given by the dates of the observations alone. In fact, by correct modern computation the numerical difference between synodic arc and time is not constant, and to assume that it is can only be an approximation. But it is precisely the use of such an approximation, *which would be unnecessary were there any reliable way of finding synodic arc independently of synodic time*, that I believe shows that something like the procedure developed here was used by the Babylonians themselves for deriving the parameters of the ephemerides. Underlying the constant difference between synodic time and synodic arc are two more fundamental principles, both also approximations. The first is that in the planetary theory the sun is taken to move uniformly; that is, only the mean motion of the sun, the "mean" sun, is considered. Whether this use of the mean sun ultimately, although indirectly, underlies its use in Ptolemy's planetary theory I do not know, but it would not surprise me if it does. The second, formulated by van der Waerden (1957), who called it the "Sonnenabstandsprinzip," the solar-distance principle, is that the heliacal phenomena take place at fixed elongations from the sun, characteristic of each phenomenon. This is not strictly true, as the elongations are taken with respect to the mean not the true sun, and are subject to periodic variations—much larger by modern computation than the small variations implied by the ephemerides—but it is this approximation, and it is just that, that makes the planetary theory of the ephemerides possible at all. In this way, the mean motion of the sun, the mean sun, returning to the same elongation from a planet in a given synodic time, measures the synodic arc, which is otherwise unknown.

Consider first the mean synodic times and arcs of heliacal phenomena determined by counting the number Π of synodic phenomena of the same kind occurring in a nearly integral number of zodiacal rotations Z of the phenomena in a nearly integral number of solar years Y. Since each phenomenon takes place at a specific, characteristic elongation of the planet from the mean sun, in the synodic time between successive phe-

nomena of the same kind, a faster inferior planet, with a limited elongation on either side of the sun, must return to the specific elongation, and
the mean sun must overtake a slower superior planet and return to the
specific elongation. The number of phenomena Π is therefore equal to
the difference between the number of zodiacal rotations of the sun Y and
of the phenomenon Z, which is not necessarily the same as the number of
rotations of the planet, and the relation between Π, Z, and Y may be
expressed as

$$i\Pi = Y - Z \quad \text{or} \quad Y = i\Pi + Z$$

$$\text{where } i = \begin{cases} 1 & \text{for Saturn} \\ 1 & \text{for Jupiter} \\ 2 & \text{for Mars} \\ 1 & \text{for Venus} \\ 0 & \text{for Mercury} \end{cases} \tag{1}$$

Since Π is the number of phenomena, of synodic periods, in Y years, the
integer quotient $i = (Y - Z)/\Pi$ is the number of complete years in the
planet's synodic period, that is, the synodic period of Mars is greater than
2 years, of Saturn, Jupiter, and Venus greater than 1 year, and of Mercury
greater than 0 and less than 1 year. The fractional part of the synodic
period is Z/Π times the length of the year.

These relations, along with a definition of the solar year, the annual
motion of the sun, suffice to determine the mean synodic arcs and times
of the planets, that is, the mean intervals of distance and time between
successive occurrences of the same phenomenon. As used in the planetary
ephemerides, the length of the year in synodic months $y = 12;22,8m$
where $m = 30^{\tau}$—the unit τ now called by the Sanskrit term "tithi"—
an approximation to the lunar calendar month of 29 or 30 days, which
would otherwise have to be determined month by month. We now
define one rotation of the zodiac as $6,0°$ and one year as $12;22,8m = (6,0 + e)^{\tau} = 6,11;4^{\tau}$ where the "epact" $e = 11;4^{\tau}$ is the excess of the solar
year over the "lunar year" of 12 months, $12m = 6,0^{\tau}$. In the case of the
sun, which completes $6,0°$ in $(6,0 + e)^{\tau}$, these two quantities are equiva-
lent and differ only in their units, tithis or degrees, in the ratio $(6,0 + e)/6,0 = 1 + e/6,0$, that is, we may define the year as the motion of the sun
through $6,0°$ of longitude or in $(6,0 + e)^{\tau}$ of time. Further, *provided that the
sun move uniformly*, the numerical relations between the time ΔT in which
the sun describes any arc $\Delta\Lambda$ and the numerical difference $\Delta T - \Delta\Lambda$ are
given by

$$\varDelta T = \frac{6,0+e}{6,0}\,\varDelta\varLambda = \varDelta\varLambda + \frac{e}{6,0}\,\varDelta\varLambda, \tag{2a}$$

$$\varDelta\varLambda = \frac{6,0}{6,0+e}\,\varDelta T = \varDelta T - \frac{e}{6,0+e}\,\varDelta T, \tag{2b}$$

$$\varDelta T - \varDelta\varLambda = \frac{e}{6,0}\,\varDelta\varLambda = \frac{e}{6,0+e}\,\varDelta T. \tag{2c}$$

We have noted that the mean motion of the sun measures the synodic arc described in a given synodic time. Since the return of the sun to a specific elongation with respect to a planet defines the synodic arc and synodic time, the mean synodic motion of the sun $\overline{\varDelta\varLambda}$ and the mean synodic time $\overline{\varDelta T}$ are given by

$$\overline{\varDelta\varLambda} = \frac{Y}{\varPi}6,0° = \left(i+\frac{Z}{\varPi}\right)6,0°, \tag{3a}$$

$$\overline{\varDelta T} = \frac{Y}{\varPi}(6,0+e)^{\tau} = \left(i+\frac{Z}{\varPi}\right)(6,0+e)^{\tau}. \tag{3b}$$

Since the mean synodic motion of the sun exceeds the mean synodic motion of the phenomenon by $i\,6,0°$, the mean synodic arc of the phenomenon $\overline{\varDelta\lambda}$ and the excess $\overline{\varDelta t}$ of $\overline{\varDelta T}$ over $i\,12m = i\,6,0^{\tau}$, are

$$\overline{\varDelta\lambda} = \overline{\varDelta\varLambda} - i\,6,0° = \frac{Z}{\varPi}6,0°, \tag{4a}$$

$$\overline{\varDelta t} = \overline{\varDelta T} - i\,6,0^{\tau} = \frac{Z}{\varPi}(6,0+e)^{\tau} + i\,e^{\tau}. \tag{4b}$$

And from (2c), the numerical difference c between synodic time and synodic arc is in both cases

$$c = \overline{\varDelta t} - \overline{\varDelta\lambda} = \overline{\varDelta T} - \overline{\varDelta\varLambda} = \frac{e}{6,0}\,\overline{\varDelta\varLambda}$$

$$= \frac{e}{6,0+e}\,\overline{\varDelta T} = \left(i+\frac{Z}{\varPi}\right)e, \tag{5a}$$

because $6,0°$ and $6,0^{\tau}$ are numerically the same. Finally, the numerical difference C of the mean synodic time $\overline{\varDelta T}$ and the mean synodic arc of the phenomenon $\overline{\varDelta\lambda}$ is

$$C = \overline{\varDelta T} - \overline{\varDelta\lambda} = i\,6,0 + \left(i+\frac{Z}{\varPi}\right)e = i\,6,0 + c = i\,12m + c, \tag{5b}$$

where $\overline{\varDelta T}$ is expressed in months and tithis and $\overline{\varDelta\lambda}$ in degrees.

Table 1
ACT periods, mean synodic arcs, times, and differences

Planet	Π	Z	Y	$P = \Pi/Z$
Saturn	4,16	9	4,25	28;26,40
Jupiter	6,31	36	7, 7	10;51,40
Mars	2,13	18	4,44	7;23,20
Venus	12, 0	7,11	19,11	1;40,13,55,
Mercury	25,13	8, 0	8, 0	3; 9, 7,30

	$\overline{\Delta\lambda}$	$\overline{\Delta t}$	$c = \overline{\Delta t} - \overline{\Delta\lambda}$
Saturn	12;39,22,30°	24; 6,43, 7,30$^\tau$	11;27,20,37,30
Jupiter	33; 8,44,48,	45;13,52,56,	12; 5, 8, 8,
Mars	48;43,18,29,	1,12;21,10,22,	23;37,51,52,
Venus	3,35;30	3,53;10	17;40
Mercury	1,54;12,36,38,	1,57;43,15,54,	3;30,39,15,

Table 1 gives for each planet Π, Z, and Y for the ACT periods of the ephemerides; the "period" $P = 6,0°/\overline{\Delta\lambda} = \Pi/Z$, the number of phenomena, not an integer, in one rotation of the zodiac, its application to be explained later; the mean synodic arc of the phenomenon $\overline{\Delta\lambda}$; the excess $\overline{\Delta t}$ of the mean synodic time $\overline{\Delta T}$ over $i6,0^\tau$; and the numerical difference $c = \overline{\Delta t} - \overline{\Delta\lambda}$, which we give without units here and in what follows. For Venus, $\overline{\Delta t}$ and c, which is never used, follow, not from the ACT period, but from the short period $5p = 1,39m - 4^\tau$, where p is the synodic period. Fractions ending in commas are non-terminating.

It thus follows from the principles that heliacal phenomena are (1) taken with respect to the mean sun, which moves uniformly, and (2) occur at fixed, characteristic elongations from the mean sun, that (1) the mean synodic motion of the sun $\overline{\Delta\Lambda}$ measures the mean synodic arc $\overline{\Delta\lambda}$ of the phenomenon described in the mean synodic time $\overline{\Delta T}$ and (2) the numerical difference between the mean synodic time and mean synodic arc is constant, $\overline{\Delta T} - \overline{\Delta\lambda} = C$. Since only mean values have been considered, this is hardly surprising. But it may be significantly extended, for it is evident from the ephemerides that, with some notable exceptions, the difference between true synodic time ΔT and true synodic arc $\Delta\lambda$ is also taken as constant, for this approximation is applied to find ΔT from $\Delta\lambda$ by $\Delta T = \Delta\lambda + C$, which shows that the principles and consequences just stated must also hold for true synodic arc and time. Suppose a phenomenon occurs at an elongation η from the mean sun; the next phenomenon will occur after the true synodic time ΔT in which the phenomenon has moved through the true synodic arc of the phenomenon $\Delta\lambda$ and the mean sun through the true synodic arc of the sun $\Delta\Lambda = \Delta\lambda + i6,0°$. Since the motion of the phenomenon and of the mean sun differ by $i6,0°$, the elongation is again η and the synodic motion of the mean sun $\Delta\Lambda$ in the synodic time ΔT still measures the synodic arc of the phenomenon $\Delta\lambda$.

There is, however, now an error in using a constant difference between true synodic arc and true synodic time, although it is quite small. From (5a) the mean difference c between synodic time and arc,

$$\overline{\Delta t} - \overline{\Delta\lambda} = \overline{\Delta T} - \overline{\Delta\Lambda} = \frac{e}{6,0}\,\overline{\Delta\Lambda} = \frac{e}{6,0 + e}\,\overline{\Delta T}, \tag{6a}$$

and since the true synodic arc, as any arc, is also measured by the mean motion of the Sun, the true difference, from (2c),

$$\Delta t - \Delta\lambda = \Delta T - \Delta\Lambda = \frac{e}{6,0}\,\Delta\Lambda = \frac{e}{6,0 + e}\,\Delta T. \tag{6b}$$

Subtracting (6a) from (6b) and substituting $\Delta\lambda - \overline{\Delta\lambda}$ for $\Delta\Lambda - \overline{\Delta\Lambda}$, since they are equal, the error in time in tithis $\varepsilon(t)$ in adding C to $\Delta\lambda$ to find ΔT, or c to $\Delta\lambda$ to find Δt,

$$\varepsilon(t) = \frac{e}{6,0}(\Delta\lambda - \overline{\Delta\lambda}) = 0;1,50,40(\Delta\lambda - \overline{\Delta\lambda}). \tag{7a}$$

And since, from (2b), for any arc $\Delta\Lambda = (6,0/(6,0 + e))\Delta T$, the error in arc in degrees $\varepsilon(\lambda)$ in subtracting C from ΔT or c from Δt to find $\Delta\lambda$,

$$\varepsilon(\lambda) = \frac{6,0}{6,0 + e}\,\varepsilon(t) = \frac{e}{6,0 + e}(\Delta\lambda - \overline{\Delta\lambda})$$

$$\approx 0;1,47,22(\Delta\lambda - \overline{\Delta\lambda}), \tag{7b}$$

only slightly smaller than (7a). Taking for each planet the greatest difference of $\Delta\lambda - \overline{\Delta\lambda}$ that can occur, the maximum values of $\varepsilon(t)$ and $\varepsilon(\lambda)$, for Mars, are $1;3^{\tau}$ and $1;1°$ respectively, but since $\Delta\lambda$ can be greater or less than $\overline{\Delta\lambda}$, the errors are periodic and in the course of some number of synodic periods presumably sum to close to zero. The maximum errors for Saturn are about 0;3 tithis or degrees, for Jupiter, 0;9, and for Mercury's four phenomena extend from 0;12 to 0;54. None of this applies to Venus, for which the difference between synodic time and arc is not constant.

What this means is that for the variable, true synodic times and arcs of heliacal phenomena the two principles just given, of fixed elongations from the mean sun, hold and that small periodic errors result only from the approximation of using the mean values of (6a) while the principles hold strictly for the true values of (6b). Further, the same principles may be applied to find ΔT from $\Delta\lambda$, as in the ephemerides, or $\Delta\lambda$ from ΔT, that is, $\Delta\lambda = \Delta T - C$, where ΔT is known from the dates of observed phenomena. In this way, an unknown true synodic arc $\Delta\lambda$ may be found from a known true synodic time ΔT without any measurement of location, and it is this possibility that lies at the foundation of finding the parameters of the ephemerides from the dates of phenomena recorded in the Diaries. Of course, without the use of the mean difference C, it would still be possible to find $\Delta\Lambda$ from ΔT by (2b), and then $\Delta\lambda = \Delta\Lambda - i6,0°$, but this does not appear to be necessary since the converse is not used to find ΔT from $\Delta\lambda$ in the ephemerides. The difference is, in any case, very small. Thus, for deriving the parameters of the ephemerides, we assume that from observational records are taken the dates T_1 and T_2 of successive phenomena of the same kind—my guess would be first visibility—which may then be used to find the synodic time $\Delta T = T_2 - T_1$, with ΔT taken as "lunar years" of $12m = 6,0^\tau$ and tithis. Then, from ΔT we find the excess $\Delta t = \Delta T - i6,0^\tau$, from which the synodic arc $\Delta\lambda = \Delta t - c$; and thus in what follows, by synodic time we shall generally mean the excess Δt and by synodic arc the synodic arc of the phenomenon $\Delta\lambda$.

For the superior planets, the derivations depend upon only the maximum and minimum Δt and $\Delta\lambda$, while Mercury requires either one or two additional values and Venus is not amenable to these methods because the difference between Δt and $\Delta\lambda$ is not constant. Since for the derivations here, which are theoretical, we shall assume Δt implied by the ephemerides, our procedure is in a sense circular. But there is no other way, for the ephemerides produce very closely the synodic arcs and times that are used to derive their parameters; very simply, what comes out had to go in, and very precisely so. Thus, were it possible to find $\Delta\lambda$ independently of Δt, precisely the values used here would likewise be required. How these values of Δt were originally selected from the considerable irregularities given by the observed dates of phenomena in the Diaries and related records I do not know—a detailed analysis of the problem is presented in our longer study—but I do know that only these values will do.

Since our object is the derivation of numerical parameters, we shall, of course, work numerically, in the case of Mercury at rather great length. But the only way of showing that a method of deriving the parameters is sound is to derive the parameters. Here we shall consider only the principal systems for each planet—the variants are considered in our longer study—and we begin with System B, which is the more straightforward in that the parameters can be derived from Δt directly without using $\Delta\lambda$.

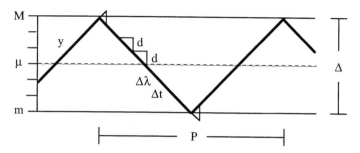

Figure 1
System B

SYSTEM B

System B is an arithmetic function of constant and numerically identical second differences for the first differences of both longitude and time, the synodic arc $\Delta\lambda$ and synodic time Δt of each phenomenon, that fall within fixed maximum and minimum limits. As pointed out by Aaboe (1964, 225), each value of $\Delta\lambda$ and Δt is a function, not of longitude, but of the number of each phenomenon in the period of Π phenomena in Z zodiacal rotations, and its relation to zodiacal longitude is indirect and in fact variable. When graphed, the function forms alternately ascending and descending branches of a "linear zigzag." A drawing is shown in figure 1 in which the graph of $\Delta\lambda$ or Δt is the function y on the sloping lines between the maximum limit M and minimum m; the range or amplitude $\Delta = M - m = \Delta\lambda_M - \Delta\lambda_m = \Delta t_M - \Delta t_m$, and the mean value $\mu = \frac{1}{2}(M + m)$ is $\overline{\Delta\lambda}$ or $\overline{\Delta t}$. The constant difference for each value of y is d such that $y_{n+1} = y_n \pm d$, where d is added on an ascending branch and subtracted on a descending. Where a limit would be crossed, the part of $y_n \pm d$, that is, of d, that falls outside the limit is turned back within, so that

at M: $y_{n+1} = M - ((y_n + d) - M) = (2M - d) - y_n$,

at m: $y_{n+1} = m + (m - (y_n - d)) = (2m + d) - y_n$.

And letting y be Δt or $\Delta\lambda$, the date and longitude of each successive phenomenon are then $t_{n+1} = t_n + \Delta t_{n+1}$ and $\lambda_{n+1} = \lambda_n + \Delta\lambda_{n+1}$. The synodic time and arc may also be computed from each other by $\Delta t - \Delta\lambda = c$, although Δt and $\Delta\lambda$ appear to operate independently in the ephemerides and may even be out of phase, presumably by error, which could not happen if c were used for conversion.

The "period" P is the number of phenomena in one rotation of the zodiac, and thus between successive identical points of the function in magnitude and direction, as M or m, and is given by

$$P = \frac{2\Delta}{d} = \frac{6{,}0^\circ}{\overline{\Delta\lambda}} = \frac{\Pi}{Z}. \tag{1}$$

It follows that a function of System B is determined by specifying M and m, which determine Δ, and P. Since $P = \Pi/Z$ is presumed known from records of the dates of phenomena through some number of years, all that are necessary are the greatest and least synodic times between successive phenomena, Δt_M and Δt_m, recoverable from the same records, to find d and completely determine the function. A similar method of derivation using $\Delta\lambda$ rather than Δt was proposed by Neugebauer (1968; HAMA, 429–30). It is of course possible to derive the function either way, since Δ and d are numerically the same for $\Delta\lambda$ and Δt, but we shall use Δt directly. Suppose now that from records of the dates of phenomena we take the maximum and minimum synodic times, Δt_{Mo} and Δt_{mo}, converted, if necessary, to an integral number of tithis. Then, as a first approximation,

$$\Delta_o = \Delta t_{Mo} - \Delta t_{mo}, \quad d_o = \frac{Z}{\Pi} 2\Delta_o = \frac{2\Delta_o}{P}. \tag{2}$$

Then d_o is rounded to d, where d is a small, preferably "regular" number, small for convenience in calculation, since d must be added or subtracted for each entry, regular so that its reciprocal will be a terminating fraction, although this last is not strictly necessary and is violated in an ephemeris for Mars. In order to maintain the proper period, $P = 2\Delta/d = \Pi/Z$, a new value of Δ is then found from

$$\Delta = \frac{1}{2}\left(\frac{\Pi}{Z} d\right) = \frac{1}{2} Pd, \tag{3}$$

and since Δ is the same for both Δt and $\Delta\lambda$, the final parameters for the limits M and m are then

$$\Delta t_M = \overline{\Delta t} + \tfrac{1}{2}\Delta, \quad \Delta t_m = \overline{\Delta t} - \tfrac{1}{2}\Delta,$$
$$\Delta\lambda_M = \overline{\Delta\lambda} + \tfrac{1}{2}\Delta, \quad \Delta\lambda_m = \overline{\Delta\lambda} - \tfrac{1}{2}\Delta. \tag{4}$$

The final values will have odd fractional places, which have nothing to do with the precision of the function, only the necessity of preserving the correct period.

Saturn: System B

All the ephemerides for Saturn, ACT 700–09, are System B, some of which use rounded parameters. For our derivation, we let

$$\Delta t_{\mathrm{Mo}} = 26^{\tau}, \quad \Delta t_{\mathrm{mo}} = 23^{\tau}, \quad \Delta_{\mathrm{o}} = 3^{\tau}.$$

Then, from (2),

$$d_{\mathrm{o}} = \frac{9}{4,16} 6^{\tau} = 0;12,39\ldots^{\tau} \approx 0;12^{\tau} = d,$$

from which, by (3),

$$\Delta = \frac{1}{2}\left(\frac{4,16}{9} 0;12^{\tau}\right) = 2;50,40^{\tau}, \quad \frac{1}{2}\Delta = 1;25,20^{\tau},$$

and the final parameters to full precision are, from (4),

$$\Delta t_{\mathrm{M}} = 25;32, 3,7,30^{\tau}, \quad \Delta\lambda_{\mathrm{M}} = 14; 4,42,30°,$$

$$\Delta t_{\mathrm{m}} = 22;41,23,7,30^{\tau}, \quad \Delta\lambda_{\mathrm{m}} = 11;14, 2,30°,$$

which agree with the ephemerides although rounded values are also used.

Jupiter: System B

Most of the ephemerides for Jupiter are System A or A$'$, although ACT 620–29 are System B and ACT 640 an inconsistent System B$'$. For the derivation of parameters, it appears to have been assumed, with some obvious rounding,

$$\Delta t_{\mathrm{Mo}} = 50^{\tau}, \quad \Delta t_{\mathrm{mo}} = 40^{\tau}, \quad \Delta_{\mathrm{o}} = 10^{\tau},$$

from which, by (2),

$$d_{\mathrm{o}} = \frac{36}{6,31} 20^{\tau} = 1;50,29\ldots^{\tau} \approx 1;48^{\tau} = d,$$

1;48 being the closest small regular number. Then from (3),

$$\Delta = \frac{1}{2}\left(\frac{6,31}{36} 1;48^{\tau}\right) = 9;46,30^{\tau}, \quad \frac{1}{2}\Delta = 4;53,15^{\tau},$$

and using the rounded $\overline{\Delta t} \approx 45;14^{\tau}$ and $\overline{\Delta\lambda} \approx 33;8,45°$, from (4),

$$\Delta t_{\mathrm{M}} = 50; 7,15^{\tau}, \quad \Delta\lambda_{\mathrm{M}} = 38; 2, 0°,$$

$$\Delta t_{\mathrm{m}} = 40;20,45^{\tau}, \quad \Delta\lambda_{\mathrm{m}} = 28;15,30°,$$

as in the ephemerides. System B', in which, because of roundings in the derivation, P and Δ differ slightly for Δt and $\Delta\lambda$, is considered in our longer study.

MARS: SYSTEM B

System B for Mars is known from a tiny fragment ACT 510, of ten lines of the column for longitude of an unknown phenomenon, containing only one complete number, and an obscure reference in the procedure text ACT 811a.11. It was correctly identified by P. Huber and first published by Aaboe (1958). For our derivation, we take as the synodic times,

$$\Delta t_{Mo} = 1,44^{\tau}, \quad \Delta t_{mo} = 40^{\tau}, \quad \Delta_o = 1,4^{\tau},$$

from which, by (2),

$$d_o = \frac{18}{2,13}\,2,8^{\tau} = 17;19,23\ldots^{\tau} \approx 17^{\tau} = d,$$

and d is in this case not a regular number (although it is small). The final values follow from (3):

$$\Delta = \frac{1}{2}\left(\frac{2,13}{18}\,17^{\tau}\right) = 1,2;48,20^{\tau}, \quad \tfrac{1}{2}\Delta = 31;24,10^{\tau}.$$

Then taking the rounded $\overline{\Delta t} \approx 1,12;21,10^{\tau}$ and $\overline{\Delta\lambda} \approx 48;43,18,30°$, from (4),

$$\Delta t_M = 1,43;45,20^{\tau}, \quad \Delta\lambda_M = 1,20;7,28,30°,$$

$$\Delta t_m = 40;57,0^{\tau}, \quad \Delta\lambda_m = 17;19,8,30°.$$

The values of $\Delta\lambda$ are those of Huber's reconstruction; there is no evidence for confirming Δt.

SYSTEM A

System B is limited by the symmetry to its mean value and is entirely determined once its period and limits have been specified. System A is adaptable to any number of different synodic arcs and times, in any order and extending over any part of the zodiac, each of which, order and extent, may be specified separately, although this advantage does not appear until more than two synodic arcs and times are used. Its principle is to divide the zodiac into zones in which the synodic arc and time are of constant length, subject to algebraic and arithmetic conditions that are

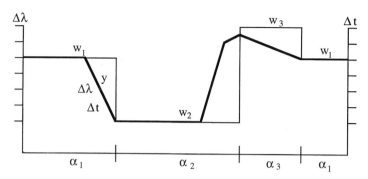

Figure 2
System A

essential to both the derivation and application of the system. A hypothetical System A of three zones is shown in figure 2. The zodiac is
divided into zones α_1, α_2, α_3 in which the length of the synodic arc or
time is w_1, w_2, w_3, called, for obvious reasons, a "step" function. The
graph of the synodic arc $\Delta\lambda$ or time Δt is the function y on the heavy line,
which coincides with w if it lies entirely in one zone or has a constant
slope if it extends through two or more zones. Given an initial value of
the longitude or time z_n, the next value will be $z_{n+1} = z_n + y_{n+1}$. If z_n
and $z_n + w_i$ lie within the same zone α_i, then $y_{n+1} = w_i$. But if $z_n + w_i$
extends into the next zone α_{i+1} by an amount x, then x must be adjusted
in the ratio w_{i+1}/w_i, that is,

$$y_{n+1} = (w_i - x) + \frac{w_{i+1}}{w_i}x = w_i + \left(\frac{w_{i+1}}{w_i} - 1\right)x, \quad z_{n+1} = z_n + y_{n+1}.$$

For Mars and Mercury, the function may extend into a third zone in the
same way. In Figure 2, the short stretch of the graph of reduced slope near
the end of α_2 extends into α_3 and α_1, so it is affected by both w_3/w_2 and
w_1/w_3. Finally, letting t or λ or be z and Δt or $\Delta\lambda$ be y, the date and longitude are $t_{n+1} = t_n + \Delta t_{n+1}$ and $\lambda_{n+1} = \lambda_n + \Delta\lambda_{n+1}$. The synodic time
or arc may also be found from the other by $\Delta t - \Delta\lambda = c$, for which reason
the range of the synodic arc and time is numerically the same,
$\Delta = \Delta\lambda_M - \Delta\lambda_m = \Delta t_M - \Delta t_m$, in common with System B. However,
the limits are not symmetrical to $\overline{\Delta\lambda}$ and $\overline{\Delta t}$, and there is an exception to
the use of c for System A_2 of Mercury.

System A is based upon the same period $P = 6{,}0°/\overline{\Delta\lambda} = \Pi/Z$ as
System B, the number of phenomena in one rotation of the zodiac, but
where in System B the period appears as the number of steps of d in the

return of the function to the same value, $P = 2\Delta/d$, in System A it is the sum of the synodic arcs per zone, $P = \Sigma(\alpha_i/w_i)$. We shall call the summation of the zones α_i and of the synodic arcs w_i in each zone, $\Sigma(\alpha_i/w_i)$, the "system"—it is often called the "generating function", the resulting true synodic arcs $\Delta\lambda$ or times Δt, shown in figure 2 by the heavy line, the "function"; and $P = \Sigma(\alpha_i/w_i) = \Pi/Z$ the "period." There are three conditions essential to System A concerned with maintaining the period:

$$1. \ \sum \frac{a_i}{w_i} = \frac{6{,}0°}{\overline{\Delta\lambda}} = \frac{\Pi}{Z} = P, \quad 2. \ \Sigma\alpha_i = 6{,}0°, \quad 3. \ \frac{w_{i+1}}{w_i} = \frac{p}{q},$$

where p/q is a ratio of small, "regular" numbers, composed only of the prime factor 2, 3, 5. The first condition is the fundamental definition of System A, while the second and third specify the conditions for α_i and w_i. The meaning of the condition for α_i is a metrical specification of the division of the zodiac into 6,0 parts—other units, including units of time, could likewise be used. The condition for w_i is more complex. Each time a boundary between zones is passed, a correction w_{i+1}/w_i is applied, and as the function passes through successive zones and returns to its original value, the reciprocal of w_{i+1}/w_i will likewise be applied, either at once in a two-zone system or through a product of ratios in a system of more than two zones. In order for both w_{i+1}/w_i and its reciprocal to be terminating sexagesimal fractions, it must be reducible to a ratio of numbers with the prime factors 2, 3, 5, a consequence of the sexagesimal system, called "regular" numbers, which are used constantly in Babylonian mathematics to replace division with multiplication by a reciprocal. Although w_i and w_{i+1} themselves may contain other prime factors, they must be contained in common. Were w_{i+1}/w_i not reducible to a ratio of regular numbers, it or its reciprocal would not be a terminating sexagesimal fraction and would have to be rounded or truncated in order to be used with convenience. But multiplication by a rounded or truncated number would introduce an error each time a boundary is crossed, which would immediately violate the period, affect every subsequent value of the function, and eventually accumulate to notable errors. In fact, w_{i+1}/w_i not only fulfills the condition of small regular numbers—there are exceptions, although not in the principal systems—but is also almost always of the form $(n \pm 1)/n$. The parameters of the principal systems of Saturn, Jupiter, and Mars are shown in table 2.

The distribution of the phenomena in each zone requires further examination, the principal studies of which are van der Waerden (1957) and Aaboe (1964), which we follow along with HAMA. If we consider

Table 2
Parameters of principal systems of superior planets

System	Zone	Limits	α_i	w_i	w_{i+1}/w_i $= p/q$	Π $\quad$ Z $P = \Pi/Z$	
Saturn A	1	♌ 10° → ♓ 0°	3,20°	11;43, 7,30°	6/5˙	4,16	9
	2	♓ 0 → ♌ 10	2,40	14; 3,45	5/6	28;26,40	
Jupiter A	1	♊ 25 → ♐ 0	2,35	30	6/5	6,31	36
	2	♐ 0 → ♊ 25	3,25	36	5/6	10;51,40	
Jupiter A′	1	♋ 9 → ♏ 9	2, 0	30	9/8	6,31	36
	2	♏ 9 → ♑ 2	53	33;45	16/15	10;51,40	
	3	♑ 2 → ♉ 17	2,15	36	15/16		
	4	♉ 17 → ♋ 9	52	33;45	8/9		
Mars A	1	♋ 0 → ♍ 0	1, 0	30	4/3	2,13	18
	2	♍ 0 → ♏ 0	1, 0	40	3/2	7;23,20	
	3	♏ 0 → ♑ 0	1, 0	1,0	3/2		
	4	♑ 0 → ♓ 0	1, 0	1,30	3/4		
	5	♓ 0 → ♉ 0	1, 0	1,7;30	2/3		
	6	♉ 0 → ♋ 0	1, 0	45	2/3		

only mean arcs, the Π phenomena will be distributed, not successively, into the mean intervals, also called "steps" (*Schritte*), $\bar{\delta} = 6{,}0°/\Pi = \overline{\Delta\lambda}/Z$, and thus the mean synodic arc $\overline{\Delta\lambda} = Z(6{,}0°(/\Pi) = Z\bar{\delta}$. However, the phenomena are not distributed uniformly throughout the zodiac, but occur in greater numbers and are more closely spaced where $\Delta\lambda$ and Δt are small, and in lesser numbers and further apart where $\Delta\lambda$ and Δt are large. Since the total number of phenomena $\Pi = ZP$, and $P = \Sigma(\alpha_i/w_i)$, the number of phenomena π_i in each zone α_i and the true interval δ_i between them are given, and related to each other, by

$$\pi_i = Z\frac{\alpha_i}{w_i} = \frac{\alpha_i}{\delta_i}, \quad \delta_i = \frac{w_i}{Z} = \frac{\alpha_i}{\pi_i}.$$

Just as the mean synodic arc $\overline{\Delta\lambda} = Z\bar{\delta}$, so, provided that the synodic arc remain within the same zone α_i, the true synodic arc $\Delta\lambda = w_i = Z\delta_i$. And in general, as the synodic arc extends into successive zones, the true synodic arc $\Delta\lambda = Z\delta$, that is, as each zone is passed, $\Delta\lambda$ will contain Z intervals of length $\delta_i, \delta_{i+1}, \delta_{i+2}$, which is the maximum extent reached.

TWO-ZONE SYSTEM

Conditions 1–3 above, which, as we have said, define System A, along with the difference c between synodic arc and time, are sufficient to derive the parameters from the true synodic time, just as was done for System B. From the records of the true dates of phenomena, assume initial values of the minimum and maximum synodic time Δt_{mo} and Δt_{Mo}. Then as a first estimate

$$w_{1\mathrm{o}} = \Delta\lambda_{\mathrm{mo}} = \Delta t_{\mathrm{mo}} - c, \quad w_{2\mathrm{o}} = \Delta\lambda_{\mathrm{Mo}} = \Delta t_{\mathrm{Mo}} - c.$$

We now solve the simultaneous linear equations from conditions 1 and 2,

$$\frac{\alpha_{1\mathrm{o}}}{w_{1\mathrm{o}}} + \frac{\alpha_{2\mathrm{o}}}{w_{2\mathrm{o}}} = P, \quad \alpha_{1\mathrm{o}} + \alpha_{2\mathrm{o}} = 6,0°.$$

If necessary, $\alpha_{1\mathrm{o}}$ and $\alpha_{2\mathrm{o}}$ are then rounded to α_1 and α_2, and new values of w_1 and w_2 found by simultaneous linear equations from conditions 1 and 3,

$$\frac{\alpha_1}{w_1} + \frac{\alpha_2}{w_2} = P, \quad \frac{w_2}{w_1} = \frac{p}{q},$$

where p/q is a ratio of small regular numbers. I need hardly point out that the solution of simultaneous linear equations would pose no problem for the Babylonians. The corresponding synodic times are then

$$w_1(t) = w_1(\lambda) + c, \quad w_2(t) = w_2(\lambda) + c.$$

SATURN: SYSTEM A

There are no known ephemerides of System A for Saturn, which is known only from the procedure texts ACT 801.3–8 and 802 and two dateless series of longitudes published as DCL texts A–B. The period, system, and mean synodic arc and time are as follows:

$$\sum \frac{\alpha_i}{w_i} = \frac{3,20}{11;43,7,30} + \frac{2,40}{14;3,45} = \frac{4,16}{9} = 28;26,40$$

$$\overline{\Delta\lambda} = 12;39,22,30°, \quad \overline{\Delta t} = 24;6,43,7,30^{\tau},$$

$$c = 11;27,20,37,30 \approx 11;30.$$

In order to derive the parameters, we assume

$$\Delta t_{\mathrm{mo}} = 23^{\tau}, \quad \Delta t_{\mathrm{Mo}} = 26^{\tau}, \quad \Delta_{\mathrm{o}} = 3^{\tau},$$

the same values used for System B, from which, by subtracting $c = 11;30$,

$$w_{1o} = \Delta\lambda_{mo} = 11;30°, \quad w_{2o} = \Delta\lambda_{Mo} = 14;30°, \quad \Delta_o = 3°.$$

From conditions 1 and 2, we have the simultaneous equations

$$\frac{\alpha_{1o}}{11;30} + \frac{\alpha_{2o}}{14;30} = 28;26,40, \quad \alpha_{1o} + \alpha_{2o} = 6,0°,$$

from which

$$\alpha_{1o} = 3,21;2,13,20°. \quad \alpha_{2o} = 2,38;57,46,40°.$$

We now round α_{1o}, α_{2o}, and w_2/w_1 in accordance with conditions 2 and 3,

$$\alpha_1 = 3,20, \quad \alpha_2 = 2,40°, \quad \frac{w_2}{w_1} = \frac{14;30}{11;30} = \frac{29}{23} \approx \frac{6}{5},$$

and from conditions 1 and 3, we solve the simultaneous equations

$$\frac{3,20}{w_1} + \frac{2,40}{w_2} = 28;26,40, \quad \frac{w_2}{w_1} = \frac{6}{5},$$

to find

$$w_1 = 11;43,7,30°, \quad w_2 = 14;3,45°,$$

the final parameters of System A. However, as John Britton pointed out to me, correctly, $29/23 \approx 5/4$, which curiously had escaped me, so there is clearly an inconsistency in the two steps of the derivation, meaning either that this derivation is incorrect or that the inconsistency was deliberate. From the unrounded value of c, the corresponding synodic times are

$$w_1(t) = 23;10,28,7,30^\tau, \quad w_2(t) = 25;31,5,37,30^\tau,$$

although there is no textual evidence for the use of these numbers.

JUPITER: SYSTEM A

Unlike Saturn, Jupiter is well represented by System A ephemerides ACT 600–08 and in addition the dateless longitudes DCL text D. The period, system, and mean synodic arc and time, are

$$\sum \frac{\alpha_i}{w_i} = \frac{2,35}{30} + \frac{3,25}{36} = \frac{6,31}{36} = 10;51,40,$$

$$\overline{\Delta\lambda} = 33;8,44,48\ldots° \quad \overline{\Delta t} = 45;13,52,56\ldots^\tau$$

$$c = 12;5,8,8\ldots \approx 12;5,10 \approx 12.$$

The derivation of the parameters for Jupiter is very straightforward. We assume that

$$\Delta t_{mo} = 42^{\tau}, \quad \Delta t_{Mo} = 48^{\tau}, \quad \Delta_o = 6^{\tau},$$

while System B uses 40^{τ} and 50^{τ} respectively. Subtracting $c = 12$,

$$w_{1o} = \Delta\lambda_{mo} = 30°, \quad w_{2o} = \Delta\lambda_{Mo} = 36°, \quad \Delta_o = 6°,$$

from which $w_2/w_1 = 6/5$, satisfying condition 3 so that $w_1 = w_{1o}$ and $w_2 = w_{2o}$. From conditions 1 and 2, we have the simultaneous equations

$$\frac{\alpha_1}{30} + \frac{\alpha_2}{36} = 10;51,40, \quad \alpha_1 + \alpha_2 = 6,0°,$$

from which the final parameters are directly

$$\alpha_1 = 2,35°, \quad \alpha_2 = 3,25°.$$

From $c = 12;5,10$, the corresponding synodic times are

$$w_1(t) = 42;5,10^{\tau}, \quad w_2(t) = 48;5,10^{\tau},$$

as in the ephemerides. It is notable that for Jupiter $\Delta\lambda_m$ and Δt_m are 2 units greater and $\Delta\lambda_M$ and Δt_M 2 units less in System A than in System B, while for Saturn they are about the same. One might explain the difference for Jupiter by noting that the extrema of System B are seldom reached so the wider limits of $10°$ and 10^{τ} are necessary to keep most of the function close to the narrower limits of System A of $6°$ and 6^{τ}, while in the case of Saturn the very small range of less than $3°$ and 3^{τ} makes such a distinction of no practical purpose. By modern calculation, the range of $\Delta\lambda$ is from about $30°$ to $37.3°$ and of Δt from 41^{τ} to 52^{τ}; hence System A is closer for $\Delta\lambda$ and System B for Δt.

FOUR-ZONE SYSTEM: JUPITER, SYSTEM A'

System A with two zones is as constrained as System B in that it is completely determined by conditions 1–3 once the period and the maximum and minimum synodic arcs or times have been specified. The great advantage of System A is realized only when additional zones with different synodic arcs and times are introduced, which may be of any order and length, provided conditions 1–3 are not violated. The simpler form, however, is the introduction of transitional zones, as in the three four-zone systems for Jupiter, of which the most important, and presumably the basic one, is System A' found in ephemerides ACT 609–14. In all three, the two zones of System A are separated by transitional zones $w_2 = w_4 = 33;45°$. System A$'$ is as follows:

$$\sum \frac{\alpha_i}{w_i} = \frac{2,0}{30} + \frac{53}{33;45} + \frac{2,15}{36} + \frac{52}{33;45} = \frac{6,31}{36} = 10;51,40.$$

The period, and thus $\bar{\Delta\lambda}$, $\bar{\Delta t}$, and c are the same as System A. The transitional zones α_2 and α_4 are nearly equal, suggesting that they were formed by dividing a zone equal to their sum, $1,45°$, as closely as possible to integer degrees. Therefore, to investigate the origin of System A$'$, consider the three-zone system

$$\sum \frac{\alpha_i}{w_i} = \frac{2,0}{30} + \frac{1,45}{33;45} + \frac{2,15}{36} = \frac{6,31}{36} = 10;51,40.$$

Note that conditions 2 and 3 are fulfilled with the following relations:

$$\alpha_1 = \tfrac{1}{2}(\alpha_2 + \alpha_3), \quad \alpha_1 = 2,0°, \quad \alpha_2 + \alpha_3 = 4,0°,$$

$$\frac{w_3}{w_1} = \frac{6}{5}, \quad \frac{w_2}{w_1} = \frac{9}{8}, \quad \frac{w_3}{w_2} = \frac{16}{15}.$$

The transitional w_2 must be formed by partitioning the ratio w_3/w_1 into two ratios of regular numbers; in this way $33;45°$ was selected, rather than, say, $33°$ because the ratio $9/8$ is the very next ratio of small, regular numbers less than $6/5$, while $33°$ would give $11/10$, which is not regular. And $32°$, although also in a ratio of regular numbers, $16/15$, was presumably excluded because it is farther than $33;45°$ from $\bar{\Delta\lambda} \approx 33;8°$. The simple relations between the zones, reducing α_1 from $2,35°$ in System A to $2,0°$, is also for purely algebraic purposes, for after letting $\alpha_1 = \tfrac{1}{2}(\alpha_2 + \alpha_3) = 2,0°$, the system follows from conditions 1 and 2, that is,

$$\frac{2,0}{30} + \frac{\alpha_2}{33;45} + \frac{\alpha_3}{36} = 10;51,40, \quad \alpha_2 + \alpha_3 + 2,0° = 6,0°,$$

from which

$$\alpha_2 = 1,45°, \quad \alpha_3 = 2,15°.$$

Then α_2 is divided closely into the two transitional zones of $53°$ and $52°$. From $c = 12;5,10$, the corresponding synodic times for the four-zone system are

$$w_1(t) = 42;5,10^\tau, \quad w_3(t) = 48;5,10^\tau, \quad w_2(t) = w_4(t) = 45;50,10^\tau.$$

Systems A$''$ and A$'''$ are considered in our longer study, as is Jupiter's six-zone system.

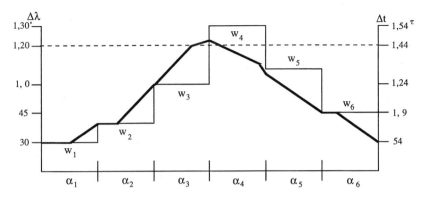

Figure 3
Mars. System A

SIX-ZONE SYSTEM: MARS, SYSTEM A

System A for Mars is known in fragmentary ephemerides *ACT* 500–03 and in dateless longitudes (DCL texts G–J, the last also ACT 504, DCL text K). The system, in which the zodiac is divided into six equal zones of two signs each, is shown in figure 3, and we give here the period, the system, and the mean arc and time with the zones reordered from ACT and HAMA to begin with the slowest.

$$\sum \frac{\alpha_i}{w_i} = \frac{1,0}{30} + \frac{1,0}{40} + \frac{1,0}{1,0} + \frac{1,0}{1,30} + \frac{1,0}{1,7;30} + \frac{1,0}{45} = \frac{2,13}{18} = 7;23,20$$

$$\overline{\Delta\lambda} = 48;43,18,29,46\ldots° \approx 48;43,18,30°$$

$$\overline{\Delta t} = 1,12;21,10,22,33\ldots^\tau \approx 1,12;21,10,30^\tau$$

$$c = 23;37,51,52,47\ldots \approx 23;37,52 \approx 24$$

The rounded values are used in ephemerides, and $c = 24$ will be used in our analysis to convert synodic time to arc.

    System A for Mars, with six equal zones, is simpler than it appears and may be derived from a symmetrical system based upon the minimum and maximum synodic times with an adjustment to fit the period. We assume that from observational records of true dates—perhaps with considerable adjustment, as shown by the round numbers for $\Delta\lambda$—the minimum and maximum synodic times are

$$\Delta t_{\text{mo}} = 54^\tau, \quad \Delta t_{\text{Mo}} = 1,44^\tau, \quad \Delta_o = 50^\tau,$$

from which, subtracting $c = 24$, the synodic arcs,

$\Delta\lambda_{\mathrm{mo}} = 30°$, $\Delta\lambda_{\mathrm{Mo}} = 1,20°$, $\Delta_{\mathrm{o}} = 50°$.

Note that the maximum values agree with System B while the minimum values are about 14 units greater, so that the range Δ, a round number, is smaller for System A than for System B, as was also true of Jupiter. Yet the range is so great—more than eight times that of Jupiter—that some number of transitional zones between the extrema are a necessity. Therefore, assuming a system of six equal zones of $1,0°$, for a minimum arc of $30°$ and a maximum of $1,20°$, we must find appropriate values of w_i, so we initially assume a symmetrical system with a slow zone of $30°$, two transitional zones of $1,0°$, and three zones, including the maximum, yet to be determined:

$$\sum \frac{\alpha_i}{w_i} = \frac{1,0}{30} + \frac{1,0}{w_2} + \frac{1,0}{1,0} + \frac{1,0}{w_4} + \frac{1,0}{1,0} + \frac{1,0}{w_6}.$$

Since $\Delta\lambda_{\mathrm{Mo}} = 1,20°$ is greater than $\alpha_4 = 1,0°$, we must find the value of w_4 that will give $\Delta\lambda_{\mathrm{Mo}} = 1,20°$, which will occur if $\Delta\lambda$ either (1) begins at the beginning of w_4 and ends within w_5 or (2) begins within w_3 and ends at the end of w_4. (We shall use conditions of this kind repeatedly in deriving the parameters of Mercury; here they occur but once.) We may use either case, but taking (2), we have

$$\frac{w_4}{w_3} = \frac{\alpha_4}{\alpha_4 + w_3 - \Delta\lambda_{\mathrm{Mo}}} = \frac{1,0}{1,0 + 1,0 - 1,20} = \frac{3}{2}, \quad w_4 = \tfrac{3}{2}w_3 = 1,30°,$$

Using case (1) would give the same result. Hence, for case (2) $w_4 = 1,30°$ will give the required

$$\Delta\lambda_{\mathrm{M}} = w_4 + \left(\frac{w_3}{w_4} - 1\right)(w_4 - \alpha_4) = 1,30° - \tfrac{1}{3}30° = 1,20°.$$

Now, using the same ratio of $3/2$ with respect to $w_1 = \Delta\lambda_{\mathrm{mo}} = 30°$, we find $w_2 = w_6 = 45°$ and fill in the symmetrical system as follows:

$$\sum \frac{\alpha_i}{w_i} = \frac{1,0}{30} + \frac{1,0}{45} + \frac{1,0}{1,0} + \frac{1,0}{1,30} + \frac{1,0}{1,0} + \frac{1,0}{45} = \frac{22}{3} = 7;20.$$

The result is quite interesting in itself since $\Pi/Z = 22/3$ forms the period $Y = 2\Pi + Z = 47$, that is, 47 years, the goal-year period of Mars used for the prediction of the dates of phenomena, as it is nearly an integral number of months, 22 synodic periods equal 581 months plus about 1 day. Since we have thus far made no assumption about the period, it may appear remarkable that a standard period simply falls out of the symmetrical system, but a sum of small-number ratios of α_i/w_i, and all here are either $\frac{3}{2}$ or $\frac{4}{3}$ and their reciprocals, must give a small-number ratio for Π/Z. The result is also evidence that the observations necessary to find the parameters need not extend over a period longer than, say, the goal-year periods themselves—all that are required are Δt_M and Δt_m—and further that the system could have been developed, although not necessarily used, for the goal-year period and then adjusted to the longer and more accurate ACT period.

And that is precisely what we must do now, that is, we must change the values of one or more w_i—α_i is fixed—so that the period P is increased by 0;3,20, from 7;20 to 7;23,20. No adjustment was necessary for Saturn or Jupiter since each algebraic step in deriving α_i and w_i maintained condition 1, $\Sigma(\alpha_i/w_i) = P$. However, in finding w_i for Mars with α_i given, condition 1 was not applied, and now it must be under three special conditions: (1) The ratios of w_{i+1}/w_i must be small, regular numbers. (2) Neither $w_1 = 30°$ nor $w_4 = 1,30°$ can be changed since these are required for the correct maximum and minimum $\Delta\lambda$. (3) The symmetry must be disturbed as little as possible. The adjustment must have been done by trial, and one cannot, to maintain symmetry, alter either $w_2 = w_6 = 45°$ or $w_3 = w_5 = 1,0°$ equally, for the former gives $44;4,53\ldots°$ and the latter $58;22,42\ldots°$, both violating condition 1. It is necessary to alter two transitional zones, and the optimal choice was made in altering w_2 to $40°$ and w_5 to $1,7;30°$, at least for keeping the ratios of w_{i+1}/w_i small and regular. In this way, decreasing w_2 to $(4/3)w_1 = 40°$ increases P by 0;10 to 7;30, and, maintaining symmetry as far as possible, increasing w_5 to $(3/4)w_4 = 1,7;30°$ decreases P by 0;6,40 to 7;23,20 exactly. We thus have the final system and period as given before and the ratios w_{i+1}/w_i given in table 2. However, the maximum synodic arc has become

$$\Delta\lambda_M = w_4 + \left(\frac{w_5}{w_4} - 1\right)(w_4 - \alpha_4) = 1,30° - \tfrac{1}{4}30° = 1,22;30°,$$

in excess of $\Delta\lambda_{Mo} = 1,20°$ underlying the symmetric system, which, however, was doubtless a rounding. What this shows, I assume, is that the conditions necessary to keep w_{i+1}/w_i a ratio of small, regular numbers and maintain the period take precedence over the accuracy of any specific value of the function. Finally, the minimum and maximum synodic times follow from adding $c = 23;37,52$,

$$\Delta t_m = 53;37,52^{\tau}, \quad \Delta t_M = 1,46;7,52^{\tau},$$

the former in good agreement with $\Delta t_{mo} = 54^{\tau}$, the latter about 2^{τ} higher than $\Delta t_{Mo} = 1,44^{\tau}$ owing to the increase in the function of $\Delta\lambda_M$.

But since the implied Δt_{mo} and Δt_{Mo} reflect the rounded values of $\Delta\lambda_{mo}$ and $\Delta\lambda_{Mo}$, the difference may not be significant. A method of deriving the parameters for Mars from the number of phenomena in each zone, proposed by Aaboe (1980), is discussed on our longer study.

Mercury

The most difficult, and impressive, accomplishment in Babylonian planetary theory, second only to the lunar theory, is that for the heliacal phenomena of Mercury. In the case of Saturn, Jupiter, and Mars, the same systems, with auxiliary rules, were used for all phenomena, but for Mercury the four heliacal phenomena, first and last visibility in the morning and evening, behave so differently and, compared to the superior planets, are so irregular and asymmetrical, that four entirely different functions were developed with the condition that they have (very nearly) the same period. The four functions have been divided into two systems, for first and last visibility respectively, each with its own procedure for finding the adjacent, following phenomenon. The systems, summarized in table 3, are

System A_1: 3 zones (ACT 300–05)   System A_2: 4 zones (ACT 300a–b)

Γ, First visibility in the morning Σ, Last visibility in the morning

Ξ, First visibility in the evening Ω, Last visibility in the evening

Unlike the other planets, the stationary points of Mercury, perhaps because they lie so close to Ω and Γ and are often not visible, are not considered. The "exact" period, found only for Ξ but closely approximated in the other three, and its mean synodic arc and time are:

$\Pi = 25,13 \quad Y = Z = 8,0 \quad P = 3;9,7,30$

$\overline{\Delta\lambda} = 1,54;12,36,38,32,45\ldots.^{\circ}$

$\overline{\Delta t} = 1,57;43,15,54,7,57\ldots^{\tau}$

$c = 3;30,39,15,35,12\ldots \approx 3;30,39 \approx 4$

The ephemerides for System A_1 use the rounded $c = 3;30,39$ for conversions between $\Delta\lambda$ and Δt. There are no procedure texts explaining the conversion for System A_2, and although no consistent rule can be derived from the two surviving ephemerides, in which only integer dates are given, it is clear that nothing as simple as a constant difference is used, since $\Delta t - \Delta\lambda$ for Σ lies between 0 and 6 and for Ω between 0 and 10. It is probably significant that the simple rule applies to the first visibilities of

Table 3
Parameters of systems A_1 and A_2 of Mercury

Zone		Original limits	α_i	Adjusted limits	α_i	w_i	w_{i+1}/w_i $= p/q$	Π	Z
								$P = \Pi/Z$	
Γ	1	♌ 0° → ♒ 0°	3, 0°	♌ 1° → ♉ 16°	2,45°	1,46°	4/3	44,33	14,8
	2	♒ 0 → ♊ 0	2, 0	♉ 16 → ♊ 0	2,14	2,21;20	2/3	3;9,7,38,…	
	3	♊ 0 → ♌ 0	1, 0	♊ 0 → ♌ 1	1, 1	1,34;13,20	9/8		
[Π]	1	♏ 0 → ♓ 0	2, 0	♎ 26 → ♓ 10	2,14	1,46;40	9/10	25,13	8,0
	2	♓ 0 → ♋ 0	2, 0	♓ 10 → ♋ 6	1,56	1,36	5/3	3;9,7,30	
	3	♋ 0 → ♏ 0	2, 0	♋ 6 → ♎ 26	1,50	2,40	2/3		
Σ	1	♋ 0 → ♎ 0	1,30	♋ 0 → ♎ 0	1,30	1,47;46,40	6/5	20,23	6,28
	2	♎ 0 → ♉ 0	1,30	♎ 0 → ♉ 6	1,36	2, 9;20	3/4	3;9,7,25,…	
	3	♉ 0 → ♈ 0	1,30	♉ 6 → ♈ 5	1,29	1,37	4/3		
	4	♈ 0 → ♋ 0	1,30	♈ 5 → ♋ 0	1,25	2, 9;20	5/6		
Ω	1	♋ 0 → ♉ 0	3, 0	♋ 0 → ♉ 0	3, 0	1,48;30	10/9	11,24	3,37
	2	♉ 0 → ♓ 0	1, 0	♉ 0 → ♓ 0	1, 0	2, 0;33,20	9/10	3;9,7,27,…	
	3	♓ 0 → ♉ 0	1, 0	♓ 0 → ♉ 0	1, 0	1,48;30	5/4		
	4	♉ 0 → ♋ 0	1, 0	♉ 0 → ♋ 0	1, 0	2,15;37,30	4/5		

System A_1 and the more complex relation to the last visibilities of System A_2. Since it seems odd that $\Delta\lambda$ would be known independently of Δt only for last visibilities, there was probably another reason, which may be an empirical correction of $\Delta t = \Delta\lambda + c$ to Δt found from observation. Or the reason may have something to do with bringing Σ and Ω in System A_2 into agreement with the same phenomena computed by "pushes" in System A_1, which are not considered here although they are taken up in our longer study. In our conversions for deriving the parameters, we shall use the integer $c = 4$ for both systems.

The functions for Mercury's heliacal phenomena are continuous in that they assign $\Delta\lambda$ and Δt between phenomena everywhere in the zodiac even though there are regions in which an entire morning or evening phase is omitted because the planet does not reach sufficient elongation from the sun to be visible. In such cases a morning or evening appearance, Γ or Ξ, cannot be observed, and likewise the following disappearance, Σ or Ω. Nevertheless, in a method of calculation that proceeds through successive phenomena, it is necessary to compute the longitude and time of each one without omission, even if not visible, in order to proceed to the next. Hence in the ephemerides all such omitted appearances are computed and said to "pass by," the same term used to describe omitted appearances in the Diaries, and the same is true of disappearances, whether found by pushes in System A_1 or independently in System A_2. The procedure text ACT 801.1–2 specifies for System A_1 that Γ "will pass by" from Υ 10° to ϑ 20° and Ξ from $\mathfrak{m}$ 30° to $\mathfrak{m}$ 5°, and the disappearances are said to "pass by, similarly" without specifying exactly where, although they must be at a greater longitude. Using the pushes, Neugebauer has estimated that Σ "will pass by" from Υ 24° to $\mathrm{I\!I}$ 5° and Ω from $\triangle$ 18° to $\mathfrak{m}$ 30°. What these limits are supposed to be for System A_2 is not known, and they are not necessarily the same, but we shall assume that they are similar, as appears from ACT 300a. In our figures for each system, the invisible regions are shown by a dashed line in the graph of the function.

Method of Analysis and Derivation: "Stepping" through the Function

For Saturn, Jupiter, and Mars, the length of each zone α_i exceeds the synodic arc $\Delta\lambda$, and thus directly $w_i = \Delta\lambda < \alpha_i$; the one exception is $\Delta\lambda_M$ of Mars, which extends through more than one zone, so that it was necessary to find both w_i and w_{i+1}. In the case of Mercury, for ten out of the fourteen zones in the four systems $w_i > \alpha_i$, which means that $\Delta\lambda$ must extend into a following zone and cannot be directly identified with w_i, nor can w_i be found directly from $\Delta\lambda$. However, when each of the functions is graphed, certain values of $\Delta\lambda$ appear, either for linear stretches or at inflection points of the graph, which occur where $\Delta\lambda$ first reaches into a following zone. These were first noted by Aaboe (1958), and it is our assumption that it was integer forms of these values of $\Delta\lambda$, found from Δt, that were used to derive w_i, along with further conditions, both observational and algebraic, for specifying the length of the zones α_i. For the other planets, we used only the extrema of $\Delta\lambda$ and Δt, but for Mercury we must use the values implied by the function itself, which always include the extrema and either one or two additional values.

What we wish to find are the successive values of w_i implied by each $\Delta\lambda$, which itself follows from Δt. In order to do this, we use a technique called "stepping" through the function, which we do one α_i and w_i at a time. In the few cases where $\alpha_i > w_i$, $\Delta\lambda$ may be contained in a single zone uninfluenced by following zones, so directly $w_i = \Delta\lambda$, just as for the other planets. If, however, $\alpha_i < w_i$, and thus $\Delta\lambda$ extends through two or more zones, our problem is to find the ratio w_{i+1}/w_i, if necessary rounded to a ratio of small, regular numbers, and from that w_{i+1}, such that w_i and w_{i+1} will produce $\Delta\lambda$. We did this once for Mars, and now we shall do it repeatedly for Mercury. There are two cases to consider depending upon where $\Delta\lambda$ begins and ends, which are illustrated in figure 4.

Case 1: $\Delta\lambda$ begins at the beginning of α_i and ends within α_{i+1}. Thus, we consider the effect of w_{i+1}/w_i on the part of $\Delta\lambda$ within α_{i+1}, that is,

$$\Delta\lambda = w_i + \left(\frac{w_{i+1}}{w_i} - 1\right)(w_i - \alpha_i) = \alpha_i + \frac{w_{i+1}}{w_i}(w_i - \alpha_i),$$

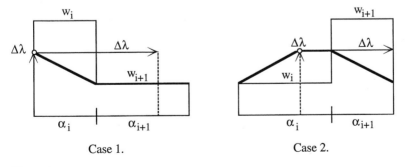

Case 1.   Case 2.

Figure 4
Mercury. Derivation of w_1

from which

$$\frac{w_{i+1}}{w_i} = \frac{\Delta\lambda - \alpha_i}{w_i - \alpha_i}, \quad w_{i+1} = \left(\frac{w_{i+1}}{w_i}\right)w_i. \tag{1}$$

Case 2: $\Delta\lambda$ begins within α_i and ends at the end of α_{i+1}. Here we initially consider the effect of w_i/w_{i+1} on the part of $\Delta\lambda$ within α_i, as though $\Delta\lambda$ ran backward from the end of α_{i+1} to within α_i, that is,

$$\Delta\lambda = w_{i+1} + \left(\frac{w_i}{w_{i+1}} - 1\right)(w_{i+1} - \alpha_{i+1}) = \alpha_{i+1} + w_i - \frac{w_i}{w_{i+1}}\alpha_{i+1}.$$

Note that, except for the ratio, the unknown w_{i+1} has dropped out of the right side of the equation, while the known w_i remains. It follows that

$$\frac{w_{i+1}}{w_i} = \frac{\alpha_{i+1}}{\alpha_{i+1} + w_i - \Delta\lambda}, \quad w_{i+1} = \left(\frac{w_{i+1}}{w_i}\right)w_i. \tag{2}$$

To save the curious reader the trouble of finding the required values for these many calculations, we shall show them numerically.

In order to apply this method, we shall first examine the function given by the ACT system to extract the values of $\Delta\lambda$ in linear stretches and at inflection points of the graph of the function. The precise values of some of these $\Delta\lambda$s are merely artifacts of the function, but most, in an integer form, provide the synodic arcs, following from their implied synodic times, that determine the successive values of w_i. In the case of Mercury, like that of Mars, the lengths of the zones α_i were assumed in advance, but whereas for Mars these were simply six equal zones of $1,0°$, those for Mercury show a greater diversity determined, I assume, on the basis of observational records of the approximate locations by zodiacal sign of the phenomena from which $\Delta\lambda$ and Δt begin. But the initial assumptions for the lengths of the zones are still simple multiples of $1,0°$ or $1,30°$, coinciding with integral zodiacal signs and corresponding approximately to the distances between the locations of linear stretches or inflection points in the graphs of $\Delta\lambda$. In addition, three of the four systems show adjustments to the lengths of the zones in order to approximate the exact period.

The reason for these adjustments is that, like the other planets, Mercury must maintain the three essential conditions of System A:

1. $\sum \dfrac{\alpha_i}{w_i} = \dfrac{\Pi}{Z} = P,$ 2. $\Sigma\alpha_i = 6,0°,$ 3. $\dfrac{w_{i+1}}{w_i} = \dfrac{p}{q},$

where p/q is a ratio of small, regular numbers. Condition 2 has been applied in selecting the length of the zones and condition 3 in finding the successive values of w_i. However, as was also true of Mars, in which α_i and w_i were likewise determined independently, condition 1 has not yet been fulfilled. Therefore, after the initial values of α_i and w_i have been found, condition 1 is applied in a series of adjustments, changing α_i by not more than 15°, one-half a zodiacal sign, or as little as 1°, or changing w_i by a smaller amount, in each case maintaining conditions 2 and 3, in order to approximate the exact period $P = 3;9,7,30$, which is reached only for Ξ while the other systems are exceedingly close.

To begin the procedure of stepping through the function, one must have an initial w_i. In all four systems there is a w_i lying between 1,46° and 1,48;30°, and in Γ, Ξ, Ω $\alpha_i > w_i$, so $w_i = \Delta\lambda$ directly, but in Σ we shall assume a w_i within this range, and we shall take these as w_1. We have renumbered the zones of Ξ from their order in ACT and HAMA so that w_1 falls in this range. For didactic reasons we shall examine the systems in the order in which we succeeded in working them out—although we shall not trouble the reader with our failed efforts or with alternative calculations that produce the same results—in this way going from the more straightforward to the more complex cases in which the final system depends upon adjustments related to the period. For this reason, we begin with what one normally takes as the last of the systems, System A_2, Ω.

SYSTEM A_2, Ω, LAST VISIBILITY IN THE EVENING

The graph of the function for Ω is shown in figure 5 and the parameters are given in table 3. The system and period,

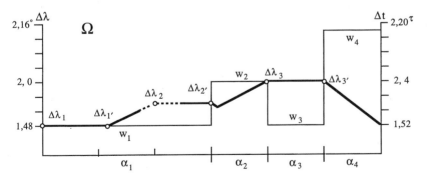

Figure 5
Mercury. System A_2. Ω

$$\sum \frac{\alpha_i}{w_i} = \frac{3,0}{1,48;30} + \frac{1,0}{2,0;33,20} + \frac{1,0}{1,48;30} + \frac{1,0}{2,15;37,30}$$

$$= \frac{11,24}{3,37} = 3;9,7,27,55\ldots,$$

and the true synodic arcs at the inflection points,

$$\Delta\lambda_1 = \Delta\lambda_{1'} = 1,48;30°, \quad \Delta\lambda_2 = \Delta\lambda_{2'} = 1,54;30°,$$

$$\Delta\lambda_3 = \Delta\lambda_{3'} = 2,0;30°.$$

The small dip, of $0;36,40°$, at the beginning of α_2 is an artifact of the function—$\Delta\lambda$ extends far into α_3 but has not yet reached α_4—and another artifact, too small to illustrate, is that $\Delta\lambda_3$ begins $0;3,20°$ before α_3. These are of no consequence for the derivation or application of the system.

Letting $c = 3;30$, from $\Delta t = \Delta\lambda + c$, the implied synodic times are

$$\Delta t_1 = \Delta t_m = 1,52^\tau, \quad \Delta t_2 = \Delta t_\mu = 1,58^\tau, \quad \Delta t_3 = \Delta t_M = 2,4^\tau.$$

Then, by subtracting $c = 4$, we find $\Delta\lambda$ for the derivation,

$$\Delta\lambda_1 = \Delta\lambda_{1'} = 1,48°, \quad \Delta\lambda_2 = \Delta\lambda_{2'} = 1,54°, \quad \Delta\lambda_3 = \Delta\lambda_{3'} = 2,0°.$$

Note that the differences in Δt and $\Delta\lambda$ are exactly 6^τ and $6°$, and we have minimum, mean, and maximum synodic times and arcs, which is hardly obvious from the resulting values of w_i; something similar turns up in the initial assumption of Δt and $\Delta\lambda$ for Σ. It appears to be assumed in advance that there are four zones divided as $\alpha_1 = 3,0°$ and $\alpha_{2,3,4} = 1,0°$. The reason for selecting these divisions seems to be that they produce nearly equal stretches of constant $\Delta\lambda_1 < \Delta\lambda_2 < \Delta\lambda_3$ with nearly equal transitions between them, all about $1,0°$, which I assume was more or less given by the observational records of the distribution of Δt in the zodiac. Such a distribution of α_i can indeed do this, provided that $w_1 < w_2 < w_4$ and $w_3 = w_1$, both of which conditions turn out to be fulfilled. For this system alone all zones begin at the beginning of zodiacal signs, with no adjustments to the length of the zones, and with α_1 at ♋ $0°$.

Now, since $\alpha_1 = 3,0° > \Delta\lambda_1 = 1,48°$, we may let $w_1 = \Delta\lambda_1 = 1,48°$. And taking $\Delta\lambda_2$ to begin within α_1 and end at the end of α_2, with $w_1 = 1,48°$ and $\Delta\lambda_2 = 1,54°$, we find from (2),

$$\frac{w_2}{w_1} = \frac{1,0}{1,0 + 1,48 - 1,54} = \frac{10}{9} = 1;6,40, \quad w_2 = 2,0°.$$

Next, taking $\Delta\lambda_{2'}$ to begin at the beginning of α_2 and end within α_3, with $w_2 = 2,0°$ and $\Delta\lambda_{2'} = 1,54°$, from (1),

$$\frac{w_3}{w_2} = \frac{1,54 - 1,0}{2,0 - 1,0} = \frac{9}{10} = 0;54, \quad w_3 = 1,48°.$$

Then, letting $\Delta\lambda_3$ begin at the beginning of α_3 and end within (actually at the end of) α_4, with $w_3 = 1,48°$ and $\Delta\lambda_3 = 2,0°$, from (1),

$$\frac{w_4}{w_3} = \frac{2,0 - 1,0}{1,48 - 1,0} = \frac{5}{4} = 1;15, \quad w_4 = 2,15°.$$

Finally, to show that the function can return consistently to its beginning, letting $\Delta\lambda_{3'}$ begin at the beginning of α_4 and end within α_1, with $w_4 = 2,15°$ and $\Delta\lambda_{3'} = 2,0°$, from (1),

$$\frac{w_1}{w_4} = \frac{2,0 - 1,0}{2,15 - 1,0} = \frac{4}{5} = 0;48, \quad w_1 = 1,48°.$$

Thus, by the assumption of minimum, mean, and maximum values of Δt and thus $\Delta\lambda$, we have the system and period:

$$\sum \frac{\alpha_i}{w_i} = \frac{3,0}{1,48} + \frac{1,0}{2,0} + \frac{1,0}{1,48} + \frac{1,0}{2,15} = \frac{19}{6} = 3;10.$$

Here again the result is itself interesting, for as with the symmetrical system for Mars, we have reached a very short period, in this case the shortest useful period for Mercury, of 19 synodic cycles in 6 years + ~10 days. Since the three required synodic times Δt_M, Δt_μ, Δt_m could likewise be found within a short period—perhaps not of 6 years but surely of 13 or 20 years—we see that here too short periods of observation suffice to establish something close to the ACT system, which now requires only an adjustment for the period. Now, the exact period is taken to be 3;9,7,30, and the six-year period gives 3;10, in excess by 0;52,30. There are different ways of making the adjustment, and in the other systems for Mercury we shall see changes in α_i. But here, as α_i is in every case 1,0° or 3,0°, the adjustment could only have been made in w_i, as for Mars. Since the period is to be reduced, w_i must be increased, and so, presumably by trial, if w_1 is raised by 0;30° from 1,48° to 1,48;30°, and w_2, w_3, w_4 recomputed from the ratios of w_{i+1}/w_i just found, we shall have exactly the values of the ACT system, and the resulting period, $P = \Pi/Z = 11,24/3,37 = 3;9,7,27,55\ldots$, is now close enough to 3;9,7,30 to serve for all practical purposes. Thus, the slightly different period for Ω, as for the other systems for Mercury, is not an independent parameter, but simply as close an approximation to 3;9,7,30 as could reasonably be reached from the values of α_i and w_i derived from the observed Δt and computed $\Delta\lambda$.

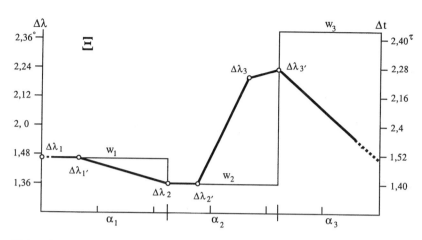

Figure 6
Mercury. System A_1. Ξ

System A_1: Ξ, First Visibility in the Evening

The function for Ξ is graphed in figure 6 and the parameters are shown in table 3. The system and period are

$$\sum \frac{\alpha_i}{w_i} = \frac{2,14}{1,46;40} + \frac{1,56}{1,36} + \frac{1,50}{2,40} = \frac{25,13}{8,0} = 3;9,7,30 \quad \text{(exactly)},$$

and the true synodic arcs at the inflection points,

$$\varDelta\lambda_1 = \varDelta\lambda_{1'} = 1,46;40°, \quad \varDelta\lambda_2 = \varDelta\lambda_{2'} = 1,36°,$$

$$\varDelta\lambda_3 = 2,20°, \quad \varDelta\lambda_{3'} = 2,23;20°.$$

The distinction of $\varDelta\lambda_3$ and $\varDelta\lambda_{3'}$ is again an artifact of the function, and it appears that the derivation began with a yet higher value. Let us assume that from observational records of true dates, evidently with some rounding,

$$\varDelta t_1 = 1,50^\tau, \quad \varDelta t_2 = 1,40^\tau, \quad \varDelta t_3 = 2,30^\tau,$$

and from $\varDelta\lambda = \varDelta t - 4$, we find for the derivation,

$$\varDelta\lambda_1 = 1,46°, \quad \varDelta\lambda_2 = 1,36°, \quad \varDelta\lambda_3 = 2,26°.$$

Let the zodiac initially be divided into three equal zones, $\alpha_{1,2,3} = 2,0°$, corresponding roughly to the distances between the locations of $\varDelta\lambda_1$, $\varDelta\lambda_2$, and $\varDelta\lambda_3$ in the figure.

To begin, since $\alpha_1 > \varDelta\lambda_1$ and $\alpha_2 > \varDelta\lambda_2$, we may estimate

$$w_1 = \varDelta\lambda_1 = 1,46°, \quad w_2 = \varDelta\lambda_2 = 1,36°,$$

of which the former will be adjusted. To find w_3, taking $\Delta\lambda_3$ to begin within α_2 and end at the end of α_3, with $w_2 = 1,36°$ and $\Delta\lambda_3 = 2,26°$, from (2),

$$\frac{w_3}{w_2} = \frac{2,0}{2,0 + 1,36 - 2,26} = \frac{12}{7} = 1;42,51\ldots \approx 1;40 = \frac{5}{3}, \quad w_3 = 2,40°.$$

This is the only ratio for all the planets that is not as $(n \pm 1)/n$, and it is as $(n + 2)/n$. Then to adjust w_1, with $w_3 = 2,40°$ and $\Delta\lambda_3 = 2,26°$, from (1),

$$\frac{w_1}{w_3} = \frac{2,26 - 2,0}{2,40 - 2,0} = \frac{13}{20} = 0;39 \approx 0;40 = \frac{2}{3}, \quad w_1 = 1,46;40°.$$

It follows that $w_2/w_1 = 1,36°/1,46;40° = 9/10$.

We have now reached the final values of w_i with the system and period

$$\sum \frac{\alpha_i}{w_i} = \frac{2,0}{1,46;40} + \frac{2,0}{1,36} + \frac{2,0}{2,40} = \frac{25}{8} = 3;7,30,$$

a very short although none-too-accurate period of 25 synodic cycles in 8 years − ~26 days, that falls short of 3;9,7,30 by 0;1,37,30. Here, rather than altering w_i, since no convenient values correct the period adequately, we increase the period by increasing α_1 by 15°, one-half a zodiacal sign, a "step" in later Greek astronomy, which we take from α_2 and α_3, now dividing the zodiac as

$$\alpha_1 = \tfrac{3}{5}(\alpha_2 + \alpha_3), \quad \alpha_1 = 2,15°, \quad \alpha_2 = 1,55°, \quad \alpha_3 = 1,50°.$$

Were these taken as the initial values of α_i, exactly the same values of w_i could be derived, but I think it far more likely that the initial values were all 2,0° and that these are adjusted. A shift of 15° to correct the period is also made for Γ. The system and period are now

$$\sum \frac{\alpha_i}{w_i} = \frac{2,15}{1,46;40} + \frac{1,55}{1,36} + \frac{1,50}{2,40} = \frac{10,5}{3,12} = 3;9,3,45,$$

which falls short of 3;9,7,30 by 0;0,3,45. Shifting 1° from α_1 to α_2, presumably by trial, gives the final system and the exact period 3;9,7,30, in fact, the only system for Mercury to reach the exact period. A consequence of the reduction of α_3 is to reduce $\Delta\lambda_3$ from 2,24° to 2,20° and $\Delta\lambda_{3'}$ from 2;26,40° to 2,23;20°, which also reduces Δt_3 accordingly by 3$^\tau$ or 4$^\tau$, and although the locations of computed values of $\Delta\lambda$ and Δt are shifted by as much as 15°, the specific values at each location differ little.

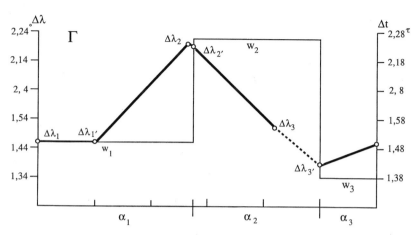

Figure 7
Mercury. System A_1. Γ

System A_1: Γ, First Visibility in the Morning

The function for Γ is shown in figure 7 and its parameters are given in table 3. The system and period are

$$\sum \frac{\alpha_i}{w_i} = \frac{2,45}{1,46} + \frac{2,14}{2,21;20} + \frac{1,1}{1,34;13,20} = \frac{44,33}{14,8} = 3;9,7,38,29\ldots,$$

and the true synodic arcs at the inflection points,

$$\Delta\lambda_1 = \Delta\lambda_{1'} = 1,46°, \qquad \Delta\lambda_2 = 2,19;30° \qquad \Delta\lambda_{2'} = 2,18;53,20°,$$

$$\Delta\lambda_3 = 1,50;50°, \quad \Delta\lambda_{3'} = 1,38;22,30°.$$

Because w_2 is so much greater than w_3, and α_3 is so short, there is a large difference between $\Delta\lambda_3$, which begins in α_2 and ends at the end of α_3, and $\Delta\lambda_{3'}$, which begins at the beginning of α_3 and ends in α_1. Curiously, $\Delta\lambda_3$ occurs just at the beginning of the region of invisibility for Γ and $\Delta\lambda_{3'}$ just at the end. Either $\Delta\lambda_3$ or $\Delta\lambda_{3'}$ is an artifact of the function, and either may be used for the derivation, as we shall see, although with $\Delta\lambda_{3'}$ it is necessary to work backward from w_1 to w_3, which may seem an odd way of proceeding.

Assume that from observational records of true dates the synodic times are

$$\Delta t_1 = 1,50^\tau, \quad \Delta t_2 = 2,24^\tau, \quad \Delta t_3 = 1,54^\tau, \quad \text{or} \quad \Delta t_{3'} = 1,42^\tau.$$

Note that Δt show a continuous decline from Δt_2 to $\Delta t_{3'}$, which suggests that $\Delta\lambda_3$ is the artifact of the computed function. The corresponding synodic arcs, from $\Delta\lambda = \Delta t - 4$, are

$$\Delta\lambda_1 = 1{,}46°, \quad \Delta\lambda_2 = 2{,}20°, \quad \Delta\lambda_3 = 1{,}50°, \quad \text{or} \quad \Delta\lambda_{3'} = 1{,}38°.$$

For the original length of the zones, we assume $\alpha_1 = 3{,}0°$, $\alpha_2 = 2{,}0°$, $\alpha_3 = 1{,}0°$, which are close to the distances between $\Delta\lambda_1$, $\Delta\lambda_2$, and $\Delta\lambda_{3'}$ in the figure.

Since $\alpha_1 > \Delta\lambda_1$, we let $w_1 = \Delta\lambda_1 = 1{,}46°$, and taking $\Delta\lambda_2$ to begin within α_1 and end at the end of α_2, with $w_1 = 1{,}46°$, $\Delta\lambda_2 = 2{,}20°$, and $\alpha_2 = 2{,}0°$, from (2),

$$\frac{w_2}{w_1} = \frac{2{,}0}{2{,}0 + 1{,}46 - 2{,}20} = \frac{1{,}0}{43} = 1{;}23{,}43{,}15\ldots$$

$$\approx 1{;}20 = \frac{4}{3}, \quad w_2 = 2{,}21{;}20°.$$

Then, letting $\Delta\lambda_3$ begin within α_2 and end at the end of α_3, with $w_2 = 2{,}21{;}20°$, $\Delta\lambda_3 = 1{,}50°$, and $\alpha_3 = 1{,}0°$, again from (2),

$$\frac{w_3}{w_2} = \frac{1{,}0}{1{,}0 + 2{,}21{;}20 - 1{,}50} = \frac{1{,}30}{2{,}17} = 0{;}39{,}24{,}57\ldots$$

$$\approx 0{;}40 = \frac{2}{3}, \quad w_3 = 1{,}34{;}13{,}20°.$$

Hence $w_1/w_3 = 9/8$. It follows that $\Delta\lambda_{3'}$ is given by

$$\Delta\lambda_{3'} = \alpha_3 + \frac{w_1}{w_3}(w_3 - \alpha_3) = 1{,}0° + \frac{9}{8}34{;}13{,}20° = 1{,}38{;}30°,$$

and with $w_3 = 1{,}34{;}13{,}20°$ and $\Delta\lambda_{3'} = 1{,}38{;}30°$, we may confirm from (1) that

$$\frac{w_1}{w_3} = \frac{1{,}38{;}30 - 1{,}0}{1{,}34{;}13{,}20 - 1{,}0} = \frac{9}{8} = 1{;}7{,}30, \quad w_1 = 1{,}46°.$$

Of course, this step is circular, but it does show the consistency of the system. If we wish to use $\Delta\lambda_{3'} = 1{,}38°$ as an initial value, we may work backward from w_1 to w_3, that is, since $\Delta\lambda_{3'}$ begins at the beginning of α_3 and ends within α_1, with $w_1 = 1{,}46°$ and $\Delta\lambda_{3'} = 1{,}38°$, from (2),

$$\frac{w_3}{w_1} = \frac{1{,}0}{1{,}0 + 1{,}46 - 1{,}38} = \frac{15}{17} = 0{;}52{,}56{,}28\ldots$$

$$\approx 0{;}53{,}20 = \frac{8}{9}, \quad w_3 = \frac{8}{9}w_1 = 1{,}34{;}13{,}20°.$$

I do not know how likely this last, backward, step is, but evidently it can be done, and it does seem reasonable to begin with the one $w_i < \alpha_i$ and work in both directions. In principle, one can compute in either direction, both here and in the other functions, although I will not tire the reader with the alternatives.

We now have the final values of w_i, but with the period

$$\sum \frac{\alpha_i}{w_i} = \frac{3,0}{1,46} + \frac{2,0}{2,21;20} + \frac{1,0}{1,34;13,20} = \frac{11,15}{3,32} = 3;11,2,15\ldots,$$

which exceeds 3;9,7,30 by the large amount of 0;1,54,45. The first change would appear to be a shift of 15°, one-half a zodiacal sign, as we also did for Ξ, from α_1 to α_2 to give

$$\sum \frac{\alpha_i}{w_i} = \frac{2,45}{1,46} + \frac{2,15}{2,21;20} + \frac{1,0}{1,34;13,20} = \frac{22,15}{7,4} = 3;8,54,54\ldots,$$

and the period is now too small by 0;0,12,36. Again, it would be possible to derive the same values of w_i with these as the initial values of α_i, but I think that unlikely. Shifting 1° from α_2 to α_3 will give the final system and period $P = \Pi/Z = 44,33/14,8 = 3;9,7,38,29\ldots$, with an error of +0;0,0,8,29, which, although the largest of any of the systems, was evidently considered satisfactory. These changes can produce differences of no more than about 1° and 1^τ in computing $\Delta\lambda$ and Δt, although, as for Ξ, the locations of specific values can be shifted by as much as 15°.

System A$_2$: Σ, Last Visibility in the Morning

The function for Σ is graphed in figure 8 and the parameters are listed in table 3. The system and period are

$$\sum \frac{\alpha_i}{w_i} = \frac{1,30}{1,47;46,40} + \frac{1,36}{2,9;20} + \frac{1,29}{1,37} + \frac{1,25}{2,9;20} = \frac{20,23}{6,28}$$

$$= 3;9,7,25,21\ldots,$$

and the true synodic arcs at the inflection points,

$\Delta\lambda_1 = \Delta\lambda_{1'} = 1,51;20°, \quad \Delta\lambda_2 = 2,3;46,40°, \quad \Delta\lambda_{2'} = 2,1°,$

$\Delta\lambda_3 = \Delta\lambda_{3'} = 1,39;40°, \quad \Delta\lambda_4 = 1,58;15°, \quad \Delta\lambda_{4'} = 2,1;56,40°.$

Unlike the four-zone system for Ω, which had three significant values of $\Delta\lambda$, Σ has four, that is, there are two nearly equal maxima given by $\Delta\lambda_2$

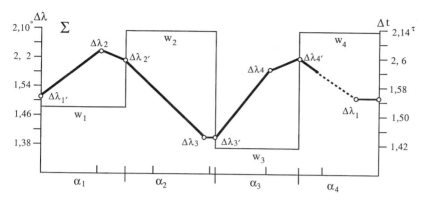

Figure 8
Mercury. System A_2. Σ

and $\varDelta\lambda_4$. Let us assume from observational records of true dates that

$$\varDelta t_1 = \varDelta t_\mu = 1{,}56^\tau, \quad \varDelta t_2 = \varDelta t_M = 2{,}8^\tau,$$

$$\varDelta t_3 = \varDelta t_m = 1{,}44^\tau, \quad \varDelta t_4 = 2{,}4^\tau.$$

Note that, as for Ω, we have a minimum, mean, and maximum $\varDelta t$ with equal differences, that is, $\varDelta t_1$ is a mean between $\varDelta t_2$ and $\varDelta t_3$, with differences of $\pm 12^\tau$, while for Ω $\varDelta t_2$ was a mean between $\varDelta t_3$ and $\varDelta t_1$, with differences of $\pm 6^\tau$. The difference of 4^τ between $\varDelta t_2$ and $\varDelta t_4$ is essential to the final values of the system. (In fact, the difference is far too small, for modern computation of Σ shows that the primary maximum $\varDelta t_2$ should be $2{,}12^\tau$ and the secondary maximum $\varDelta t_4$ $1{,}58^\tau$, so the correct difference is not 4^τ but 14^τ.) The synodic arcs, taken here as $\varDelta\lambda = \varDelta t - 4$, are

$$\varDelta\lambda_1 = 1{,}52^\circ, \quad \varDelta\lambda_2 = 2{,}4^\circ, \quad \varDelta\lambda_3 = 1{,}40^\circ, \quad \varDelta\lambda_4 = 2{,}0^\circ.$$

We shall initially assume four equal zones $\alpha_{1,2,3,4} = 1{,}30^\circ$, which are close to the final distribution and correspond approximately to the distances between the locations of $\varDelta\lambda_{1,2,3,4}$.

Now, Σ is the only system for which no $\alpha_i > w_i$, so no w_i appears directly as $\varDelta\lambda$, and we must therefore assume some initial value. And since all the other systems have one $w_i = \varDelta\lambda$ lying between $1{,}46^\circ$ and $1{,}48{;}30^\circ$, we shall initially assume for Σ that $w_1 = 1{,}48^\circ$, just as we originally took for w_1 of Ω, although there directly from $\varDelta\lambda$. Next, since $\varDelta\lambda_2$ is taken to begin within α_1 and end at the end of α_2, with $w_1 = 1{,}48^\circ$ and $\varDelta\lambda_2 = 2{,}4^\circ$, from (2),

$$\frac{w_2}{w_1} = \frac{1,30}{1,30 + 1,48 - 2,4} = \frac{45}{37} = 1;12,58\ldots$$

$$\approx 1;12 = \frac{6}{5}, \quad w_2 = 2,9;36°.$$

And taking $\Delta\lambda_3$ to begin in α_2 and end at the end of α_3, with $w_2 = 2,9;36°$ and $\Delta\lambda_3 = 1,40°$, we find from (2),

$$\frac{w_3}{w_2} = \frac{1,30}{1,30 + 2,9;36 - 1,40} = \frac{3,45}{4,59} = 0;45,9,1\ldots$$

$$\approx 0;45 = \frac{3}{4}, \quad w_3 = 1,37;12°.$$

Letting $\Delta\lambda_4$ begin in α_3 and end at the end of α_4, with $w_3 = 1,37;12°$ and $\Delta\lambda_4 = 2,0°$, from (2),

$$\frac{w_4}{w_3} = \frac{1,30}{1,30 + 1,37;12 - 2,0} = \frac{1,15}{56} = 1;20,21,25$$

$$\approx 1;20 = \frac{4}{3}, \quad w_4 = 2,9;36°.$$

Finally, to return to w_1, taking $\Delta\lambda_1$ to begin in α_4 and end at the end of α_1, with $w_4 = 2,9;36°$ and $\Delta\lambda_1 = 1,52°$, from (2),

$$\frac{w_1}{w_4} = \frac{1,30}{1,30 + 2,9;36 - 1,52} = \frac{3,45}{4,29} = 0;50,11,9\ldots$$

$$\approx 0;50 = \frac{5}{6}, \quad w_1 = 1,48°.$$

What remains appear to be three small adjustments. We now have the system

$$\sum \frac{\alpha_i}{w_i} = \frac{1,30}{1,48} + \frac{1,30}{2,9;36} + \frac{1,30}{1,37;12} + \frac{1,30}{2,9;36} = \frac{1,25}{27} = 3;8,53,20,$$

the period of which, 85 synodic cycles in 27 years − ~12 days, is short by 0;0,14,10. Since the period must be increased, reduce w_3 from 1,37;12° to 1;37° and recompute the remaining w_i by the ratios just found. The result is

$$\sum \frac{\alpha_i}{w_i} = \frac{1,30}{1,47;46,40} + \frac{1,30}{2,9;20} + \frac{1,30}{1,37} + \frac{1,30}{2,9;20}$$

$$= \frac{5,6}{1,37} = 3;9,16,42\ldots,$$

which is now too long by 0;0,9,12, but note that these are the final values of w_i. To reduce the period, shift 1° between two zones, as was done for Ξ, in this case from α_3 to α_2, to give

$$\sum \frac{\alpha_i}{w_i} = \frac{1,30}{1,47;46,40} + \frac{1,31}{2,9;20} + \frac{1,29}{1,37} + \frac{1,30}{2,9;20}$$

$$= \frac{20,23}{6,28} = 3;9,7,25,21\ldots,$$

which is in fact the final period. These changes can have only a minute effect on the computed $\Delta\lambda$ and Δt. But the system also shifts 5° from α_4 to α_2, the two zones with equal w_i, which therefore cannot affect the period. Why? The reason appears to be the difference of 4° between $\Delta\lambda_2$ and $\Delta\lambda_4$. Since this is the only effect of shifting 5° from α_4 to α_2, the only plausible reason is to raise $\Delta\lambda_2$ and lower $\Delta\lambda_4$ in keeping with a difference, presumably of 4° in $\Delta\lambda$ following from 4^τ in Δt, although the resulting differences are still not this great. That a matter of 4^τ should be of concern, when a similar difference was overlooked or actually introduced in the adjustment of the zones for Ξ—and in fact the correct difference in the primary and secondary maxima is 14^τ—may seem strange, but I can see no other motivation for the change.

Conclusion to the Analysis of Mercury and General Conclusion

The reader who has passed through, or "passed by," the multitude of calculations for deriving the parameters of Mercury may be pleased to know that there are but five steps:

1. Selection of the maximum, minimum, and either one or two additional, true synodic times Δt from the observational records of the dates of phenomena.

2. Computation of the true synodic arcs $\Delta\lambda$ from the true synodic times Δt by $\Delta\lambda = \Delta t - c$ where $c = 4$.

3. Determination of the lengths and locations of the zones α_i to the nearest 1,0° or 1,30°, based upon observational records of the locations of phenomena by zodiacal sign.

4. Derivation of the ratios w_{i+1}/w_i and the successive values of w_i through applying case 1 or case 2 as required to the true synodic arcs $\Delta\lambda$ and the zones α_i.

5. Adjustment of α_i, for Γ, Ξ, Σ, and of w_i, for Σ, Ω, while maintaining the ratios of w_{i+1}/w_i, to adjust the period as closely as practical to $P = \sum (\alpha_i/w_i) = 3;9,7,30$.

When set out in this way, the procedure may not seem too complex, and in any case I can see no simpler or more direct way of reaching the final parameters. Each initial Δt and α_i must be found from observational records of the dates and locations of the phenomena, each $\Delta \lambda$ found from Δt, and each w_i then found from $\Delta \lambda$ and α_i, which makes Mercury the most empirical, the most directly dependent upon observation, of all the planets for the derivation of its parameters, although it is possible that the synodic times of Venus are even more directly empirical.

Of the five steps, I regard 3–5 as more secure, but 1 and 2 as less secure. Step 3, a partitioning of the zodiac into 2, 3, 4 or 6 zodiacal signs, multiples of $1,0°$ or $1,30°$, is entirely in keeping with the observational records, specifying only the zodiacal sign of phenomena. Step 4, the most complex, is, as far as I can tell, the most direct way to derive w_{i+1}/w_i and w_i from $\Delta \lambda$ and α_i; it must be done one step at a time, there is no simpler method. If anything, it is too simple and direct, for our own derivations may well be only a small part, the final part, of a much longer procedure involving the trial and adjustment of different values of $\Delta \lambda$. The various adjustments in step 5, which are not as numerous as they may seem, make sense of the peculiar length of the zones, differing from 2, 3, 4, or 6 zodiacal signs by $\pm 1°$, or by $5°$, $10°$, or $15° \pm 1°$, and also show why the exact period $P = 3;9,7,30$ is only closely approximated in three of the four systems by variants, which should therefore be regarded as exceedingly good approximations rather than independent, inconsistent periods. (Admittedly, for Σ the shift of $5°$ from α_4 to α_2 is not an adjustment to the period, but to $\Delta \lambda_2$ and $\Delta \lambda_4$.) But step 2, the derivation of $\Delta \lambda$ from the observational record of Δt by $\Delta \lambda = \Delta t - c$ where $c = 4$, is less certain, at least for System A_2, by which Σ and Ω are computed. For in System A_2, $\Delta t - \Delta \lambda$ is not constant but varies, for Σ from 0 to 6 and for Ω from 0 to from 10, as far as the relation is known from the ephemerides ACT 300a–b. It is possible that the greater complexity of the conversion from $\Delta \lambda$ to Δt in System A_2 is an empirical correction for reconciling the computed Δt, although not $\Delta \lambda$, with observed Δt in different parts of the zodiac, or with the dates of Σ and Ω computed with pushes in System A_1, and had no role in the derivation of the parameters. For it is hard to believe that for Mercury's last visibilities alone $\Delta \lambda$ would be known independently of Δt. Hence for the conversions I have used the same constant $c = 4$ as for

System A_1 because I can see no alternative to the use of some constant, and if this was not 4, it was something very close.

The most interesting question concerns step 1, the selection of the required values of Δt from the observed true synodic times. This is also the most complex question, and it is taken up in detail in our longer study in which we also consider the empirical basis of the derivations for all the planets. Here I will only summarize the results. For System A_1, first visibilities Ξ and Γ, the selected values are close to those given by observation, with the functions representing the entire range of the phenomena quite well, particularly for Ξ, both according to the ancient observations and by modern theory. For System A_2, last visibilities Σ and Ω, the surviving ancient observations are themselves very poor, and, although the general contours of the functions follow the phenomena according to modern computation somewhat better, System A_2 is notably inferior to System A_1. But it is for Σ and Ω that the derivations are based upon minimum, mean, and maximum values of Δt with differences of 12^τ for Σ and 6^τ for Ω, with an additional secondary maximum for Σ, all of which appears more schematic and less empirical than Ξ and Γ, as though recognizing some inherent imprecision in the dates of last visibilities, which may also have something to do with the inconsistency of the conversion between $\Delta\lambda$ and Δt in System A_2.

Much the same holds for comparison of the other planets with observation, that is, the ancient observations of heliacal phenomena in the Diaries are not very accurate—errors of two or three days in the synodic times are common and some errors are much larger—and yet the functions are better than one would expect, indeed, better than the parameters themselves would indicate, in *more or less* following the zodiacal distribution of synodic times according to modern theory. The emphasis on "more or less" is important. It appears to me that Babylonian planetary theory was *intended* as an approximation to phenomena too complex and irregular to compute with absolute precision—in part due to the inaccuracy of the observations, in part to the theory—and the theory itself based upon a number of deliberate approximations for the purpose of computation. First, heliacal phenomena take place with respect to the true sun, as is obvious for first and last visibilities, so the use of the mean sun is itself an approximation. Then, elongations, although falling within some range for each phenomenon, are not fixed, so this too is an approximation. Even with the phenomena occurring at fixed elongations from the mean sun, the use of a constant difference between synodic arc and time is an approximation. That the same functions with the same parameters are used for all phenomena of each superior planet (except for Mars's retrograde arc) is an approximation. The preselected zones α_i of Mars and Mercury are approximations. The adjustment of the zones of Mercury to fit the correct period is an approximation. The restriction of w_i so that successive values will be in ratios of regular numbers, in order to preserve the period, is an approximation. The derivation of the limits of System B so that $2\Delta/d = P$, likewise to preserve the period, is an approximation. And on and on, some larger and some smaller, but all cumulative in their effect as approximations. The same must hold true for the selection of the values of Δt used for the derivations from the synodic times given directly by observation, for the selection is itself an approximation to something more complex and irregular, in part due to errors of observation.

However, it is not specifically the errors in the observations that determine the parameters and functions or account for their limitations as approximations to the true phenomena, for a *selection* of Δt from modern computation of phenomena would give similar, although not identical, parameters and functions, from which only the selected values of Δt would appear accurately, and the rest of the function would still be an approximation. In fact, the most important limitation of the systems is theoretical rather than empirical. For Mercury in particular the computation of the synodic arc or time can be considered as the integration by summation of a function of two or at the most three discontinuous steps through one synodic period. The language here is modern, but the method is ancient, at least in mathematical astronomy. In principle, the Babylonian methods are startlingly modern, analogous to integrations of the equations of motion. If the zodiac were divided into 60 or 120 or 360 zones with very close but still discontinuous $\Delta\lambda$ and Δt in each, and w_{i+1}/w_i still a ratio of regular numbers, which could hardly remain small, it would be possible to integrate the motion and time through the synodic period with very great precision, but the computation would be horrendous. The Babylonians understood the behavior of their discontinuous functions perfectly well, and I have no doubt that they were aware of the possibility of using a great number of zones and of the obstacle of carrying out the calculation. Hence for this reason too, they must have seen their own theory as an approximation, in this case to a completely impractical computation.

There are three further parameters yet to be derived. One is the alignment of each function with the zodiac, analogous to finding the direction of an apsidal line, the second is the epoch, an initial position at a given time, and the third the subdivision of the synodic arc and time, that is, the arc and time between adjacent phenomena in a single synodic period. These are treated in detail in our longer study, and here we merely summarize what we have found. The alignment with the zodiac was taken from the locations of the phenomena by zodiacal sign in the observations of the sort recorded in the Diaries, that is, particular synodic times occur in particular parts of the zodiac, and these can be used to locate the zones of System A and the maximum and minimum synodic arcs and times in both systems. However, System B has its complications in that, because it is a function not of longitude but of the number of each phenomenon in the period, the zodiacal locations are not fixed but have periodic variations about a mean position. Great precision was evidently neither intended nor possible. Thus, in System A the zones of Mars are each two zodiacal signs with the slow zone beginning at ♋ 0°, and the original limits of the zones of Mercury in Table 3 are also integral zodiacal signs. For Saturn the fast zone begins at ♓ 0° and for Jupiter at ♐ 0°, with the length of the zones determined algebraically. The midpoints of the slow zones for the superior planets are: Saturn ♏ 20°, Jupiter ♍ 12;30°, Mars ♌ 0°, although the locations for Saturn and Jupiter result from adjusting one zone to begin at the beginning of a zodiacal sign and the original intention was probably Saturn ♏ 15°, Jupiter ♍ 15°. In System B, the mean locations of the minimum synodic arcs and times, $\Delta\lambda_m$ and Δt_m, and the variations on either side are: Saturn ♏ 10° ±5° and other close values—♏ 15° is probably intended—Jupiter ♍ 15° ±6;45°, Mars ♌ 20° ±29;30°, although there is a variant for Saturn of ♐ 15°, and the great variation for Mars may be a reason that ephemerides for System B are hardly known.

The epoch may be found from the solar-distance principle, that each phenomenon takes place at a fixed, characteristic elongation from the mean Sun. Thus, for the observed time of any phenomenon, the longitude of the phenomenon and of the planet λ_p, is equal to the longitude of the mean sun λ_s, presumed known by computation as a function of time, plus the fixed elongation η, $\lambda_p = \lambda_s + \eta$, and no measurement of position is required. For System B, λ_p must also be associated with some value of $\Delta\lambda$ in the function, a matter of some complexity and uncertainty. That the planetary ephemerides are in general not connectable suggests that the epochs may have been periodically redetermined, requiring only the date of a single phenomenon—my guess would be first visibility— since no other parameters are changed. The ephemerides imply variations in these elongations, which are artifacts of computation far smaller than their true variations according to modern theory, but the elongation for determining the epoch must be a mean value since it is then locked in as a mean about which the implied elongations vary.

Finding the elongations, however, from the surviving sources is not exactly easy or secure. By convention, we take elongations of morning phenomena, Γ first visibility and Φ first station, to be negative, at a smaller longitude than the mean sun, and of evening phenomena, Ψ second station and Ω last visibility, to be positive, at a greater longitude than the mean sun. In the case of Mars, the procedure text ACT 811a gives the change of elongation $\Delta\eta$ in the intervals $\Omega \rightarrow \Gamma$, $\Gamma \rightarrow \Phi$, $\Phi \rightarrow \Omega$, from which, if it is assumed that Ω and Γ are symmetrical to conjunction with the mean sun, an assumption we also make for the other planets, the elongations of Ω and Γ are $\pm 15°$ and of Φ $-2,0°$. Ψ is computed from Φ and does not have a fixed elongation. The elongations of Mars were first noted by van der Waerden (1957) as evidence for the solar-distance principle. For Saturn and Jupiter it is necessary to use procedure texts for the subdivision of the synodic arc and time and dateless longitude templates in System A—there is no equivalent for System B. From a tedious analysis of inconsistent sources we have found the mean elongations to be, *approximately*: Saturn, Ω and Γ $\pm 14°$, Ψ and Φ $\pm 2,2°$; Jupiter, Ω and Γ $\pm 11°$, Φ $-1,53°$, Ψ $+1,58°$; each subject to small variations between the fast and slow zones. The departures of Ψ and Φ from $\pm 2,0°$ exactly are probably artifacts of computation rather than empirical, and thus for the superior planets the underlying assumption appears to be that stations take place $\pm 120°$ from the sun, trine aspect, a convention reported by Vitruvius (9.1.11) and Pliny (2.59), as noted in HAMA (411 n.11). The subdivision of the synodic time δt between phenomena is found empirically as rather crude mean values. The subdivision of the synodic arc $\delta\lambda$ is found from the time δt in which the mean sun moves a distance $\delta\lambda_s = \delta t \cdot v_s^{\circ/\tau}$, the mean velocity of the sun, and the change of elongation $\delta\eta$ by $\delta\lambda = \delta\lambda_s - \delta\eta$. The values of $\delta\lambda$ are then adjusted so that their sum in any zone will equal the synodic arc w_i, which precludes recomputing them exactly without assuming specific adjustments, although one can come quite close. For Mercury, by an even more tedious statistical analysis using the ephemerides directly, taking $\delta\lambda$ from the ephemerides and computing $\delta\lambda_s = \delta t \cdot v_s^{\circ/\tau}$, we find $\delta\eta = \delta\lambda - \delta\lambda_s$ and, again assuming symmetry to conjunction, we find that the mean values of the elongations are *approximately*: System A_1, Σ and Ξ $\pm 14°$, Ω and Γ $\pm 13;30°$; System A_2, Σ and Ξ $\pm 15°$, Ω and Γ $\pm 14°$, with a notable variation, particularly for System A_2, although still far smaller than the true variation by modern theory. It should, however, be noted that the differences may not be significant, that is, all elongations could be assumed to be 14° or even 15° for finding the epoch, and the results for System A$_2$ are very insecure since ACT 300a has at most sixteen restorable lines for Σ and Ξ and any errors in time cannot be controlled.

The method we have set out for deriving the parameters of the ephemerides, although applied in various ways under specific conditions, both algebraic and for preserving the period, is consistent in its empirical foundation, the dates of heliacal phenomena and location by zodiacal sign, exactly the observational records in the Diaries and related collections. The dates are used to determine the maximum and minimum synodic times and, for Mercury, one or two additional synodic times, and the difference between synodic time and arc is then taken as constant, as in the ephemerides. It appears to us that this relation is the very foundation of Babylonian planetary theory, and the reason is that while the time of phenomena was specified by observation quite precisely to the day—although the observations in the Diaries are not nearly so accurate—longitude was estimated only roughly, by zodiacal sign or by beginning or end of zodiacal sign, and thus synodic arc could not be found by direct observation independently of synodic time with sufficient precision to take as the basis of planetary theory. It is for this reason that time is the principal independent variable, and the mean motion of the sun is the measure of synodic arc, both in the derivation of parameters and implicitly throughout Babylonian planetary theory. This is of some interest since taking time as the independent variable is the principle underlying the use of mean motion, a linear function of time, with corrections for computing positions of planets and satellites, whether in the tables of Ptolemy or Kepler or Leverrier. And time is also the independent variable for the variation of parameters in the perturbation theory of modern celestial mechanics, so it can be seen that, either directly or indirectly, the principle first used by the Babylonians has had profound and enduring consequences.

ABBREVIATIONS

ACT   Neugebauer, O. *Astronomical Cuneiform Texts*

ADT   Sachs, A., and H. Hunger. *Astronomical Diaries and Related Texts from Babylonia*

DCL   Aaboe, A., and A. Sachs. *Some Dateless Computed Lists of Longitudes*

HAMA   Neugebauer, O. *A History of Ancient Mathematical Astronomy*

JCS   *Journal of Cuneiform Studies*

References

Aaboe, A.
1958. On Babylonian Planetary Theories. *Centaurus* 5, 209–77.
1964. On Period Relations in Babylonian Astronomy. *Centaurus* 10, 213–31.
1980. Observation and Theory in Babylonian Astronomy. *Centaurus* 24, 14–35.

Aaboe, A., A. J. Sachs.
DCL. Some Dateless Computed Lists of Longitudes of Characteristic Planetary Phenomena from the Late Babylonian Period. *JCS* 20 (1966), 1–33.

Kugler, F. X.
1907–10. *Sternkunde und Sterndienst in Babel* I–II. Münster in Westfalen.

Neugebauer, O.
ACT. *Astronomical Cuneiform Texts.* 3 vols. London, [1955], rpr. New York, 1983.
HAMA. *A History of Ancient Mathematical Astronomy.* 3 pts. New York, 1975.
1954. Babylonian Planetary Theories. *Proc. Am. Phil. Soc.* 98, 60–89.
1968. The Origin of "System B" of Babylonian Astronomy. *Centaurus* 12, 209–14.

Neugebauer, P. V.
1938. Tafeln zur Berechnung der jährlichen Auf- und Untergänge der Planeten.
Astronomische Nachrichten 264, 313–22.

Sachs, A. J.
1948. A Classification of the Babylonian Astronomical Tablets of the Seleucid Period.
JCS 2, 271–90.
1952. A Late Babylonian Star Catalog. *JCS* 6, 146–50.
1974. Babylonian observational astronomy. *Phil. Trans. R. Soc, Lond.* A 276, 43–50.

Sachs, A. J., and H. Hunger.
ADT. *Astronomical Diaries and Related Texts from Babylonia.* I (1988), II (1989), III
(1996). Denkschr. d. Öster. Akad. d. Wiss. Phil.-Hist. Kl. 195, 210, 246.

Swerdlow, N. M.
1998. *The Babylonian Theory of the Planets.* Princeton.
1999. Acronychal Risings in Babylonian Planetary Theory. *Archive for History of Exact
Science* 54, 49–65.

Tuckerman, B.
1962. *Planetary, Lunar, and Solar Positions 601 b.c. to a.d. 1 at Five-day and Ten-day
Intervals.* Mem. Am. Phil. Soc. 56.

van der Waerden, B. L.
1957. Babylonische Planetrechnung. *Vierteljahrsschrift der Naturforschenden Gesellschaft
in Zürich* 102, 39–60.
1966. *Die Anfänge der Astronomie. Erwachende Wissenschaft II.* Groningen.

10

A CLASSIFICATION OF ASTRONOMICAL TABLES
ON PAPYRUS

Alexander Jones

1 SUMMARY

After Mesopotamia, Egypt is proving the second most important source of
archeologically recovered astronomical texts from antiquity. The principal
modern collections of papyri from Greco-Roman Egypt are mostly the
result of excavations and purchases from the decades surrounding the turn
of the century.[1] But although the heyday of the recovery of papyri lagged
only one or two decades behind that of the cuneiform tablets of Mesopotamia, it was only with O. Neugebauer's studies of astronomical papyri
from the late 1940s to the late 1980s that their historical importance began
to be appreciated.

Various causes, however, have long stood in the way of a systematic
search for such texts. They were spread thinly among numerous, widely
scattered collections. Inventories of unpublished papyri were scarce. And
until a body of representative type-specimens had accumulated, even to
know an astronomical fragment when one saw it often called for a degree
of familiarity with astronomy and its history that a papyrologist could not
be expected to possess. The still fundamental 1959 edition of horoscopes
by Neugebauer and H. B. van Hoesen showed how much we might learn
by studying even a category of documents that were meager in technical
content but that could be hunted out in the world's collections with some
claim to thoroughness because they were easy to recognize.[2] But astronomical papyri in the strict sense of the term came to light slowly and
sporadically, and the sixty or so that Neugebauer knew gave the impression of a confusing variety of formats and contents of text with few
instances of repetition.

In 1990, through the kindness of the Egypt Exploration Society, I
obtained access to the trove of Greco-Roman papyri excavated between
1896 and 1906 by B. P. Grenfell and A. S. Hunt at the site of ancient
Oxyrhynchus. Even after sixty-six published volumes containing more
than 4,000 papyrus texts, the unpublished collection kept at the Ashmolean

Museum remains enormous, and a card inventory categorizes tens of thousands of the more substantial fragments by type or subject. It became swiftly apparent that this collection included unpublished fragments of astronomical texts and tables far outnumbering the entire hitherto published corpus. When tiny bits preserving only tabular ruling or traces of one or two numerals had been excluded, there remained more than a hundred new astronomical fragments (and about seventy new horoscopes), which I undertook to edit.[3]

Such a sudden increase in the number of documents available for study cannot help bringing with it a change in the way we study them. Where Neugebauer perforce treated each papyrus as an original problem of interpreting data of unknown meaning (and many of his articles are small classics of this genre), it is now often possible to identify the contents of a papyrus through its resemblance to known examples or its conformity to well established conventions. In fact, we are now entering the stage of scholarship, very stimulating for the modern student, where enough digested matter exists to be a real help in understanding new texts, but not so much that the new texts turn out to be mere repetitions of what has gone before.

Tables account for by far the majority of the known astronomical papyri, and among the insights to be gained from the new Oxyrhynchus fragments is the recognition of recurring, well-defined varieties and formats of table. In this paper I present a classification and nomenclature that I hope will be useful for the future identification, description, and analysis of astronomical tables. An exercise of this kind can never be definitive: the next hundred papyri to come to light may transform a present anomaly into a standard format. I have taken the opportunity to draw attention to instances where comparison with analogous texts has made it possible to extend or correct interpretations by Neugebauer and others (including myself) of previously published papyri; I have also slipped in some *obiter dicta* about the original applications of the various kinds of table and the underlying methods of computation.[4]

2   GENERAL DESCRIPTION OF THE DOCUMENTS

For convenience one often refers to the documents obtained from Egyptian sites collectively as papyri, but in fact other media are represented. The one that potentially most closely resembles papyrus in its uses, parchment, was rarely used for astronomical purposes. Ostraca were cheap but gave little space for writing, and so were most suitable for short instructional texts, numerical jottings, and horoscopes. On the other hand, wooden tablets provided a handy and sturdy medium for tables that had to be consulted often, and we have examples of almanacs in this form.

Papyrus was manufactured in square or oblong sheets, and these could be used in several ways. Usually sheets would be joined to form a roll, which was by preference inscribed on the side that had its reed fibres running horizontally, although frequently the back side was later used for another text. The practice of binding sheets into a codex, which in general became common only in late antiquity, is often attested for astronomical tables—not surprisingly, given the awkwardness of a roll for ready reference. Finally, odd scraps of papyrus, sometimes cut from discarded documents, could hold shorter texts, tables, and especially horoscopes.

No known astronomical papyrus antedates the Macedonian conquest of Egypt, and among the few Hellenistic ones there are no tables or texts pertaining to tables.[5] Our earliest astronomical table is probably *P. Oxy.* 4175, an ephemeris covering parts of the sixth and seventh regnal years of Augustus (24 B.C.), and astronomical papyri in general start appearing with some frequency in the first century of our era, reaching a peak in the second and third centuries, and thereafter the numbers taper off. The latest datable tables, again ephemerides, are from the second half of the fifth century. Great care must be taken in drawing inferences from the (present) lack of known tables from outside these limits, or the apparent flow and ebb between them, for at least in its broad outlines this pattern of distribution seems to parallel the chronological distribution of non-astronomical papyri. Most of the astronomical texts and tables (including all the ones from Oxyrhynchus) are written in Greek, but there is a significant minority in Demotic Egyptian.

Greco-Roman papyri have been discovered in numerous Egyptian sites between the Nile delta (where the water table is too high for their survival) and the southern frontier, with especially rich finds in the towns of the Fayyum and in Oxyrhynchus, the capital of the ancient Oxyrhynchite nome. So far as we can tell from the scattering of astronomical tables and horoscopes, astrology was to be found throughout this area; the very high number of astronomical texts from Oxyrhynchus is perhaps just a measure of the overwhelming bulk of papyri of all kinds that were retrieved there. Papyri in many other modern collections were purchased from dealers, and are of uncertain provenance. The circumstance that the great part of our Roman-era papyri came from the ancient rubbish heaps of the towns means that the specific sites of astronomical or astrological activity can seldom be identified. We do know from ostraca and papyri found there that Egyptian temples in the Fayyum towns of Narmouthis (Medinet Madi) and Tebtynis were centers of astrological practice. But in general we are not dealing with archives from just one or two institutions with long-running scribal traditions, as was the case with the Babylonian astronomical tablets, so that it is not surprising to find less uniformity in the details of layout of the papyrus documents, written as they were during a span of more than five centuries.

The broadest categories of astronomical papyri are prose texts and tables. Very few of the texts are concerned with theoretical astronomy or observation, the majority being "procedure texts" directed to the production or use of the tables. The chief and almost unique goal of the astronomy represented in the Roman-era papyri was prediction: to find the zodiacal positions of the sun, moon, and planets and the disposition of the heavens at a given time in the past or future, and to determine the dates and circumstances of certain recurring phenomena such as eclipses and planetary appearances and disappearances. In turn these data were required for astrological prognostications and interpretations.[6] We may further classify many of the tables under two general headings: "almanacs" that register dated positions of the heavenly bodies and other data at regular intervals,[7] and "primary tables" that supplied the means by which the data in the almanacs as well as in the horoscopes were in the first instance computed. Both the almanacs and the primary tables occur in certain recurring varieties that are the subject of the ensuing classification. In addition, we have several isolated tables, some of them of unknown purpose, that do not as yet lend themselves to categorization.

With the exception of one or two fragments of theoretical treatises at one extreme and the personal horoscopes at the other, the astronomical papyri do not fit easily into the conventional dichotomy of papyrology into literary and documentary texts; hence they are usually assigned to a vague catch-all heading of "subliterary papyri," which embraces technical writings, but also magic and other ephemera, and in general those impersonal and unofficial texts that were valued in antiquity more for their content than for the precise means of expression. The handwriting (at least in the Greek papyri) is in accord with this ambivalent status: seldom a formal bookhand, still more rarely a true documentary cursive, but typically an intermediate "semiliterary" or "semidocumentary" hand comprising mostly disconnected letters but some characteristically documentary letter forms. Many of the numerical tables are carefully, even beautifully, executed, with neat rows and columns of entries in a red tabular framework. This kind of writing is often much more difficult to date on paleographical grounds than verbal texts, so that it is fortunate that a high proportion of the astronomical tables can be dated reliably from their contents.

3   NOTATIONS IN THE PAPYRI

The most common information contained in the tables consists of dates and positions of heavenly bodies. This section describes various ways in which these are expressed.

3.1   Dates

The standard specification of a date is by year, month, and day. Years in the tables are almost always the regnal years of the current Roman emperor, counted such that the entire calendar year during which his accession occurred is his year 1. Rather than the elaborate official formulae of the documentary papyri, astronomical tables usually name the relevant emperor in a short form before or beside the first of a series of regnal years. Normally only the names of emperors who were still alive at the end of their accession year occur in astronomical datings. The schematic regnal canon underlying this reckoning, which is essentially the one preserved in medieval copies of Ptolemy's *Handy Tables*, diverged in other ways from the historical realities of political chronology; for example Commodus's regnal years were counted retroactively from the accession year of his father Marcus Aurelius, and likewise Caracalla's from that of Septimius Severus. The regular institution of co-emperors from the late third century on resulted in a great complication of civil dating formulae; the horoscopes from this time mostly eschew these conventions by reckoning years from the accession of Diocletian, but it is not yet clear whether the same practice was followed in tables.

Months are usually identified by name, or abbreviated name, but sometimes by an ordinal number counted from the first month of the year. The two calendars frequently encountered in the papyri, which we conventionally refer to as the "Egyptian" and "Alexandrian," used the same twelve months of thirty days supplemented by additional "epagomenal" days at the end of the year. In the Egyptian calendar year there were always five epagomenals. Near the beginning of Roman rule in Egypt, the Alexandrian calendar was adopted as the civil calendar as a reform of the Egyptian, adding a sixth epagomenal to every fourth year.[8] Consequently an identical year number, month name, and day number in the two calendars referred to two dates, the interval between which grew by one day every four years from the institution of the reform. This ambiguity would not have caused difficulty if the Alexandrian calendar had completely superseded the Egyptian, as indeed it did in most contexts. Unfortunately, however, astronomers persisted in employing the Egyptian calendar because its invariable 365-day year simplified computations involving time intervals. There did exist verbal formulae to specify the calendar used in a dating ("according to the ancients" = Egyptian, "according to the Greeks" = Alexandrian), but these were never used in the astronomical tables. We have to deduce which calendar is meant in a given table from the agreement of the corresponding astronomical data with recomputation by modern theory, from the regularity of time intervals between recorded dates (which can reveal the presence or absence of intercalations), or, failing that, from the known practice in other tables of the same type. Since the Alexandrian year always began on August 29 or 30, a given Alexandrian month and day will always correspond (within one day) to the same Julian calendar date, whereas the counterpart of a given Egyptian date falls progressively earlier in the Julian calendar.

Day numbers ought, of course, to range from 1 to 30. Occasionally (on analogy with longitudes?) a scribe has written day 0, instead of the 30th of the preceding month. A given day number applies, without qualification, to the interval from sunrise to sunset, and sometimes to the ensuing night. The more common designation of a night is by an expression of the time "night of the xth into the $x + 1$th," which in some tables is represented simply by the numerals for the preceding and following day. Dates unaccompanied by an indication of time normally refer to some standard time of day, often the sunset ending the day in question (for computational purposes this is seldom distinguishable from 6 equinoctial hours past noon).

The usual "civil" expression for time of day is "hour x of day (or night)," where the hours are seasonal hours, that is, twelfths of the interval from sunrise to sunset or sunset to sunrise, counted from 1 to 12. The designation can in principle refer to any time during the hour in question, an ambiguity complicated by the occasional use of fractions. In certain tables, on the other hand, one encounters sexagesimal fractions following the day number (see the next subsection for the way these are written). These fractions can sometimes be meaningfully translated into times reckoned from the beginning of the day—whatever that may be. But in other cases, such as where the dates are of first or last visibilities of a planet, the time in question must be sunset or sunrise according the nature of the phenomenon, and the fractions are merely part of the "bookkeeping" involved in computing the dates.

3.2 Positions

Most of the recorded positions are longitudes along the ecliptic, reckoned in degrees. (Except in the latest tables, these are always sidereal longitudes, although this is not important for our present purposes.) The longitude is regularly broken down into zodiacal sign, degrees (counted from 0 to 29 or from 1 to 30), and, optionally, one or more places of sexagesimal fractions. The zodiacal sign may be named in full or abbreviated to its first letters. Alternatively, it may be specified by its ordinal number (1 through 12) counted from Virgo = 1; the reason for this convention may be that the sun is in Virgo at the beginning of the Alexandrian year. The equivalences are as follows:

1 Virgo	5 Capricorn	9 Taurus
2 Libra	6 Aquarius	10 Gemini
3 Scorpio	7 Pisces	11 Cancer
4 Sagittarius	8 Aries	12 Leo

The sexagesimal notation is most frequently encountered in expressing longitudes, but occurs throughout Greek mathematical astronomy (and, with the significant exception of Ptolemy's *Harmonica*, nowhere else in Greek). It is used only for fractions, the integer part of a quantity being written in the normal way. The fractional places are separated from the integer part and from each other by spaces. Zero in a fractional place or as the whole number is written with a special symbol that usually is written $\overline{\circ}$.

Celestial latitudes seem to have been of very slight interest to the practitioners of the astronomy in the papyri. The argument of latitude of the moon, however, was apparently of enough astrological significance in its own right to call for frequent computation. It is always measured from the northern limit of the moon's orbit, sometimes in degrees, but more often in special units called steps (*bathmoi*) such that 1 step = 15°.

4 PRIMARY TABLES

4.1 Epoch Tables

The chief problem in predicting the day-to-day motion of the heavenly bodies is to take account of their cyclic variations in speed. One approach to this problem is to predict the successive occurrences of one or more characteristic stages in the anomalistic cycle, and then to bridge the intervening intervals by interpolation or by applying a standard pattern of changing daily motion. We may name a table that lists epochs, that is, the dates and positions of one of the sun, moon, or planets at these characteristic moments, an "epoch table." Each epoch takes up one line of the table. Parallel columns of data may include the following:

(T).   The date of the event. The Egyptian calendar is used in all the known epoch tables except one.

(ΔT).   The interval of time (in days) between the preceding and current epoch.

(D).   The number of accumulated intercalary days between equivalent dates in the Egyptian and Alexandrian calendars is sometimes tabulated in a column preceding the Egyptian calendar date, to allow ready conversion of dates to and from the Alexandrian calendar.

(B).   The longitude of the relevant heavenly body.

(ΔB).   The interval of longitude between the preceding and current epoch. This column is not attested in any known fragment, but probably was present in those epoch tables that had a column for ΔT.

The epoch dates so far attested in epoch tables for the sun and moon are, respectively, summer solstices and dates of minimum daily motion. For the planets, in all known cases except *P. Oxy.* 4157a (for which see below), the epochs are the so-called "Greek letter" phenomena of their synodic cycles. In this paper we will use Neugebauer's notations for these, which are the following:

Mercury and Venus

Γ   First morning visibility

Φ   Morning station

Σ   Last morning visibility

Ξ   First evening visibility

Ψ   Evening station

Ω   Last evening visibility

Mars, Jupiter, and Saturn

Γ   First morning visibility

Φ   First (morning) station

Θ   Acronychal rising (opposition)

Ψ   Second (evening) station

Ω   Last evening visibility

The known examples of epoch tables are listed in table 1. It is worth remarking that before the discovery of the Oxyrhynchus fragments this format was represented only by a single specimen. All the epoch tables share this property, that the intervals in time and position from one tabulated epoch to the next are computed by arithmetical procedures of varying complexity. They are thus of considerable interest because they bring us closest to some of the methods and theoretical assumptions underlying the predictive astronomy of the papyri. The lunar tables are part of a skillfully crafted arithmetical system for computing lunar longitudes and arguments of latitude, which I call the "Standard Scheme."[9] Most planetary tables were computed using the "ACT" schemes familiar from Babylonian astronomy, in most cases adapted only minimally to accommodate the use of a calendar working with real days instead of the idealized lunar calendar of the Babylonian tables.[10]

As an illustration of one of the possible arrangements of an epoch table, table 2 presents part of an epoch table for Mercury's first morning appearance (Γ), computed for the years Septimius Severus 19–22 (= A.D. 210/211–213/214) according to System A_1. The columns for D and T are taken from $P. Oxy.$ 4153, with some restorations in the days and sexagesimal fractions. The other columns are entirely restored, and it is not certain that ΔT and ΔB were originally present. The months are represented in this translation by roman numerals in place of the abbreviated names in the papyrus. Table 3 shows a restoration of part of $P. Oxy.$ 4161, a table of Saturn's Γ for Tiberius 9 through Gaius 2 (= A.D. 22/23–37/38) computed according to System B. The five columns contain, respectively, the regnal year, the ordinal number of the month, the day (with no fractions), the ordinal number of the zodiacal sign, and the integer number of degrees.

Remarks on Individual Texts

$P. Oxy.$ 4148. Summer solstices. Only the dates are tabulated, because the longitude is assumed to be the same on all epochs. This table is discussed in Jones 1997b.

$P. Lund$ 35a. Lunar epochs. Published with photograph in Knudtzon and Neugebauer 1947. The traces visible in columns i and x, not identified by the editors, are the remains of a column for D.

$P. Oxy.$ 4149–4151. Lunar epochs. $P. Oxy.$ 4149 is the only known epoch table to give dates in the Alexandrian calendar. $P. Oxy.$ 4150 and 4151 both have part of the Standard Scheme template on the other side of the papyrus (in the case of 4151, this is published under a separate number, 4164). In

Table 1
Epoch tables

Text	Body	Years	Event	D	T	B	Other columns	Computation
P. Oxy. LXI 4148	Sun	161–237	Summer solstice	Yes	Yes	No		Constant year
P. Lund V 35a	Moon	58–108	Slowest motion	Yes	Yes	Yes		Standard Scheme
P. Oxy. LXI 4149	Moon	96–166	Slowest motion	No	Yes	Yes		Standard Scheme
P. Oxy. LXI 4150	Moon	187–198	Slowest motion	?	Yes	Yes		Standard Scheme
P. Oxy. LXI 4151	Moon	217–254	Slowest motion	Yes	Yes	Yes	Ω (vestigial)	Standard Scheme
P. Oxy. LXI 4152	Mercury	?	Ξ, Σ	?	?	Yes		System A_1, A_2
P. Oxy. LXI 4153	Mercury	206–215	Γ	Yes	Yes	?		System A_1
P. Oxy. LXI 4154	Mercury	?	Ω	?	Yes	?		System A_2
P. Oxy. LXI 4155	Mercury	?	$\Omega, \Gamma, [\Sigma]$	?	?	Yes	$T_\Gamma - T_\Omega,$ $T_\Sigma - T_\Gamma$	System A_1, A_2
P. Oxy. LXI 4156a	Mercury	?	Ω	?	Yes	Yes		System A_2
P. Oxy. LXI 4156b	Mercury	?	Γ	?	Yes?	Yes		?
P. Oxy. LXI 4157	Venus	?	?	?	Yes	?	ΔT	?
P. Oxy. LXI 4157a	Venus	106–121	Inferior conjunction	?	Yes	?	ΔT?	?
P. Oxy. LXI 4158	Mars	271–325	Φ, Θ?	?	Yes	?	ΔT	System A
P. Oxy. LXI 4159	Mars	225–235	Γ	?	Yes	?		System A
P. Oxy. LXI 4159a	Mars	102–133	Ψ	?	Yes	Yes		System A variant?
P. Oxy. LXI 4160 + P. Berol. inv. 16511	Jupiter	57–80	Γ?, Φ	No	Yes	Yes		System A′ variant
P. Oxy. LXI 4160a	Jupiter	6–13	Γ	?	Yes	?	ΔT	System B
P. Oxy. LXI 4161	Saturn	23–92	Γ	No	Yes	Yes		System B

Table 2

Epoch table for Mercury: *P. Oxy.* 4153, lines 12–22, with restored columns

D	T			ΔT	B	ΔB
	19	I	23;2	119;53,20	♋ 16;13,20°	118;13,20°
		V	8;51,10	105;49,10	♏ 0;22,30°	104;9,10°
		IX	6;38,40	117;47,30	♒ 26;30°	116;7,30°
59	20	I	8;42	127;3,20	♋ 1;53,20°	125;23,20°
		IV	22;43,40	104;1,40	♎ 14;15°	102;21,40°
		VIII	15;8,40	112;25	♒ 5°	110;45°
		XII	29;22	134;13,20	♊ 27;33,20°	132;33,20°
	21	IV	6;36,10	102;14,10	♍ 28;7,30°	100;34,10°
		VII	24;16,10	107;40	♐ 14;7,30°	106°
		XII	15;2	140;45,50	♊ 3;13,20°	139;5,50°
	22	III	20;28,40	100;26,40	♍ 12°	98;46,40°

some lines of 4151 the numerals purporting to be the lower order sexagesimal places of the epoch longitudes are in fact arguments of latitude (Ω) in steps, an intrusion suggesting an error in copying from an original that had a separate column for this quantity.

P. Oxy. 4152. "Greek letter" phenomena of Mercury. The epochs for Σ preserved on this fragment are actually mirror-reversed ink offsets on the back of the papyrus, transferred from a lost part of the same set of tables.

P. Oxy. 4155. "Greek letter" phenomena of Mercury. The remains of this table are in several small pieces. One group can be restored as a series of epochs of Ω, the other as epochs of Γ; these seem to have been in separate parts of the manuscript rather than laid out in parallel columns. In each group, there are slight traces of an unidentified column of sexagesimals to the left, then a column containing whole numbers, and then a column giving the epoch longitudes. The middle column turns out to give the number of days from the tabulated epoch to the next phase of Mercury; that is, for Ω it contains the number of days from Ω to Γ, and for Γ it contains the days from Γ to Σ. On the basis of the latter, we are able to reconstruct a sequence of longitudes of Σ.

P. Oxy. 4157–4157a. Epochs for Venus. The tables for Venus that have come to light so far are frustratingly fragmentary, and little can yet be said about the basis of computation. The evidence for the Babylonian treatment of this planet is notoriously poor, and seems to offer no help with these tables. In 4157a, a preserved heading identifies the epochs as "morning conjunctions," that is, (as the dates reveal) inferior conjunctions. This is the unique

Table 3

Epoch table for Saturn: *P. Oxy.* 4161, lines 1–16, with restorations

T		B		
9	8	12	7	11
10	8	26	7	24
11	9	10	8	8
12	9	24	8	21
13	10	8	9	5
14	10	23	9	19
15	11	7	10	3
16	11	21	10	17
17	12	5	11	1
18	12	19	11	14
19	13	2	11	27
21	1	11	12	10
22	1	24	12	23
Gaius				
1	2	4	1	5
2	2	20	1	17

exception to the rule that planetary epochs are observable "Greek-letter phenomena." Conjunctions may have been preferred for Venus in order to disentangle this planet's comparatively uniform pattern of longitudinal motion from the irregularities of its visibility phenomena.

P. Oxy. 4160 + P. Berol. inv. 16511. Γ and Φ of Jupiter. The longitudes and dates are computed by a previously unattested System A style scheme related to the Babylonian System A′ for Jupiter. The newly discovered Berlin fragment will be published in a forthcoming article by W. Brashear and A. Jones.

P. Oxy. 4161. Γ of Saturn. As noted above, the months and zodiacal signs in this table are represented by ordinal numbers. This is so far the only epoch table that uses this practice, which is common in almanacs.

4.2 Templates

"Templates" (cf. table 4) are tables that describe the progress of a heavenly body on the days following an epoch event. Such tables were evidently used in conjunction with epoch tables in order to compute the positions of the sun, moon, and planets on arbitrary dates. The majority of known templates count days continuously from one occurrence of an epoch event to the next occurrence of the same event (e.g., from Γ to Γ), but some planetary templates describe the planet's progress only during part of its synodic cycle (e.g., from Ψ to Ω). Like the epoch tables, most templates are computed according to arithmetical procedures. Here, however, the dependence on Babylonian astronomy is less direct; for although the representations of lunar daily progress by a linear zigzag function (as in the Standard Scheme) and of planetary motion by linear and second-order arithmetical sequences are already attested in cuneiform texts, it is in the papyri that we first find such sequences tabulated as a standard pattern of increments in longitude, to be added to an arbitrary epoch position. It is also noteworthy that, whereas all the lunar templates that we have are essentially copies of the same table, the other templates show considerable variations in structure and parameters, even when two pertain to the same body.

The columns of a template may include:

(T'). The number of days since epoch, increasing by steps of 1 day from 1, but generally recorded only at multiples of 5.

(B'). The increase in longitude since epoch in degrees.

(B). For the sun, the actual longitude is recorded instead of B', because the longitude is assumed to be always the same at epoch.

(ΔB). The increase in longitude since the preceding day.

(Ω'). The moon's increase in argument of latitude since epoch in steps.

Table 5 presents *P. Oxy.* 4166 as an illustration of a template; the whole of the first column for T' and parts of the others are restored. The planet in question is Saturn, with the leftmost three columns giving the end of the interval of motion from Ψ to Ω as a second-order sequence, and the remaining two giving the end of the shorter interval from Ω to Γ as a linear sequence. T' is not counted continuously from one interval to the next. Between Ψ and Ω, ΔB increases by constant $0;0,4°$. If we extrapolate backward, we find that for day 0 (which was probably not actually written in the table), $\Delta B = 0;0,23,27,16°$ and $B = 0°$ exactly. From Ω to Γ, ΔB is a constant $0;6,40°$ (not tabulated in the papyrus), and

Table 4
Templates

Text	Body	Epoch	Complete period?	T'	B'	ΔB	Other columns	Computation
P. Oxy. LXI 4162	Sun	Perigee	Yes	Yes	No	No	B	Trigonometric
P. Oxy. LXI 4163	Sun	Summer solstice	Yes	Yes	No	No	B	Linear sequences
PSI XV 1493	Moon	Slowest motion	Yes	Yes	Yes	No	Ω'	Standard Scheme
P. Oxy. LXI 4150	Moon	Slowest motion	Yes	Yes	Yes	?	Ω'	Standard Scheme
P. Oxy. LXI 4164	Moon	Slowest motion	Yes	Yes	Yes	Yes		Standard Scheme
P. Oxy. LXI 4164a	Moon	Slowest motion	Yes	Yes	Yes	Yes		Standard Scheme
P. Carlsberg 32 (demotic)	Mercury	Γ	?	Yes	Yes	Yes		2nd order sequence
P. Oxy. LXI 4217	Mercury	Γ	No	?	Yes	No		2nd order sequence
P. Oxy. LXI 4217a–c	Mercury	Ξ, Γ	No	?	Yes	No		2nd order sequence
P. Oxy. LXI 4165	Mars	Ψ	No	Yes	Yes	Yes		2nd order sequence
P. Berol. inv. 21236	Mars	Γ?	?	Yes	Yes	No		?
P. Oxy. LXI 4165a	Jupiter	Γ	?	Yes	Yes	Yes		2nd order sequence
PSI XV 1492	Saturn	Γ	Yes	Yes	Yes	Yes		2nd order and linear
P. Oxy. LXI 4166	Saturn	Ψ, Ω	No	Yes	Yes	Yes		2nd order and linear
E.E.S. inv. 79/1 (1)b	?	?	?	Yes	Yes	Yes		2nd order sequence

Table 5
From a Template for Saturn: *P. Oxy.* 4166 with restored columns

T'	ΔB	B'	T'	B'
90	0;6,23,27,16°	5;8,10,54,0°		0;46,40°
	0;6,27,27,16°	5;14,38,21,16°		0;53,20°
	0;6,31,27,16°	5;21,9,48,32°		1;0,0°
	0;6,35,27,16°	5;27,45,15,48°	10	1;6,40°
	0;6,39,27,16°	5;34,24,43,4°		1;13,20°
95	0;6,43,27,16°	5;41,8,10,20°		1;20,0°
	0;6,47,27,16°	5;47,55,37,36°		1;26,40°
	0;6,51,27,16°	5;54,47,4,52°		1;33,20°
	0;6,55,27,16°	6;1,42,32,8°	15	1;40,0°
	0;6,59,27,16°	6;8,41,59,24°		1;46,40°
100	0;7,3,27,16°	6;15,45,26,40°		1;53,20°
	0;7,7,27,16°	6;22,52,53,56°		2;0,0°
	0;7,11,27,16°	6;30,4,21,12°		2;6,40°
	0;7,15,27,16°	6;37,19,48,28°	20	2;13,20°
	0;7,19,27,16°	6;44,39,15,44°		2;20,0°
105	0;7,23,27,16°	6;52,2,43,0°		2;26,40°
	0;7,27,27,16°	6;59,30,10,16°		2;33,20°
	0;7,31,27,16°	7;7,1,37,32°		2;40,0°
	0;7,35,27,16°	7;14,37,4,48°	25	2;46,40°
	0;7,39,27,16°	7;22,16,32,4°		2;53,20°
110	0;7,43,27,16°	7;29,59,59,20°		3;0,0°
				3;6,40°
				3;13,20°
				3;20,0°

on day 0 again $B = 0°$. The total progress in the two intervals is 7;30° (effectively) and 3;20° respectively, both quantities being parameters associated with the "slow zone" of System A in Babylonian texts. It appears likely that this template was designed to be used with epochs for each of Saturn's "Greek letter" phenomena computed by System A, on the condition that the epoch longitude was in the slow zone. There would have been an analogous template associated with the fast zone.

Remarks on Individual Texts
P. Oxy. 4162. Template for the sun. This is the only template that is not based on arithmetical methods. The longitudes are computed trigonometrically from the Hipparchus-Ptolemy solar model, or something very like it.

Day 0 corresponds to the sun's passage of its perigee, at longitude Sagittarius 13;30°, which is equivalent to Ptolemy's perigee adjusted to the convention that the vernal equinoctial point is at Aries 8° instead of Aries 0°. This table and the next are discussed in Jones 1997b.

P. Oxy. 4163. Template for the sun. The scribe used the vacant space around the end of the table to copy a procedure text concerning the computation of syzygies, with an example for A.D. 259/260. Day 0 is summer solstice, at Cancer 0°. The longitudes in the preserved portion of the table are computed using long stretches of linear motion.

PSI 1493. Template for the moon. Unpublished; an edition with commentary by Neugebauer was prepared in 1964 for inclusion in volume 15 of the *Papiri della Società Italiana*, which has yet to appear. A partial translation was provided by Neugebauer 1975, 823. Like the other lunar templates *P. Oxy.* 4150, 4164, and 4164a, this table belongs to the Standard Scheme, for which see Jones 1996a. 4150 and 4164 are on the other sides of Standard Scheme epoch tables.

P. Oxy. 4217 and 4217a–c. Templates for Mercury. These are actually series of templates for Mercury's intervals of visibility, capable of bridging any interval of a whole number of days within certain limits. Because the columns progressively increase or decrease in length by 1, the general appearance of the table is triangular or trapezoidal. (The intervals of invisibility were probably represented by linear motion.)

P. Berol. inv. 21236. Template for Mars. Published in Neugebauer and Brashear 1976, where it was identified as an "ephemeris," that is, daily longitudes associated with specific dates. In fact it is clearly a template, with the conventional column for ΔT at five-day intervals, and an initial longitude of 0° exactly on day 0 (the longitude for day 1 should be read 0;34°, not 0;31°). The heading "Cancer" shows that the template is meant to be used with epochs in this sign—that is, in the slowest zone of the ecliptic for this planet—and the longitudes fit the beginning of the interval of motion starting with Γ.

PSI 1492. Template for Saturn. Unpublished, the prospects for publication being the same as for *PSI* XV 1493 above. Part of the table was translated in Neugebauer 1975, 791.

P. Oxy. 4166. Template for Saturn. The template is written on the back of a document. In the space between two columns of the document, in the same hand as the table, are a few lines of jottings or calculations involving the parameters of the template.

E.E.S. inv. 79/1 (1)b. Published in Jones 1998c. A badly damaged table containing a second-order sequence of sexagesimals in the format of a template. The astronomical meaning of the numbers has not been identified.

Table 6

Manuscripts of Ptolemy's *Handy Tables*

Text	Codex?	Tables preserved (in part) .
P. Oxy. LXI 4167	No	Oblique ascensions; anomaly table; planetary latitudes; stations; phases; greatest elongations
P. Oxy. LXI 4168	Yes	Declinations & lunar latitude; solar & lunar mean motions; parallax; planetary anomaly; unidentified supplementary tables
P. Oxy. LXI 4169	Yes	Planetary anomaly
P. Oxy. LXI 4170	Yes	Anomaly table; planetary mean motions
P. Oxy. LXI 4171	Yes	Oblique ascensions

4.3   Handy Tables and Other Kinematic Tables

Under this head we group various tables that do not strictly conform to a single format, but that, in Ptolemy's phrase, "exhibit the uniform circular motion" of kinematic models for the heavenly bodies through mean motion tables and anomaly tables. This is of course Ptolemy's own approach. And since Ptolemy published his mean motion and anomaly tables as part of a larger collection of tabular material (at first interspersed among the chapters of the *Almagest*, then independently as the *Handy Tables*) that by and large defined for a millenium of later astronomers what should go into a set of tables, we also include here any papyrus tables that are related to the other parts of the Ptolemaic corpus.

Fragments from several copies of Ptolemy's *Handy Tables* survive on papyrus (table 6). Of the major divisions of the collection as it is known to us from Byzantine manuscripts and the manuals of Ptolemy and Theon, almost all are represented on papyrus, including sections such as the tables for planetary "Greek-letter" phenomena that one suspects were seldom used. Most of these ancient manuscripts were codices, and the "look" of the page is often close to the medieval copies.

Remarks on Individual Texts

P. Oxy. 4167. This is the only known copy of the *Handy Tables* in roll form. It was a handsome manuscript, almost certainly in several rolls, for the 30 meters or so required for the full set of tables would have been unmanageable unless divided into smaller sections. The numerous extant fragments come from scattered parts of the *Handy Tables*: the spherical astronomy toward the beginning; tables for planetary latitudes, stations, greatest elongations, and visibilities towards the end; and a small fragment of one of the anomaly tables (it is not clear whether the one for the sun and moon or part of the planetary set). Aside from particular readings, the divergences from the medieval text are comparatively slight: omitted index columns in the oblique ascensions, and a superfluous column of zeros in the planetary latitudes. The fragments of 4167 were found in several folders of the unpublished Oxyrhynchus papyri, interspersed with many other bits of tables—almost certainly the remnants of an astrologer's library discarded *en masse*. Among the other tables in this archive were arithmetically-based epoch tables (4153–4155, 4159) and templates (4165, 4217a–c) as well as other kinematic tables (4169, 4173a) and almanacs covering dates from the early third to the first decade of the fourth century A.D.

P. Oxy. 4168. The tables preserved from this codex are, taken singly, almost identical to their counterparts in the medieval copies, but there were some differences in the sequence of tables. For example, the parallax tables begin on the verso of the table for solar declination and lunar latitude, whereas in the medieval manuscripts the eclipse tables and others come between. Other bits in the same hand, and presumably from the same manuscript, do not belong to the *Handy Tables* as we know it; one of these was part of a geographical table of uncertain purpose and structure.

P. Oxy. 4170. Seemingly part of a page from a codex; but if so, the arrangement of the tables was peculiar. On one side is the *second* page of the planetary mean motions (25-year intervals for Mars, Venus, and Mercury); on the other, the last page of an unidentified anomaly table.

While the *Handy Tables* clearly had a wide diffusion, Ptolemy's earlier versions of the tables in the *Almagest* have not so far turned up on papyrus (nor have any fragments of the text of the *Almagest* been identified). There are, however, papyrus tables that are based on the astronomical parameters set out in the *Almagest*, but presented in a form that is not known to be due to Ptolemy, as well as tables that present apparently non-Ptolemaic data within the distinctive structure of certain of Ptolemy's tables (table 7). Leaving aside the improbable hypothesis that these were versions Ptolemy himself published but that did not survive through the medieval manuscript tradition, either the resemblances are due to Ptolemy's adhering to established tradition, or they reflect others' attempts to improve on Ptolemy. In my opinion the latter is the more likely explanation for all but the chronological tables.

Remarks on Individual Texts

P. Ryl. 522/523. (Photograph of 522 in plate 4 of *P. Ryl.* III.) One side of this fragment of a codex page, published as *P. Ryl.* 522, contains part of the first two columns from a table of noteworthy cities, essentially the one of the *Handy Tables* but with variants from the (itself rather unstable) medieval text.

Table 7
Other tables related to Ptolemy's tables

Text	Codex?	Tables preserved (in part)
P. Ryl. III 522/523	Yes	Noteworthy cities; oblique ascensions
P. Lond. III 1278	Yes	Geographical table; geographical latitudes and longest daylight; right ascensions and equation of time; oblique ascensions; planetary mean motions, with column of unknown significance
P. Oxy. LXI 4173	Yes	Planetary mean motions
P. Oxy. LXI 4173a	Yes	Anomaly? and others?
P. Oxy. LXI 4172	No	Planetary phases?
P. Oxy. I 35	No	Regnal canon with lengths of reigns but no totals, Augustus to Decius (A.D. 249)
P. Oxy. XXXI 2551	Yes	Regnal canon with lengths of reigns but no totals, Cyrus to Philip (A.D. 243); procedure text
P. Ryl. I 27	No	Regnal canon with lengths of reigns and totals (eras Augustus and Commodus), Commodus to Gallus (A.D. 250); procedure text

The columns on this side were of approximately 45 lines. The other side (P. Ryl. 523) has the end of a table of oblique ascensions, based on the same parameters as the one in the *Handy Tables* but with minor numerical deviations in the minutes. The table was laid out in a different format from the *Handy Tables*, in such a way that each page contained the ascensions for the thirty degrees of one zodiacal sign for three *climata*; and there is no column for the number of equatorial degrees rising in one seasonal hour. The preserved part is from the last page (Pisces, *climata* 4–6), so that in the original codex this was the recto.

P. Lond. 1278. Published in Neugebauer and T. C. Skeat 1958. Fragments from two codices, or one codex written by two scribes. In the first hand are parts of (1) a geographical table; (2) a table relating maximum length of daylight to geographical latitude; (3) a table for right ascension, sine of right ascension, and equation of time normed for the Era Nabonassar; and (4) oblique ascensions and rising times of 10° arcs for the seven *climata*. Although all these tables are derived from the parameters of the *Almagest*, none seems to have formed part of the original *Handy Tables* as described by Ptolemy in his introduction to that work. Nevertheless versions of (1), (2), and part of (3) are found in some of the medieval copies of the *Handy Tables*. In the second hand is a fragment from a table of mean motions for Saturn, laid out in the format of the *Handy Tables*, with 25-year intervals counted from the Era

Philip. From the table of mean motions at 25-year intervals we have a column for the planet's mean motion in longitude reckoned from the apogee, as in the *Handy Tables* but with one more fractional place, and a column for the planet's mean motion in longitude reckoned from Aries 0°, which is not found among Ptolemy's tables. On the other side of the fragment is the table of mean motions at 1-year intervals, with a column for Saturn's mean motion in anomaly, again to one more fractional place than in the *Handy Tables*, and a mysterious column with quantities increasing by 155;23,30 per line.

P. Oxy. 4173. Tables of mean motions of Regulus and the five planets for the Egyptian months and days. The calendrical structure of the table is that of the *Handy Tables*, but the mean motions are given to one more sexagesimal place, and as far as they go they are in exact agreement with the parameters of the *Almagest*. The table is also un-Ptolemaic in presenting only one mean motion for each planet: for Saturn, Jupiter, and Mars, the mean longitude; for Venus and Mercury, the mean anomaly.

P. Oxy. 4173a. Tables in a hand similar or identical to that of 4173; there is no other evidence that they were parts of the same manuscript. Most of the fragments are from tables that were very similar in structure to the anomaly tables of the *Handy Tables*, with a double column of arguments ascending from 1° and descending from 360° at 1° intervals (30° to a page), and several columns of dependent functions. The numbers in these columns do not match any of Ptolemy's anomaly tables, and I have not succeeded in identifying them.

P. Oxy. 4172. The tables in the papyrus look, at least superficially, like the planetary visibility tables of the *Handy Tables*; but there is no specific match. I am very unsure of this identification.

There remain a few tables—surprisingly few—that point to the existence of traditions of kinematic modeling that owed little, if anything, to Ptolemy (table 8).

Table 8
Other kinematic tables

Text	Codex?	Tables preserved (in part)
P. Oxy. LXI 4174	Yes	Lunar mean motion
P. Oxy. LXI 4174a	No	Solar mean motions (longitude, anomaly, latitude)
P. Oxy. LXI 4220	Partly	Lunar mean motions; anomaly?
P. Oxy. LXI 4223	No	Sinusoidal function (anomaly?)
P. Oxy. LXI 4224	No	Sinusoidal function (anomaly?)

Remarks on Individual Texts

P. Oxy. 4174. A rather crude table. On one side is a table of lunar mean longitudes, in integer degrees, for the beginnings of a series of Alexandrian years. On the other side are supplements of lunar mean motion in longitude for the single days of a year, again without fractions.

P. Oxy. 4174a. Mean motions of the sun, tabulated for the days of a calendar month and for equinoctial hours. Three independent motions are given, which correspond approximately to periods of $365\frac{1}{4}$, $365\frac{1}{2}$, and $365\frac{1}{8}$ days. A passage in Theon of Smyrna (ed. Hiller, p. 172) allows us to identify these as the sun's mean motion in longitude, anomaly, and latitude according to a theory of unknown provenance.

P. Oxy. 4220. Fragments from, apparently, two manuscripts. The two fragments of the first manuscript, written on one side of the papyrus only, belong to a table for the mean motion of the moon's apogee in sidereal longitude for Egyptian years, and a table for the mean elongation of the moon from the sun for single days. The other fragment, from a codex, is from what seems to be an anomaly table. The numerals in one of the preserved columns reach a maximum value of 47;0 and then decrease (Venus's synodic anomaly?).

P. Oxy. 4223. A table with sets of columns of 45 lines, each comprising an index number (increasing by steps of 1) and a corresponding quantity with one sexagesimal fractional place. At index number 96, a maximum of 7;30 is attained, after which the numbers decrease, evidently approaching zero at or near index 180. The pattern suggests a simple table of equations.

P. Oxy. 4224. The table is roughly like 4223 in structure, with an index column and a column of corresponding quantities with two fractional places. A maximum, 19;24,19, seems to occur at index 98, after which the numbers decrease with more or less constant second differences. Extrapolation suggests again that zero was reached near index 180.

5   ALMANACS

5.1   Ephemerides

The name "Ephemeris" was applied already in antiquity to a table that listed the positions of the heavenly bodies at intervals of one day, and it is desirable to restrict the term to this meaning.[11] Ephemerides (table 9) are laid out according to a civil calendar, either the Alexandrian or the Roman, with a set of columns for each month and one line for each day of the month. The leftmost columns of each set are used for the day numbers in the principal calendar of the ephemeris and in one or more other calendars. After these follow columns containing longitudes of the heavenly bodies and other data such as syzygies. Some ephemerides also

Table 9
Ephemerides

Text	Years	Codex?	Calendar	Almanac?	Columns in ephemeris
P. Oxy. LXI 4175	24 B.C.	No	Alexandrian	Yes	Roman, Alexandrian calendars; moon
P. Dublin TCD F.7	100	No	Roman?	No	...planets;...
P. Oxy. LXI 4176	111	No	Alexandrian	Yes	...moon; time of moon's sign-entries
P. Oxy. LXI 4177	121–127	No	Alexandrian	Yes	Alexandrian calendar; moon (only on dates of sign-entry)
P. Harris I 60	140	No	Roman	Yes	Planetary weeks; Roman, Alexandrian calendars ... moon?; ?
P. Oxy. LXI 4181	161	No	Roman?	No	Roman, Alexandrian, lunar calendars; moon; time of moon's sign-entries; ... planets
P. Oxy. LXI 4177a	245	No	Alexandrian	?	Moon; time of moon's sign-entries
P. Oxy. LXI 4178	261	No	Roman	?	Roman, Alexandrian, lunar calendars; moon; time of moon's sign-entries
P. Oxy. LXI 4184a	293	No	Alexandrian	No	Alexandrian calendar; moon ....
P. Oxy. LXI 4179	348	No	Roman	No	Roman, "Greek", Alexandrian, lunar calendars; moon; time of moon's sign-entries; syzygies; sun; planets
P. Oxy. LXI 4180	465	Yes	Alexandrian	No	...Roman, lunar calendars; evaluation of day; moon; sun; planets
P. Mich. inv. 1454	467	Yes	Roman	No	Weeks; Roman, Alexandrian, lunar calendars; moon; time of moon's sign-entries; sun; planets; evaluation of day
P. Vind. inv. G. 29370b	471	Yes	Roman	?	Weeks; Roman, Alexandrian, lunar calendars; moon; ...
P. Vind. inv. G. 29370	489	Yes	Roman	No	Weeks; Roman, "Greek", Alexandrian, lunar calendars; moon;... planets; ascending node; evaluation of day
P. Oxy. LXI 4182	?	Yes	Roman	No	Roman, Alexandrian, lunar calendars; moon;... planets; eval. of day?
P. Oxy. LXI 4183	?	Yes	Roman	No	Roman, Alexandrian, lunar calendars;...?
P. Oxy. LXI 4184	?	No	?	?	... unknown calendar; moon ....
P. Vind. inv. G. 3231	?	Yes	?	No	... moon and lunar sign-entries;... planets ...
P. Berol. inv. 21240	?	No	?	No	... sun; planets ....

have a column giving brief astrological appraisals of the auspiciousness of each day, apparently based on the data in the preceding columns. The time of day for which the longitudes in ephemerides were reckoned seems always to have been six equinoctial hours past noon.

The selection and order of the columns of data vary from one ephemeris to the next. Broadly, however, there are two kinds: ephemerides that have columns for each of the sun, moon, and five planets; and "almanac-ephemerides," in which the ephemeris proper only has a column for the lunar longitude while the progress of the planets is recorded above the ephemeris in a monthly almanac (see section 5.3 below). The almanac-ephemeris seems to be the older format. An anonymous text explaining in detail how to construct ephemerides of the later type is preserved in several medieval astronomical and astrological manuscripts.[12] It allows for the inclusion of some kinds of information, such as eclipse predictions and weather signs derived from fixed star phases, that have not so far appeared in the papyri.

Table 10 shows a hypothetical ephemeris, not based on any actual papyrus but covering the same month as *P. Oxy.* 4176. Here the first four columns give the date respectively in the Alexandrian, Roman, "Greek" (i.e., Roman, but counting the days forward as we do), and lunar (counting from conjunction). The next three columns belong to the moon, giving the zodiacal sign, degrees and minutes, and the time of crossing into a new sign in seasonal hours. The remaining six columns, giving the longitudes of the sun and the five planets, would be omitted in an almanac-ephemeris.

The five-day almanac P. Heid. inv. 34 (see 5.4 below) also has a section giving daily longitudes of the moon, but as a continuous vertical series running through several columns rather than in ephemeris format.

Remarks on Individual Texts

Dublin TCD F.7. Published with photograph in F. A. J. Hoogendijk 1982; identified as ephemeris and dated in Jones 1995 (correcting false interpretation in Jones 1991).

P. Oxy. 4177. The arrangement differs from the other almanac-ephemerides (*a*) in having the monthly almanac laid out with columns for the planets and rows for the months, as in an independent almanac, and (*b*) in omitting all lines in the ephemeris on which the moon does not enter a new zodiacal sign.

P. Harris 60. Identified as almanac-ephemeris and dated in Jones 1994b.

P. Mich. inv. 1454. Published with photograph in H. D. Curtis and F. E. Robbins 1935.

Table 10
Ephemeris for Trajan 15, Choeac (Alexandrian calendar) = November/December, A.D. 111

Alex.	Roman	Greek	Lunar	Moon	Sign-entries	Sun	Saturn	Jupiter	Mars	Venus	Mercury
1	4	28	10	♈		♐ 9;13	♓ 10;45	♉ 29;16	11;9	♎ 29;49	♏ 15;55
2	3	29	11			10;15	10;47	29;28	11;23	♏ 1;2	16;59
3	1	30	12	♉	N 1	11;17	10;49	29;40	11;37	2;15	18;5
4	Kal.	1	13			12;19	10;52	29;51	11;51	3;29	19;15
5	4	2	14	II	N 10	13;20	10;54	♒ 0;3	12;7	4;42	20;30
6	3	3	15			14;22	10;57	0;15	12;22	5;54	21;45
7	1	4	16	♋	D 9	15;24	10;59	0;27	12;39	7;8	23;2
8	Non.	5	17			16;26	11;1	0;39	12;55	8;22	24;25
9	8	6	18	♌		17;28	11;3	0;51	13;15	9;34	25;48
10	7	7	19		N 7	18;29	11;6	1;3	13;36	10;47	27;13
11	6	8	20	♍		19;30	11;9	1;15	13;54	12;0	28;41
12	5	9	21		D 11	20;32	11;12	1;28	14;14	13;14	♐ 0;9
13	4	10	22	♎		21;34	11;15	1;41	14;33	14;27	1;41
14	3	11	23			22;35	11;19	1;54	14;53	15;40	3;14
15	1	12	24	♏	N 8	23;36	11;22	2;6	15;13	16;54	4;50
16	Id.	13	25			24;39	11;26	2;19	15;33	18;8	6;28
17	19	14	26	♏	D 3	25;41	11;29	2;32	15;53	19;22	8;6
18	18	15	27			26;42	11;33	2;45	16;13	20;35	9;46

Alex.	Roman	Greek	Lunar	Moon		Sign-entries	Sun	Saturn	Jupiter	Mars	Venus	Mercury
19	17	16	28	♐	3;1	D 7	27;43	11;37	2;58	16;35	21;49	11;27
20	16	17	29		17;16		28;44	11;40	3;10	16;57	23;2	13;8
21	15	18	1	♑	1;46	D 9	29;46	11;44	3;23	17;19	24;16	14;52
22	14	19	2		16;23		♑ 0;47	11;48	3;36	17;41	25;30	16;36
23	13	20	3	♒	1;18	D 10	1;49	11;52	3;49	18;3	26;45	18;22
24	12	21	4		16;19		2;51	11;56	4;2	18;27	27;58	20;7
25	11	22	5	♓	1;18	D 10	3;52	12;0	4;16	18;50	29;12	21;53
26	10	23	6		15;56		4;53	12;4	4;30	19;13	♈ 0;26	23;41
27	9	24	7	♈	0;1	D 12	5;55	12;8	4;43	19;36	1;40	25;29
28	8	25	8		13;34		6;56	12;13	4;56	19;59	2;54	27;18
29	7	26	9	♉	26;48	6 N	7;57	12;17	5;10	20;24	4;8	29;7
30	6	27	10		9;41		8;59	12;22	5;23	20;48	5;22	♑ 0;57

P. Vind. G. 29370 and 29370b. Published with photograph in H. Gerstinger and Neugebauer 1962. For the correct dates and identification of some of the columns, see Jones 1994c.

P. Vind. G. 3231. Published in Neugebauer and P. J. Sijpesteijn 1980b. The fragment preserves the headings of an ephemeris in codex format, as can be seen by comparison with *P. Oxy.* 4180–4182.

P. Berol. inv. 21240. Published in Ioannidou 1996, 144, and plate 62. The legible columns include the longitudes of the sun, Saturn, Jupiter, Mars, and Venus.

5.2   Sign-Entry Almanacs

"Sign-entry almanacs" record the dates on which each planet makes its crossings from one zodiacal sign to a neighboring sign. This is the most frequently encountered variety of table among the papyri (table 11), and was already recognizable as a standard type before arrival of the Oxyrhynchus material. The usual format is year-by-year and, within each year, planet-by-planet in the standard Greek order Saturn, Jupiter, Mars, Venus, Mercury. This sequence puts the planets in order of decreasing longitudinal and synodic periods, so that the number of sign-entries increases from one planet to the next. Sometimes the almanac is so arranged that each column starts a new year.

Each line of the almanac has three entries: the month (which may be omitted if it is the same as for the preceding line), the day of the month, and the sign entered. The month may be named or, more frequently, identified numerically by its sequence within the months of the Egyptian or Alexandrian calendar, starting with Thoth = 1. Similarly, the zodiacal signs are often represented as numbers in order of increasing longitude, starting with Virgo = 1.

In addition to the sign-entries, other events such as phases or stations may be recorded in the almanac, in which case the sign specified is the one *within which* the event occurs. The sign occupied by each planet on the first day of the year is often recorded in this way.

Two variants of the sign-entry almanac should be noted. In one group of almanacs from Oxyrhynchus, the date of nocturnal sign-entries is given as a double date (e.g., "16 17" signifies sign-entry during the night of the 16th leading into the 17th), and a fourth column of numerals follows the zodiacal sign. The numbers in this column, which never exceed 12, represent the time of sign-entry in seasonal hours. A further peculiarity of the four-column almanacs is that the sign-entries that they contain were apparently all computed using the *Handy Tables*. These almanacs

Table 11
Sign–Entry Almanacs

Text	Years	Calendar	Medium	Remarks
Tab. Amst. inv. 1	26 B.C.–24 B.C.	Egyptian or Alex.	Boards	Years in Callippic period
P. Berol. inv. 8279 (demotic)	17 B.C.–13 B.C.	Egyptian	Roll	
P. *Laur.* IV 144	69–75	Alexandrian	Codex	
Bodl. MS Gr. Class. F 7 (P)	98–103	Egyptian	Boards	Years in era Titus
Stobart Tablets (demotic)	70–133	Alexandrian	Boards	
P. Oxy. LXI 4185	129–133	Alexandrian	Roll	
P. Oxy. LXI 4189	136–144?	Alexandrian	Codex	
P. Oxy. LXI 4186	142–146	Egyptian	Roll	
P. Oxy. LXI 4187	160/161	Alexandrian	Roll	
P. Oxy. LXI 4188	161–165	Alexandrian	Codex	
P. Oxy. LXI 4193	195–203	Alexandrian	Codex	
P. Oxy. LXI 4188a	201–208	Alexandrian	Codex	
P. Oxy. LXI 4196	218–220	Alexandrian	Codex	Four-column
P. Oxy. XLVI 3299	217–225	Alexandrian	Codex	
P. Oxy. LXI 4190	241–243	Alexandrian	Codex	Four-column
P. Oxy. LXI 4191	236–245	Alexandrian	Codex	
P. Oxy. LXI 4192	276–280	Alexandrian	Codex	Four-column
P. Oxy. LXI 4194	282–286	Alexandrian	Codex	Four-column
P. Oxy. LXI 4195	300–304	Alexandrian	Codex	Four-column
P. Oxy. LXI 4196a	?	Alexandrian	Codex	Four-column
P. Strasb. inv. gr. 1097 (ined.)	?	?	?	
P. Vind. inv. gr. 36041 + 26011 fr. n.	Perpetual	Alexandrian	Codex	
P. Oxy. LXI 4197	Perpetual	Alexandrian	Codex	
P. Oxy. LXI 4198	Perpetual	Alexandrian	Codex	

must have been a great trouble to produce—vain labor, to some extent, since the planetary tables of the *Handy Tables* cannot claim sufficient accuracy to justify searching for the precise hour of sign-entry.

Secondly, we have examples of perpetual sign-entry almanacs that give entries for an entire Goal-Year period for each planet in turn. These clearly follow the same principle of arrangement as the medieval "Almanacs," where one used modulo arithmetic to locate a given year in each planet's cycle in order to obtain the positions for that year.

Tables 12 and 13 allow us to compare the sign-entries from the same two Alexandrian years from a three-column and four-column almanac. In both, the sections ruled off below the regnal year contain the sign-entries for Saturn, Jupiter, Mars, and (in table 12) Venus, the rest of the almanac having been broken off at the bottom. Lost or illegible numerals are represented by "x," but restorations that are certain have not been marked.

Remarks on Individual Texts

Tab. Amst. inv. 1. Published with photograph in Neugebauer, P. J. Sijpesteijn, and K. A. Worp 1977. Date established in Jones 1993. The years are numbered according to the (fifth) Callippic period, as I will show elsewhere.

P. Berlin 8279. Published with photographs in Neugebauer and R. A. Parker, *EAT* v. 3, 228–232 and plates 66–73.

P. Laur. 144. Published with photograph in R. Pintaudi and Neugebauer 1978, reproduced in *P. Laur.* v. 4. Corrected edition and dating of the back in Jones 1998b.

Bodl. MS Gr. Class. F 7 (P). Published with photograph in Neugebauer 1957; corrections and dating in Neugebauer 1972.

Stobart Tablets. Published with photographs in Neugebauer and R. A. Parker, *EAT* v. 3, 225–240, and plates 74–78.

P. Strasb. inv. gr. 1097. Unpublished; see Neugebauer 1975, 788.

P. Vind. G. 36041+26011 fr. n. Published with photograph in Neugebauer and P. J. Sijpesteijn 1980a; corrections and identification as perpetual almanac in Jones 1994c.

5.3   Monthly Almanacs

A "monthly almanac" (table 14) is a tabular almanac in which columns correspond to the months of the year, and rows to the planets, or vice versa; hence there is space for recording at least one planetary position for each planet for each month. A single entry comprises a day number

Table 12
Sign-entry almanac for Gordian 5–6 (A.D. 241–243) from *P. Oxy.* 4191

month	day	sign	month	day	sign
	Year 5			Year 6	
1	1	12	1	1	1
2	1x	1			
6	1x	12	1	1	6
11	1	1	6	xx	7
1	1	5	1	5	3
6	2	6	3	14	4
			4	23	5
1	21	11	5	xx	6
6	17	10	7	6	7
7	17	11	8	xx	8
9	xx	12	9	xx	9
11	x	1	11	xx	10
12	25	2	12	29	11
1	11	1	1	8	1
2	2	2		27	2
	22	3	2	27	3
3	11	4	4	21	4
4	1	3	5	9	5
	21	4		26	6
5	17	5	6	13	7
6	x	6	8	16	8
	xx	7	9	2	9
(rest of column lost)				18	10
			10	7	11
			12	18	12
			epag.	x	1

(rest of column lost)

Table 13

Four-column sign-entry almanac for Gordian 5–6 from *P. Oxy.* 4190

month	day	sign	hour	month	day	sign	hour
		Year 5				Year 6 of Gordian	
1	1	12		1	1	1	
2	2	1	5	1	1	6	
7	7/8	12	5	6	16/17	7	6
10	30	1	5	1	1	2	
1	1	5		2	2	3	3
6	5/6	6	1	3	12/13	4	6
1	1	10		4	21/22	5	11
	23/24	11	x	5	30	6	5
5	10/11	10	x	7	8	7	11
7	10/11	11	x				
9	17	12	x		(rest of column lost)		

(rest of column lost)

within the month, the planet's position on that day, and sometimes an indication of an event (e.g., sign-entry, station, or first appearance) to which the date and position apply. Whether a planet's progress through the year is to be read across a row or down a column is a matter of convenience of layout. When a monthly almanac is laid out above a lunar ephemeris for the same year (see 5.3 above), the planets are usually assigned rows so that the planetary positions for any month can be read directly above the ephemeris entries for the same month. On the other hand, most independent monthly almanacs assign the columns to the planets, to allow a wider space for the tabular entries. In addition, we have monthly almanacs for single planets.

The monthly almanacs fall into two groups in another way. A simpler format records, for a given planet and month, either (1) the day of the month on which the planet makes a sign-entry, with the newly occupied sign identified, or (2), if no sign-entry occurs during the month, the zodiacal sign occupied throughout the month. Such almanacs are close kin to sign-entry almanacs. A more elaborate kind reports the planet's phases and stations as well as sign-entries and gives the associated longitudes in degrees and even minutes; when nothing else happens to a planet during a month, the longitude on the first of the month is given. Such tables are

Table 14
Monthly almanacs

Text	Years	Calendar	Medium	Planets	Phases & degrees?	Remarks
P. Oxy. LXI 4199	14 B.C.–6 B.C.	Egypt. or Alexandrian	Codex	Jupiter only	Yes	
PSI inv. 75D + E.E.S. 79/82(1)	48–55	Egyptian	Roll	All (columns)	Yes	
P. Berol. inv. 21226	52–56	Alexandrian	Roll	Jupiter only	Yes	Years in era Augustus
P. Nelson	101–103	Alexandrian	Roll	All (columns)	Yes	Years in era Titus
P. Tebt. II 274	107–118	Alexandrian	Roll	All (columns)	No	Years in era Titus; new moons
P. Lund V 35b	119–122	Alexandrian	Roll	All (columns)	No	
P. Oxy. LXI 4201	126–128	Alexandrian	Codex	All (columns)	Yes	
P. Oxy. LXI 4202	152–155	Alexandrian	Roll	All (columns)	Yes	
P. Oxy. LXI 4203	215/216	Alexandrian	Codex	Saturn only?	Yes	
P. Oxy. LXI 4200	?	Alexandrian	Codex	Jupiter only	Yes	Unknown era
P. Oxy. LXI 4203a	?	Alexandrian	Roll	Mercury only	Yes	

Table 15
Monthly almanac for A.D. 154/155 from *P. Oxy.* 4202

Month	Day	Saturn	Day	Jupiter	Day	Mars
I	[ ]	♍ [ ]	[ ]	♍ 6° 23rd Γ 12°	1	♐ 2;45°
II	[ ]	♍ [ ]	x	♍ xx;7°	1	♑ 0;41°
III	[ ]	♍ xx;3x°	x	♎ 0;x°	13	♒ 0;37°
IV	[ ]	♍ xx;41°	1	♎ 1;x4°	26	♓ 0;3°
V	[ ]	♍ 21;35°	xx	♎ [ ]	1	♓ 4;1x°
VI	[ ]	♍ 20;29°	[ ]	♎ [ ]	7	♈ 0;38°
VII	[ ]	♍ 17;5x°	[ ]	♎ [ ]	16	♉ 0;7°
VIII	[ ]	♍ 15;14°	1	♍ 26°	1	♉ 9;14°
IX	[ ]	♍ 12;x° Ψ	26	♍ 23;21° Ψ	3	♊ 0;4° 26th ♌ 17°
X	[ ]	♍ 13;5°	6	♍ 24;21°	14	♋ 0;14°
XI	1	♍ 14;2°	1	♍ 26;47°	23	♌ [ ]
XII	[ ]	♍ 20;57°	[ ]	♎ xx	5	♌ x;6° 2nd ♍ 0°

potentially rich with information about the methods used to compute planetary motion; unfortunately none of the specimens known at present is in a good state of preservation. Table 15 shows one year's data for Saturn, Jupiter, and Mars from an almanac of this kind.

Remarks on Individual Texts
PSI inv. 75D + E.E.S. 79/82(1). Published with partial photograph in M. Manfredi and Neugebauer 1973; new fragment in Jones 1998c.

P. Berol. inv. 21226. Published in W. Brashear and Neugebauer 1983.

P. Nelson. Published in C. A. Nelson 1970; identified in Neugebauer 1971; date established by Jones, 1998b.

P. *Tebt.* 274. Published with photograph in Neugebauer 1942, 241–242.

P. *Lund* 35b. Published with photograph in E. J. Knudtzon and Neugebauer 1947.

5.4   Five-Day Almanacs
"Five-day almanacs" (table 16) are arranged like sign-entry almanacs, by year and within each year by planet in the standard order; but instead of sign-entries, the table records the planet's longitude (degrees and minutes) at five-day intervals starting with either the first or the fifth day of each month. Many more lines are needed for a year than in a sign-entry almanac, and a common arrangement is to have a column reserved for

Table 16
Five-day almanacs

Text	Years	Codex?	Remarks
P. Oxy. LXI 4212	217/218	No	
P. Oxy. LXI 4205	257/258	Yes	
P. Oxy. LXI 4206	272–274	Yes	
P. Oxy. LXI 4207	290/291	Yes	
P. Oxy. LXI 4208	295/296	No	
P. Oxy. LXI 4209	295/296	No	Stations included
P. Oxy. LXI 4210	300–302	?	
P. Oxy. LXI 4213	301/302	No	Same MS as P. Oxy. 4210?
P. Oxy. LXI 4211	306/307	Yes	
P. Heid. inv. 34	345–349	Yes	Also lunar longitudes and syzygies

each of the planets. The name of each month is written above the six lines required for its positions. Each line will give the day of the month, the zodiacal sign (this may be omitted if unchanged from preceding lines), and the degrees and minutes of longitude. As in sign-entry almanacs, the zodiacal sign is often denoted by its number, starting from 1 = Virgo. The specimen in table 17 is from the top of an almanac laid out with two columns per planet per year, so that the preserved portion has the beginning of the second half of A.D. 273/274 for Jupiter and the beginning of the first half of the same year for Mars.

All known five-day almanacs turn out to have been computed using the *Handy Tables*. The format is probably the ancestor of the medieval Almanacs, which also give planetary longitudes at five (or ten) day intervals; but no example of a perpetual five-day almanac employing the Goal-Year periods has yet come to light on papyrus.

Remarks on Individual Texts
P. Heid. inv. 34. Published with photograph in Neugebauer 1956; dating of lunar longitudes and syzygies in Neugebauer 1975 1056–1057.

5.5   Miscellaneous Almanacs
P. Oxy. 4204 is, so far, the only known specimen of a format that has some resemblance to the more detailed kind of monthly almanac. It covers the last six months of the Alexandrian year 54/55, recording, for each planet in turn (in the standard order), the dates and longitudes of its phases, stations, and sign-entries.

Table 17
Five-day almanac for A.D. 273/274 from *P. Oxy.* 4206

Day	Sign	Longitude	Day	Sign	Longitude
Jupiter			Mars		
Month VII			Month I		
1	2	19;43°	1	10	22;24°
6	2	19;22°	6	10	25;7°
11	2	18;57°	11	10	27;44°
16	2	18;29°	16	11	0;17°
21	2	17;59°	21	11	2;48°
26	2	17;23°	26	11	5;13°
Month VIII			Month II		
1	2	16;46°	1	11	7;31°
6	2	16;8°	6	11	9;xx°
11	2	15;31°	11	11	11;41°
16	2	14;52°	16	11	13;41°
21	2	14;xx°	21	11	15;30°
26	2	13;36°	26	11	17;2°
(rest of column lost)			Month III		
			1	11	18;26°
			(rest of column lost)		

E.E.S. inv. 79/82 (2), published in Jones 1998c, presents dates of sign-entry of Mercury over at least three years, in a format related to the monthly almanacs but using columns for the calendar months and rows for consecutive years.

5.6   Syzygy Tables and Eclipse Canons
With the exception of the ephemerides, the almanacs almost never include dated positions of either the sun or the moon. In the sun's case this is probably because solar longitudes hardly vary for the same (Alexandrian) date from year to year; and the moon's swift motion and rapid changes of speed mean that tabulation at intervals of more than one day will be of little use. On the other hand, full moons and new moons are of sufficient special interest to call for their own tables.

What seems to have been a standard format for syzygy tables is represented by two examples: *P. Mich.* III 150, *P. Oxy.* LXI 4214, part of the five-day almanac P. Heid. inv. 34, and the demotic P. Vienna D 4876 (published with photograph in Neugebauer and Parker, *EAT*, v. 3, 243–

250, and plate 80A). The full moons or new moons for a year are each given a line in the table, with consecutive columns for the date (month, day, seasonal hour and fraction thereof) and longitude of the moon (zodiacal sign, degrees, minutes). Another demotic papyrus, P. Carlsberg 9 (published with photograph in Neugebauer and Parker, *EAT*, v. 3, 220–225, and plate 65) contains a table of schematic dates of lunation according to a 25 Egyptian-year calendrical cycle. This text has frequently been discussed in relation to the Egyptian lunar calendar.[13]

P. Colker is, to date, the unique example of a Babylonian-style syzygy table, with many columns of data leading up to the computation of the circumstances of a series of conjunctions or oppositions. Neugebauer 1988 provided a photograph and partial translation; a full edition is given in Jones 1997d. The second of the two partially preserved columns of sexagesimals is a sequence of values of the linear zigzag function known as "column G" of the Babylonian System B lunar theory, representing an approximation to the duration of the synodic month. The first column, together with the indications "additive" and "subtractive" between the columns, is probably a version of "column J," the correction applied to G to account for solar anomaly.

The two extant eclipse canons are laid out as prose texts rather than tables, but their standardized and repetitive order of presentation of computed data has more affinity to the tables than to the theoretical and procedure texts. P. Berlin 13146 + 13147, published in Neugebauer, Parker, and K.-T. Zauzich 1981, is a demotic papyrus, and its date (predicted eclipse possibilities from 85 to 74 B.C.) is half a century older than any of the other papyri discussed in this article. Nevertheless it is remarkably similar in character to *P. Oxy.* LXI 4137, which contains eclipse forecasts for A.D. 56 and 57. In both texts the following information is provided for a succession of eclipse possibilities: (*a*) the number of synodic months since the preceding eclipse possibility; (*b*) the date according to the current Callippic period, with Egyptian month and day; and for actual eclipses, (*c*) the time of mid-eclipse; and (*d*) the direction of obscuration. In the demotic canon there are also some past-tense remarks about planets in the sky at the time of eclipse, implying an observational component. In *P. Oxy.* 4137, the Callippic cycle year is accompanied by an Athenian calendar month, and the description of the eclipse includes the approximate magnitude and duration, "inclinations" (*prosneuseis*) of the beginning, middle, and end of obscuration, and the location of the moon with respect to fixed stars.

6   OTHER PAPYRI CONTAINING OR REFERRING TO TABLES

Although it is not the intention of this paper to provide a complete bibliographical index to the astronomical papyri,[14] it nevertheless seems
worthwhile, if only as a measure of the incompleteness of the foregoing
classification, to draw attention to those tables on papyrus that do not fit
into it. I omit fragments that are in too poor a state of preservation to offer
much hope of recognition or analysis.

P. Carlsberg 31 (demotic). Published with photograph in Neugebauer and
Parker, *EAT*, v. 3, 241–243, and plate 79A. Identified by A. Aaboe 1972 as
an auxiliary table for a cycle of eclipse possibilities.

P. Heid. inv. 4144 + P. *Mich.* III 151. The joined fragments published with
photograph in Neugebauer 1960. Interpreted in Jones 1990 as auxiliary tables
for a variant of the System A scheme for Mars. I am no longer so sure of this,
since the only similar table that has come to light, P. *Oxy.* 4216 (see below),
does not fit into my reconstruction.

P. Oslo inv. 1336 (demotic). Published with photograph in Neugebauer and
Parker, *EAT*, v. 3, 254–255, and plate 79C. A numerical table of unknown
significance.

P. *Ryl.* III 526. A numerical table of unknown significance.

P. *Oxy.* LXI 4140. In the left of the two preserved columns, large consecutive (?) whole numbers (preserved: 571 and 572); in the right, a prose description of circumstances of some sort of event involving the moon's
longitude, argument of latitude, and other unidentified elements.

P. *Oxy.* LXI 4216. A table of similar structure to P. Heid. inv. 4144 + P.
Mich. 151.

P. *Oxy.* LXI 4218. Multiples of 5;0, 3;45, and 2;30, each up to 30;0.

P. *Oxy.* LXI 4219. A numerical table of unknown significance.

P. *Oxy.* LXI 4221. A table of longitudes, in one part increasing by constant
$0;56,40°$ in Scorpio.

P. *Oxy.* LXI 4222. A table associating latitudes (?) with the twelve zodiacal
signs: zero in Libra (and Aries?), maximum 7 southerly in Capricorn.

P. *Oxy.* LXI 4225. Index numbers up to 30, with three columns of small
whole numbers exhibiting no apparent pattern.

P. *Oxy.* LXI 4226. Tables headed with names of planets; beside each of the
zodiacal signs, one or two circular spots or eight-pointed stars in black or red ink.

P. *Oxy.* LXI 4228. Columns with second-order arithmetical sequences:
(1) descending from 180 to 103;45,0; (2) ascending from 180 to 256;15,0;
(3) descending from 360 to 256;15,0.

Lastly, there are several papyri containing prose texts that refer to tables:

P. Ant. III 141. A procedure text concerning computations of lunar longitudes, with methods closely related to *P. Oxy.* 4174; see Jones 1998b.

P. Ryl. I 27. Corrections to text, and translation, in Jones 1997a Appendix 2.

PSI XV 1490. Published in M. Manfredi 1966. Concerning the construction of mean motion tables for the sun and an anomaly (?) table for the sun.

PSI XV 1491. Unpublished (see above, section 4.2, on *PSI* 1493). Discussion in Neugebauer 1962, 388; 1975, 946; and Jones 1984—probably all incorrect.

P. Oxy. LXI 4136. Text on calculating the running totals of the Standard Scheme template.

P. Oxy. LXI 4142. Instructions for the *Handy Tables*: time conversions involving seasonal and equinoctial hours, geographical longitude, and the equation of time.

P. Oxy. LXI 4143. Instructions for the *Handy Tables*: the planetary latitude tables.

7 Concluding Remarks

At our present state of knowledge it seems appropriate to treat the astronomy of the papyri as a distinct "Greco-Egyptian" tradition with links not only to the Greek theoretical astronomy of Hipparchus and Ptolemy but also backward to the Babylonians and forward to the astronomers of Byzantium, Islam, and the Latin West. The debt to Babylonian astronomy is deep both in the kinds of things that Greco-Egyptian astronomy set out to accomplish and the methods that it employed. For example, the monthly almanacs give dates of the planets' sign-entries, dates and longitudes of their phenomena, and their longitudes at the beginning of each month—items that all are regularly reported in the Babylonian almanacs.[15] Again, the planetary epoch tables on papyrus not only resemble ACT planetary tables in content, but were mostly computed using the same arithmetical procedures. On the other hand, the specific formats of the Babylonian texts do not reappear on papyrus, and many kinds of data prominent in the cuneiform texts (e.g., normal-star passages, and computations of lunar crescent visibility) are absent from the papyri. Between the two traditions we have to hypothesize a process not just of transmission but of profound transformation.

This had happened already by the beginning of the period for which we have good documentation, which roughly coincides with the beginning of Roman rule in Egypt in the last decades of the first century B.C. From this time on, there is a remarkable uniformity in the kind of astronomy practiced in the papyri, and not merely at the superficial level of long-lived formats such as the ephemeris and the sign-entry almanac. One significant event that was certainly felt in Greco-Egyptian astronomy was the advent of Ptolemy's tables after A.D. 150, which apparently for the first time popularized kinematic methods of astronomical prediction. But even these did not instantly drive arithmetical methods out of use; instead, they inaugurated a period, which lasted at least two centuries, during which arithmetical and kinematic tables subsisted side by side. How much, besides the *Handy Tables*, survived out of this hotchpotch may become clearer by comparison of the papyrus almanacs and ephemerides with their medieval counterparts.

Notes

1. See the account in E. G. Turner 1968, 17–41.

2. Neugebauer and van Hoesen 1959. The recent "*aggiornamento*," Baccani 1992, places more emphasis on the horoscopes as astrological and social documents than on their astronomical aspect.

3. Jones, *APO*. I have especially to thank J. R. Rea and R. A. Coles for their untiring assistance in making these texts accessible to me, and to D. Fowler for bringing the Oxyrhynchus material to my attention in the first place.

4. Where possible I have used the abbreviations recommended in J. F. Oates et al. 1992 for Greek papyri. Papyri in published series are given in italics, with the volume (Roman) preceding the text number at the first reference. For bibliographical consistency the new texts edited in Jones, *APO* are referred to as *P. Oxy.* LXI 4133–4300a, although only a provisional list of their contents was included in vol. 61 of *The Oxyrhynchus Papyri*. For publication data of other papyri listed in the tables, see Appendix B of Jones, *APO*.

5. For a fuller discussion of the problems of establishing the provenance and identity of the users of the astronomical papyri, see Jones 1994a.

6. The prevalence of astrology in Roman Egypt is directly witnessed by numerous personal horoscopes and fragments of astrological treatises on papyrus, and the majority of the astronomical papyri too may be counted among the astrologers' papers.

7. This general definition of "almanac" embraces the medieval Arabo-Latin Almanacs as well as the varieties of cuneiform text that A. Sachs has named "almanacs" and "normal-star almanacs."

8. The early history of the Egyptian calendar reform is disputed. A regular four-year cycle of intercalations was unquestionably followed after A.D. 8.

9. Jones 1997a.

10. Jones 1998a.

11. In the context of Babylonian astronomy the term has been applied, confusingly and imprecisely, both to the kinds of text later renamed by Sachs as "almanacs" and "normal-star almanacs" (e.g., by Kugler) and to the "ACT" numerical tables (e.g., by Neugebauer).

12. Neugebauer 1975, 1055–1056.

13. Parker 1950; Jones 1997c; Depuydt 1998.

14. A bibliographical checklist of published astronomical papyri appears as Appendix B in Jones, *APO* (bringing up to date Neugebauer 1962).

15. Sachs 1948, 277.

BIBLIOGRAPHY

Aaboe, A. 1972. Remarks on the theoretical treatment of eclipses in antiquity. *Journal for the History of Astronomy* 3, 105–118.

Baccani, D. 1992. *Oroscopi greci: documentazione papirologica*. Ricerca Papirologica 1. Messina.

Brashear, W., and O. Neugebauer. 1973. Zwei Berliner Papyri: Ein Horoskop und eine Jupiter-Tafel. *Österreichische Akademie der Wissenschaften, phil.-hist. Klasse, Anzeiger* 110, 306–312.

Curtis, H. D., and F. E. Robbins. 1935. An ephemeris of 467 A.D. *Publications of the Observatory of the University of Michigan* 6.9, 77–100.

Depuydt, L. 1998. The demotic mathematical astronomical papyrus Carlsberg 9 reinterpreted. *Egyptian Religion: the last thousand years*. Studies dedicated to the memory of Jan Quaegebeur, ed. W. Clarysse et al. Leuven, 1277–1292.

Gerstinger, H., and O. Neugebauer. 1962. Eine Ephemeride für das Jahr 348 oder 424 n. Chr. in den PER, Pap. Graec. Vindob. 29370. *Österreichische Akademie der Wissenschaften, Phil.-hist. Klasse, Sitzungsberichte* 240.2, 3–25.

Hoogendijk, F. A. J. 1982. Fragment of a Greek planetary table. *Zeitschrift für Papyrologie und Epigraphik* 48, 135–141 and plate vi.

Ioannidou, G. 1996. *Catalogue of Greek and Latin Literary Papyri in Berlin (P. Berol. inv. 21101–21229, 21911)*. Berliner Klassikertexte 9. Mainz am Rhein.

Jones, A. 1999. *APO. Astronomical Papyri from Oxyrhynchus*. Memoirs of the American Philosophical Society 233. Philadelphia.

——— 1984. A Greek Saturn Table. *Centaurus* 27, 311–317.

———— 1990. Babylonian and Greek astronomy in a papyrus concerning Mars. *Centaurus* 33, 97–114.

———— 1991. A second-century Greek ephemeris for Venus. *Archives Internationales d'Histoire des Sciences* 41, 3–12.

———— 1993. The date of the astronomical almanac Tab. Amst. inv. no. 1. *Chronique d'Égypte* 68, 178–185.

———— 1994a. The place of astronomy in Roman Egypt. In *The Sciences in Greco-Roman Society*, ed. T. D. Barnes. Apeiron 27.4. Edmonton. 25–51.

———— 1994b. An astronomical ephemeris for A.D. 140: P. Harris I.60. *Zeitschrift für Papyrologie und Epigraphik* 100, 59–63.

———— 1994c. Two astronomical papyri revisited. *Analecta Papyrologica* 6, 111–126.

———— 1995. On the planetary table, Dublin TCD Pap. F. 7. *Zeitschrift für Papyrologie und Epigraphik* 107, 255–258.

———— 1997a. Studies in the astronomy of the Roman period. I. The standard lunar scheme. *Centaurus* 39, 11–16.

———— 1997b. Studies in the astronomy of the Roman period. II. Tables for solar longitude. *Centaurus* 39, 211–229.

———— 1997c. On the reconstructed Macedonian and Egyptian lunar calendars. *Zeitschrift für Papyrologie und Epigraphik* 119, 157–166.

———— 1997d. A Greek papyrus concerning Babylonian lunar theory. *Zeitschrift für Papyrologie und Epigraphik* 119, 167–172.

———— 1998a. Studies in the astronomy of the Roman period. III. Planetary epoch tables. *Centaurus* 40, 1–41.

———— 1998b. Notes on astronomical papyri. *Zeitschrift für Papyrologie und Epigraphik*. 121, 203–210.

———— 1998c. Three astronomical tables from Tebtynis. *Zeitschrift für Papyrologie und Epigraphik*. 121, 211–218.

Knudtzon, E. J., and O. Neugebauer. 1947. Zwei astronomische Texte. *Bulletin de la Société royale des lettres de Lund* 1946–1947, part 2, 77–88 and plates 1–2.

Manfredi, M. 1966. Presentazione di un testo astronomico e discussione di un documento di Antinoe. *Atti dell' XI Congresso Internazionale di Papirologia, Milano 2–8 Settembre 1965*. Milano. 237–243.

Manfredi, M., and O. Neugebauer. 1973. Greek planetary tables from the time of Claudius. *Zeitschrift für Papyrologie und Epigraphik* 11, 101–114.

Nelson, C. A. 1970. Astronomical table. *Bulletin of the American Society of Papyrologists* 7, 35–37.

Neugebauer, O. 1942. Egyptian planetary texts. *Transactions of the American Philosophical Society* 32.2, 209–250.

———— 1956. An astronomical almanac for the year 348/349 (P. Heid. Inv. No. 34). *Det Kongelige Danske Videnskabernes Selskab, Historisk-filologiske Meddelelser* 36.4.

———— 1957. A Greek planetary table. *Chronique d'Égypte* 32, 269–272.

———— 1960. A new Greek astronomical table (P. Heid. Inv. 4144 + P. Mich. 151). *Det Kongelige Danske Videnskabernes Selskab, Historisk-filosofiske Meddelelser* 39.1.

———— 1962. Astronomical papyri and ostraca: bibliographical notes. *Proceedings of the American Philosophical Society* 106, 383–391.

———— 1971. On a fragment from a planetary table. *Zeitschrift für Papyrologie und Epigraphik* 7, 267–274.

———— 1972. The date of the planetary tables Bodleian MS. Gr. Class. F 7 (P). *Chronique d'Égypte* 93, 224–226.

———— 1975. *A History of Ancient Mathematical Astronomy*. Berlin, Heidelberg, New York.

———— 1988. A Babylonian lunar ephemeris from Roman Egypt. In *A Scientific Humanist: Studies in Memory of Abraham Sachs*, ed. E. Leichty, M. deJ. Ellis, and P. Gerardi. Philadelphia. 301–304.

Neugebauer, O., and W. Brashear. 1976. An ephemeris for Mars. *Zeitschrift für Papyrologie und Epigraphik* 20, 117–118.

Neugebauer, O., and R. A. Parker, *EAT. Egyptian Astronomical Texts.* 3 vols. 1960–1969. Providence.

Neugebauer, O., R. A. Parker, and K.-T. Zauzich. 1981. A demotic lunar eclipse text of the first century B.C. *Proceedings of the American Philosophical Society* 125, 312–327.

Neugebauer, O., and P. J. Sijpesteijn. 1980a. A new version of Greek planetary tables. *Zeitschrift für Papyrologie und Epigraphik* 37, 280–293.

———— 1980b. Fragment of an astronomical text. *Zeitschrift für Papyrologie und Epigraphik* 39, 163–164.

Neugebauer, O., P. J. Sijpesteijn, and K. A. Worp. 1977. A Greek planetary table. *Chronique d'Égypte* 52, 301–310.

Neugebauer, O., and T. C. Skeat. 1958. The astronomical tables, P. Lond. 1278. *Osiris* 13, 93–113.

Neugebauer, O., and H. B. van Hoesen. 1959. *Greek Horoscopes*. Memoirs of the American Philosophical Society 48. Philadelphia.

Oates, J. F., R. S. Bagnall, W. H. Willis, and K. A. Worp. 1992. *Checklist of Editions of Greek and Latin Papyri, Ostraca and Tablets*. 4th ed. Atlanta.

P. Ant. = *The Antinoopolis Papyri.*

P. Harris = *The Rendel Harris Papyri.*

P. Oxy. = *The Oxyrhynchus Papyri.*

P. Ryl. = *Catalogue of the Greek Papyri in the John Rylands Library, Manchester.*

PSI = *Papiri della Società Italiana.*

P. Tebt. = *The Tebtunis Papyri.*

Parker, R. A. 1950. *The Calendars of Ancient Egypt.* Studies in Ancient Oriental Civilization 26. Chicago.

Pintaudi, R., and O. Neugebauer. 1978. Pap. Laurentiana III/423: a planetary table. *Zeitschrift für Papyrologie und Epigraphik* 30, 211–218.

Sachs, A. J. 1948. A classification of the Babylonian astronomical tablets of the Seleucid period. *JCS* 2, 271–290.

Turner, E. G. 1968. *Greek Papyri: An Introduction.* Oxford.

11

THE ROLE OF OBSERVATIONS IN PTOLEMY'S
LUNAR THEORIES
Bernard R. Goldstein and Alan C. Bowen

Ptolemy's lunar theory in the *Almagest* has been described from several points of view by previous scholars, though their emphasis has been on mathematical detail. It is our primary goal to build upon these previous studies and to focus on the way Ptolemy appealed to observations as part of his argument. A secondary and related goal is to characterize Ptolemy's use of earlier observations and his attitude to earlier theories. Our analysis depends on a methodology according to which reports, quotations, and analytical accounts of earlier works are accepted as authentic evidence for the state of astronomy prior to the date of the texts in which they are embedded only under special circumstances.[1] Ideally, acceptance should, in our view, be predicated on independent confirmation, or at least support, that is demonstrated by recourse to extant texts from the antecedent period. But, should this not be possible because the relevant earlier sources are lacking, we think that acceptance must wait upon credible argument that is made case by case and looks critically to the question of the reliability of other citations in the later texts. On this basis, we argue that there is no credible evidence for a Greek geometric lunar theory prior to Ptolemy that was used to compute lunar positions. Nevertheless, there is considerable evidence for lunar theories dependent on Babylonian arithmetic schemes prior to Ptolemy and, we suggest, there is reason to believe that Ptolemy was aware of them. It has often been said that Ptolemy's first lunar theory was already available to (if not invented by) Hipparchus (second century B.C.), but our analysis of the evidence casts doubt on that claim.

Ptolemy presents two theories to account for the lunar motion in longitude which we designate the first and the second lunar theories.[2] The first was posited by Ptolemy as the definitive theory for determining the position of the Moon at syzygy (conjunction and opposition of the Moon with the Sun); the second, for determining lunar positions when the elongation from the Moon to the Sun is not equal to $0°$ or $180°$. At elongations $0°$ and $180°$ the second theory reduces to the first theory.

Ptolemy devotes books 4 and 5 of the *Almagest* to lunar theory, and he begins by presenting the periods for the Moon's motion. As is well known, the periods he cites are of Babylonian origin. Of greater interest for us is that he ascribes one long period to Hipparchus that in fact is Babylonian, claiming that Hipparchus had deduced it from eclipses chosen in light of certain criteria. Toomer has reported that he was able to identify the eclipses in question, and has argued that "Hipparchus *confirmed* (not *derived*, as Ptolemy says)" the Babylonian period relationship [Toomer's emphasis].[3] But Ptolemy mentions no source for his remark about Hipparchus, and those detailed remarks by Ptolemy concerning the constraints on eclipses that would be useful for determining the periods (*Alm.* iv.2[4]) are not explicitly related to any procedure used by Hipparchus whether for deriving or confirming the period. Rather, it seems to us more likely that Ptolemy is describing a correct mathematical procedure that might be used to find this period, but not one that was actually used as Toomer assumes. After all, the constraints are sufficiently severe that a very long list of eclipses would be needed in order to select those that met the various criteria, and there is no evidence that such a list was available in the Greco-Latin world or that anyone ever made such a selection.[5] Moreover, no specific work by Hipparchus is mentioned in Ptolemy's discussion of Hipparchus's derivation of the (Babylonian) period from the selected eclipses, and so it may be that this information came to Ptolemy *via* some intermediary source whose reliability we cannot determine. Indeed, there are relatively few instances in the *Almagest* where Ptolemy refers to a specific work by Hipparchus (none of these works survive); elsewhere Ptolemy does not indicate the source for his information about Hipparchus or his reasons for accepting it as reliable. Of course, we do not deny the *possibility* that Hipparchus verified the Babylonian period by observation. But asserting such a mere possibility is very weak and a stronger claim is unwarranted, given the evidence at hand.

The construction of the first lunar theory requires the mean motions Ptolemy derived from the period relations, a set of three lunar eclipses, and a theory for finding the eccentricity (alternatively, the radius of the epicycle) and apsidal line from such a set of lunar eclipses.[6] The construction of the second theory requires additional observations meeting certain stated criteria and a method for deriving the appropriate parameters. For the first theory Ptolemy analyzes four triples of lunar eclipses: one set that he observed himself; one set observed by the Babylonians some 850 years earlier; another set observed by the Babylonians reported by Hipparchus; and a final set observed in Alexandria and reported by Hipparchus. For the second lunar theory Ptolemy adduces one observation of his own and three by Hipparchus. None of these observations are known from any source prior to Ptolemy's *Almagest*,[7] though in several places Ptolemy suggests that he had a greater number of usable observational reports, and that the observations cited were chosen from among them (*Alm.* iv.6, v.3, and v.5).[8]

In light of our interest here in the role of observation in Ptolemy's argument as a whole, we will put aside the geometric proofs Ptolemy invokes in order to derive parameters from specified observations, and concentrate instead on the observations themselves and their cognitive force in his argument. Thus, we examine these observations in order to assess their reliability as perceived in Ptolemy's time, when the question of reliability was not so clearly reduced to the question of accuracy as it frequently is nowadays.

It is hardly surprising that Ptolemy relies on his own observations, but what is gained by his appeal to observations by his predecessors, and why does he cite specific dated observations?[9] Which previous authors of astronomical treatises had invoked such specific observational data?

Let us consider these questions in turn, starting with the last one. There are, in fact, very few observations reported in Greek or Latin texts written prior to Ptolemy and, while Ptolemy suggests that Hipparchus had collected observations of his predecessors,[10] no such work is extant. This makes it difficult to tell how Ptolemy came to make this assertion about Hipparchus. One certainly need not assume that there actually was such a systematic collection of observations: after all, Ptolemy may simply have inferred that Hipparchus did this on the basis of scattered references in the books by Hipparchus, that is, those books Ptolemy designates by title but which may have been available to him only in the form of digests. The most extensive set of dated observations from antiquity are the Babylonian Diaries that cover the period from −651 to ca. −50 (with many gaps in the extant record);[11] but, in general, the Babylonian texts that survive do not indicate the way their theories were constructed from specific observations. And, though it is sometimes supposed that Babylonian eclipses cited in the *Almagest* ultimately derive from these Diaries, the intermediate stages in the transmission have not been discovered.[12] Observational reports of astronomical phenomena (some of which are dated) occasionally appear in Greek and Latin historical and literary works written before Ptolemy's time,[13] but they are usually incidental and are never part of an astronomical argument. The same is true of most pre-Ptolemaic scientific treatises. Aristotle, for example, includes such reports in some of his works and, in at least one case, he claims to have observed the event himself ("we ourselves have observed Jupiter coinciding with one of the stars of the Twins and hiding it . . ."; *Meteor.* 343b30) without specifying the date. There are also the careful (undated) observations of the apparent size of the Sun made by Archimedes.[14] However, the earliest scientific text in Greek or Latin to report dated observations is Pliny's *Natural History*, ii.180, where eclipses are cited and associated with specific dates and specific reporters. Unlike Ptolemy, however, Pliny does not invoke these observations in order to construct a theory. In other words, Ptolemy's methods for deriving parameters from observations appear in no text prior to the *Almagest*. In fact, this is the only extant work in Greek or Latin that presents dated observations that are then used to construct an astronomical theory.

As for Ptolemy's citing specific observations, this is a defining characteristic of his concept of astronomy. In fact, the observations he adduces are the primary data or phenomena that the theory is to account for (or, in ancient terminology, "save"), and this account, a quantified geometrical model, is to be applicable to all phenomena of the same kind at any time. We note that for Ptolemy (but apparently not for his predecessors) the phenomena are the positions of a given celestial body at any time. Here again we see that Ptolemy's project is unprecedented in the extant literature. Indeed, the only basis for supposing that this project goes back to an earlier period is a set of inferences from certain passages in the *Almagest* itself. Some of these passages are discussed below, and we will argue that they do not support such inferences.

As for the question of how Ptolemy's citations of his predecessors advance his argument, this is in part a matter of the reliability that he ascribes to these observations. If one considers the period prior to Ptolemy, it would seem that for the Babylonians the reliability of observations depended on the status of the bureaucracy in whose name they were reported; whereas for some Greeks and Romans, reliability was related to the personal status of the observer/reporter. Thus, Pliny retails only astronomical observations reported by prominent military leaders and well known astronomers.

Contemporary with Pliny there is also attested the view that the basic parameters of an astronomical theory were observed. Thus, for instance, Geminus (first century A.D.) claims that the parameters for the arithmetic theory of lunar motion described in chapter 18 of his *Introduction to astronomy* were established by observation "from ancient times," without mentioning any specific observations or indicating any reservations about the results.[15] Nevertheless, though Geminus was the first to raise the idea that observations can serve as a basis for constructing theory, he did not do this by deriving parameters for models from observations as Ptolemy did a century later. Rather, his project was to explain, according to Greco-Latin standards of empirical knowledge, the Babylonian treatment of lunar motion as he found it in the numerical data that had entered the Hellenistic world during the first century B.C.[16] Thus, for Geminus the reliability of the observations fundamental to the theory was a consequence of the success of the theory.

To understand Ptolemy's appeal to observations, it is important to distinguish models from theories. A model is a geometrical representation of celestial motion, and the astronomer's task, as Ptolemy understands it, is to assign numerical parameters to a model so that it can be used to account for observed phenomena.[17] The combination of a model with its parameters is what is meant by an astronomical theory. Thus, the model has the certainty of geometry, and the parameters that are derived from observations that are themselves certain have the certainty of arithmetic, but the theory itself has only degrees of confidence. There are several reasons why this is so. First, the model may be inappropriate for the phenomena at issue; and next, the observations may not be sufficient to support a theory that can account for *all* the relevant phenomena at *any* time. Finally, there is the fact that the explanatory power of the theory (of which the model is a part) is largely a matter of its success in accounting for these phenomena.

So let us consider how Ptolemy gains confidence in the first lunar theory.[18] Ptolemy presupposes both an epicyclic model and an eccentric model (which he shows are equivalent), and applies a method requiring a set of three lunar eclipses. In *Alm.* iv.6 he presents three Babylonian eclipses (dated −720 Mar. 19/20, −719 Mar. 8/9, −719 Sep. 1/2) and three eclipses that he himself observed (dated 133 May 6/7, 134 Oct. 20/21, 136 Mar. 5/6). He then applies the same method to both sets of eclipses and obtains a value for the epicyclic radius of 5;15, very nearly (in the first case Ptolemy's result is 5;13, and in the second, it is 5;14) where the radius of the deferent is 60.[19] Thus, over a period of about 850 years, the lunar epicyclic radius had not changed. These results also allowed Ptolemy to verify and correct the lunar mean motions in longitude and anomaly.

It is our view that, though the Babylonian eclipse observations are presented before Ptolemy's own, he began with his observations because they were certain by virtue of his direct experience and his control over the observational procedures. Moreover, we think that his confidence in the theory was only enhanced (not established) by agreement of these ancient observations with his theory.[20] In other words, we propose that he accepted as reliable the Babylonian observations over which he had no direct access because of their agreement with the results of his own theory.

In *Alm.* iv.11 Ptolemy presents two additional sets of three lunar eclipses that, he says, were used by Hipparchus. (Note that Hipparchus is not mentioned in *Alm.* iv.6 where the earliest Babylonian eclipses are cited.) In this chapter Ptolemy expresses surprise that Hipparchus had found one value for the lunar eccentricity $(3144 : 327\frac{2}{3} \approx 60 : 6;15)$, and a different value for the epicyclic radius $(3122\frac{1}{2} : 247\frac{1}{2} \approx 60 : 4;46)$. Ptolemy remarks "Such a discrepancy cannot, as some think, be due to some inconsistency between the [epicyclic and eccentric] hypotheses."[21] He then shows that the data he had from Hipparchus for both sets of eclipses yield results in agreement with those he had previously demonstrated in *Alm.* iv.6.

We infer from this that Ptolemy's theory was not taken from Hipparchus, and that the goal of iv.11 is to explain the puzzling fact that Hipparchus constructed a false theory from reliable observations. Ptolemy mentions no work of Hipparchus in which he found the data given here; his analysis is clearly a reconstruction in which Ptolemy assumes that Hipparchus used the same method as he had, and he seeks to demonstrate that Hipparchus made errors in calculation. Though Ptolemy offers no argument in support of his assumption about Hipparchus's method, scholars have found one in *Alm.* iv.5, where Ptolemy says: "In this first part of our demonstrations we shall use the methods of establishing the theorem which Hipparchus, as we shall see, used before us."[22] But, as Toomer notes, this claim in iv.5 may only be an allusion to iv.11, and that passage does not really support it. Our conclusion, then, is that Ptolemy's confidence in the first lunar theory depended neither on Hipparchus's observations nor on his mathematical methods.[23]

An alternative view is that Hipparchus did not apply the same method as Ptolemy did, and that Hipparchus's method (or methods) yielded the discordant results reported in some text that Ptolemy did not identify. Jones has argued that Hipparchus invoked a Babylonian arithmetic scheme for determining solar positions (and hence lunar positions at mid-eclipse), using an arithmetic scheme for quantitative purposes while adhering to a geometric model for qualitative purposes.[24] This reconstruction thus resolves one difficulty, since there is no longer any reason to assert that Hipparchus had a table of chords and methods of trigonometry.[25] Nevertheless, we do not in general accept numerical agreement between the reconstruction and the text as a sign that the method of the ancient author has been recovered. At best, such. agreement merely establishes one possible way of reaching the desired result: it is not sufficient to eliminate other possibilities or to show that the one proposed was the way in which the result was actually obtained.[26] In this instance we also doubt that Ptolemy's source even included a procedure of any sort, for it is by no means clear that Hipparchus indicated the method used to get from a set of observations to parameters. This is an expectation we have on the basis of Ptolemy's systematic derivation of parameters from stated observations, but there does not seem to be any precedent for this in an extant text prior to the *Almagest*. Thus, we suggest that it is best to admit ignorance of the methods used by Hipparchus when there is too little evidence available to decide what they were, and we maintain that it is inappropriate to attempt fill this gap in what is known of Hipparchus by ascribing Ptolemy's methods to him.

Ptolemy's stated purpose for introducing the two sets of eclipse triples in *Alm.* iv.11 is to show that no rival theory is adequate—specifically the two theories he ascribes to Hipparchus—and that the observations used to construct those theories actually yield the parameters he derived in *Alm.* iv.6 from earlier and later eclipse triples. The first set of eclipses in iv.11 are Babylonian (dated −382 Dec. 22/23, −381 Jun. 18/19, −381 Dec. 12/13), and the second set come from Alexandria (dated −200 Sep. 22, −199 Mar. 19, −199 Sep. 11).[27] After demonstrating that the data transmitted by Hipparchus yield the parameter previously derived for the value of the lunar eccentricity (or, alternatively, the epicyclic radius), he concludes:

> Thus, the reason for the above discrepancy has come into our view as well as the fact that we can use with still more confidence (θαρροῦντες ἔτι μᾶλλον) the ratio we have demonstrated for the anomaly at lunar syzygies, since these very eclipses have been found in close agreement with our hypotheses.[28]

All in all, Ptolemy finds that his parameters for lunar motion account for lunar eclipses at about −720, −380, −200, and 135, a quite impressive result and one that rightly merits his increased confidence in his own theory.

So far we have proposed that for his first lunar theory Ptolemy depends on his own observations, which he takes to be certain, and exercises judgment in the case of reports of earlier observations. He finds at least some of the observations by his predecessors to be reliable, and his criterion for determining this is that they produce results in agreement with his theory. At the same time, to the extent that his own theory accounts for earlier observations, Ptolemy indicates that they are no longer a legitimate basis for any rival theory such as the ones Hipparchus constructed from the two sets of lunar eclipse triples. Moreover, as his treatment of Hipparchus shows, Ptolemy is quite prepared to grant that one can make and report reliable observations without being a reliable theoretician.

Let us now return to the thesis that Ptolemy adopted methods already used (if not invented) by Hipparchus. Determining mean positions for each of the planets (including the Sun and the Moon) is an essential feature of Ptolemy's procedures, for it is the first step in computing the planet's true position. So, for those who accept this thesis, there is the difficulty that Ptolemy never ascribes a mean position of the Sun or the Moon to Hipparchus. But this suggests to us that Ptolemy's theories for the Sun and the Moon were structured differently from those of Hipparchus.

As we remarked above, it is quite possible to use a geometric model to gain a qualitative understanding of variation in velocity (or, more properly, the variation in the daily displacement in longitude) while using an arithmetic scheme for the computation of positions for a given time.[29] And this is certainly consistent with the information Ptolemy provides in *Alm.* iii.1[30] about the way Hipparchus computed a solar position: Hipparchus went from the time of the preceding vernal equinox directly to the true solar position at some time following that equinox, which is quite different from Ptolemy's procedure of introducing the mean position of the Sun and a correction for its anomaly. In any case, there is no evidence that Hipparchus ever discussed the equation of time. Jones has even argued that some of the corrections Ptolemy applied to the data reported in the name of Hipparchus in *Alm.* iv.11 were to account for Hipparchus's neglect of the equation of time.[31] But without a proper understanding of the equation of time, one cannot compute correctly the time intervals between lunar eclipses or the mean position of the Moon at a given moment. So, it would seem, Ptolemy only accepted Hipparchus's observations as well as those he reported, and this entailed no commitment to Hipparchus's methods.

But beyond the question of method, it seems that Ptolemy operated in a different conceptual framework from his predecessors. In Ptolemy's account of lunar motion there are apparent phenomena, true phenomena, and mean phenomena. Now, an apparent phenomenon is a determinate celestial position that is observed by someone at a particular location on Earth, whereas a true phenomenon is a celestial position that would be seen by an observer at the center of the Earth. The differences between apparent and true phenomena are due to parallax. A mean phenomenon is a celestial position computed on the basis of mean motion in longitude only, and the differences between true and mean phenomena are to be analyzed quantitatively, using a geometric model, as periodic deviations from the mean. There is, however, no evidence that these distinctions were all in place prior to the *Almagest*, despite the fact that elements of the framework—for example, various senses of mean motion—had been applied both by Babylonian and Greek astronomers before Ptolemy.

To take another case in point, Geminus makes rather different distinctions from the ones Ptolemy introduces. For, as Geminus would have it in *Intro. ast.* i.19–22, the phenomena (i.e., such qualitative features of planetary motion as stations or variations in daily displacement in longitude) display irregularities that are to be explained by appealing to true motions that are smooth and circular in conformity with the nature of the divine (and not directly observable).[32] This is obviously not the same as Ptolemy's distinction of true and apparent phenomena. And, what is more, Geminus does not share Ptolemy's distinction of mean and true phenomena. For, in *Intro. ast.* xviii he proposes that certain celestial phenomena, that is, the observed variations in daily lunar displacement in longitude, can be analyzed as periodic departures from a mean daily displacement according to an arithmetic scheme.[33]

The originality of Ptolemy's treatment of the data is also evident in Ptolemy's second lunar theory. To construct this theory Ptolemy does not adduce any Babylonian observations; rather, he computes its parameters from a total of four observations: one observation of his own and three by Hipparchus. Ptolemy remarks that the size of the lunar epicycle seems to be greater at quadrature than at syzygy (*Alm.* v.2), and presents two observations to determine the maximum correction to the mean position of the Moon at quadrature (*Alm.* v.3). The first observation (dated 139 Feb. 9) of this pair is his own and the second (dated −127 Aug. 5), Hipparchus's. Both yield the same result, 7;40°, for the maximum correction. Ptolemy also introduces a distinction between mean and true epicyclic apogee and, to determine this variation, he turns to a pair of observations of Hipparchus (dated −126 May 2 and −126 July 7) reported in *Alm.* v.5.

It is important to notice that the reports of Hipparchus's observations used in constructing the second lunar theory include solar and lunar positions. Though the reports use verbs of seeing, Ptolemy treats these positions as computed by Hipparchus. Ptolemy then ignores these computed values, and relies only on what he takes to be the observed quanta, namely, the differences between the reported solar and lunar positions.[34] Ptolemy computes solar positions from his theory for the times in question,[35] and they do not agree with the values reported for Hipparchus. This indicates that Ptolemy does not regard Hipparchus's solar theory to be identical with his own, despite the contrary impression one gets from *Alm.* iii.4. Moreover, Ptolemy's disregard of Hipparchus's lunar positions here is consistent with the demonstration in *Alm.* iv.11 that Hipparchus's lunar theory is wrong.

This treatment of observational reports seems rather strange because Ptolemy takes what is observed to be computed, and considers values inferred from the observed data to be the observed quanta. As Britton has persuasively suggested, Ptolemy does this because he assumes that Hipparchus made his observations in the same way as he did, that only the elongations from the Sun to the Moon were measured whereas the longitudes were computed.[36] It also seems odd that Ptolemy takes the elongation associated with Hipparchus's observation in *Alm.* v.3, derives a lunar correction from it (using his own values for the true position of the Sun and the mean position of the Moon at the time of the observation), deems it to be reliable, and then uses this correction to exemplify his own theory. But, in effect, he supposes that the observed quantity is reliable because it agrees with his theories.

In addition to these peculiarities, what is really remarkable is that in *Alm.* v.5 Ptolemy constructs a theoretical component of his lunar theory without citing any of his own observations; rather, he appeals to elongations associated with Hipparchus's observations. Curiously, Ptolemy does not call attention to this, but it reveals a new aspect of his understanding of the relationship between theory and observation. Our explanation is that Ptolemy has two criteria for the reliability of obervations made by others. According to his first, an observation is reliable if it is in agreement with his theory; whereas, according to his second, it is reliable if it comes from a reliable observer, that is, from an observer who has made a number of reliable observations and is thus proven to be reliable himself. But, once Ptolemy deems an observer reliable, he considers it legitimate to use the observer's observations (or inferences from them) to construct theory.[37]

Nevertheless, Ptolemy's confidence in his second lunar theory depended to a much greater degree on his own observations of lunar elongations than on Hipparchus's reliability as an observer. These eight observations of elongations were all made between December 138 and

July 139, and no other observations by Ptolemy are recorded in the *Almagest* during this interval. The observations concern apparent positions of the Moon: to compare them with positions derived from the second lunar theory, Ptolemy corrects the true position of the Moon (derived from the theory) to its apparent position by computing the lunar parallax in longitude. In this way, these observations confirm Ptolemy's second lunar theory together with his parallax theory, taken as a whole. In one case, Ptolemy reports an elongation from the Sun to the Moon on 139 Feb. 23: he computes the true position of the Sun, adds the observed elongation to it, and remarks that this "is also the position it [the Moon] should occupy according to our hypotheses."[38] In other words, this observed elongation enhanced Ptolemy's confidence in his lunar theory, although he offers no details of his computations. In another case Ptolemy presents an observation he made of Mercury on 139 May 17/18. He first gives the position of Mercury at Gem $17\frac{1}{2}^{\circ}$ with respect to Regulus, and adds that at the same time Mercury was "$1\frac{1}{6}^{\circ}$ to the rear of the Moon's center." He then goes on to compute the true and apparent positions of the Moon at that time, and adds that "from this [computation] too we find that Mercury's longitude was Gem $17\frac{1}{2}^{\circ}$ (since it was [observed] $1\frac{1}{6}^{\circ}$ to the rear of the Moon's center)."[39] Clearly, there is no reason to include the elongation from Mercury to the Moon for the purpose of establishing the position of Mercury, because it is already determined in relation to that of the star, Regulus. Hence, the aim in observing the elongation from Mercury to the Moon is to use the position of Mercury (or, more properly, the position of Regulus) to confirm the second lunar theory together with the parallax that corrects the true lunar position to its apparent position. Now, there are many questions one might well ask about the details of Ptolemy's computations of the lunar positions as well as about the accuracy of the observed elongations. Our interest on this occasion, however, is not in these matters but with the use Ptolemy makes of these observations to confirm and thereby increase confidence in his second lunar theory.

In this account we have only discussed the observations Ptolemy uses to construct and increase confidence in his lunar theory,[40] and we have shown that he was responsible for a number of innovations. We have not, for example, discussed the unprecedented way in which Ptolemy constructed theories, nor have we addressed his use of theories to devise tables of a sort that is attested nowhere in the earlier extant ancient literature.[41] Our principal conclusion is that there are good reasons to maintain that in his lunar theory Ptolemy is indebted to his predecessors only for models and for some of the observations used to quantify them, and that he did not rely on his predecessors for their theories. Indeed, we have argued that Ptolemy assessed the reliability of previous observers, using as his criterion that their observations produce agreement with a theory based on his own observations.

ACKNOWLEDGMENT

We are most grateful to J. P. Britton for his comments on an earlier draft of this paper.

NOTES

1. For an indication of our methodology, see Bowen and Goldstein (1991), especially p. 235. A more detailed treatment is forthcoming. Suffice it to say that our analysis of Ptolemy's method of presentation is not intended to address his process of discovery.

2. Cf. Neugebauer (1969), 191 ff. We are not persuaded by the proposal in some modern accounts, e.g., Petersen (1969), that Ptolemy had three theories; rather, we hold that the second theory is presented in two stages (cf. Neugebauer 1975, 84 ff.).

3. Toomer (1984), 176n10.

4. Toomer (1984), 176; cf. 178.

5. Such lists of eclipse reports have been found among Babylonian documents (see Britton 1989, 4, 5, and 49nn9–10; Aaboe et al. 1991), but their mere existence does not imply that they were available to the Greeks or that they were ever used for this purpose. In any case, the citation of a few Babylonian eclipse reports in the *Almagest* does not warrant supposing that the entire list of eclipses was translated into Greek. Moreover, if Hipparchus simply borrowed Babylonian period relations, he did not need any mathematical procedure to discover them. As for Ptolemy, he concludes his discussion with the remark that "we devised [procedures] to check these values in simpler and more practical ways" (Toomer 1984, 179). On Ptolemy's description of this procedure, cf. Neugebauer (1975), 73, where it is noted that Ptolemy does not give the dates of the eclipses used by Hipparchus; that Ptolemy's own procedure for the determination of the mean lunar anomaly does not rest on the selection of two pairs of eclipses; and that "it seems as if [Ptolemy's] analysis is meant as an explanation of the fact that Hipparchus's parameters were not final but required certain corrections . . . ." In other words, there is no reason to suppose that Ptolemy reports a procedure that anyone had actually used.

6. Ptolemy's procedure is described in Neugebauer (1975), 73 ff.

7. There is one case where an eclipse mentioned in the *Almagest*—not used for determining the model for lunar longitude—is also described in a Babylonian text, but there are significant differences between the two accounts (cf. Britton 1992, 58 ff.).

8. See Toomer 1984, 190, 223–4, and 227, respectively.

9. Since Ptolemy's lunar theory is better in several respects than the observations cited, it is amusing to reflect that his modern reputation might have been enhanced by omitting the observational data.

10. See especially Ptolemy's introduction to the planetary theory in *Alm.* ix.2; Toomer (1984), 421.

11. Sachs 1974, 48; see also Sachs and Hunger 1988–96.

12. For an analysis of early observational reports in the *Almagest* and other ancient Greek texts, see Goldstein and Bowen (1991), and the references cited therein.

13. For eclipses, see, e.g., Herodotus, *Hist.* vii.37; Thucydides, *Hist.* ii.28; Livy, *Ab urbe cond.* xliv.37.

14. Cf. Shapiro (1975).

15. Geminus, *Intro. ast.* xviii.4; Aujac (1975), 94. For an analysis of this chapter, see Bowen and Goldstein (1996).

16. Some authors have maintained that these Babylonian data were already available in the second century B.C. (e.g., Jones 1991a, 117 ff.), but this view is primarily an inference from the data Ptolemy ascribes to Hipparchus. If one considers only the extant sources from the period before Ptolemy, one finds a paucity of Babylonian data prior to the first century B.C.—largely related to the calendar, length of daylight, and star risings—and an impressive amount of Babylonian data beginning in the first century B.C. See Goldstein (1997).

17. According to Ptolemy, the task of the harmonic theorist is similar: cf. Bowen (1999).

18. Ptolemy seems to have used his first lunar theory to enhance his confidence in his solar theory. For, he computes the longitudes of the Moon at the middle of lunar eclipses (when the Moon is in opposition to the Sun) from his solar theory: the solar theory yields the true longitude of the Sun, and the true longitude of the Moon at that time is found simply by adding 180° to this solar longitude. So, had the solar theory been seriously deficient, no coherent lunar theory could have been based on it. Ptolemy hints at this when he argues in *Alm.* iii.1 (Toomer 1984, 136) that there is no second solar inequality because it would destroy the agreement of his first lunar theory with the observed times of eclipses: cf. Britton (1992), 40.

19. See Toomer (1984), 197 and 202, respectively.

20. See Ptolemy, "*On the* Kriterion *and* Hegemonikon," in Huby and Neal 1989, 201 *et passim*. In *An Outline of Empiricism*, Galen (129 to ca. 200) reports on sources of knowledge used by the Empiricists. For one such source of knowledge, Galen introduces the term "autopsia" meaning "one's own perception" (Walzer and Frede, trans., 1985, 25). Another source of knowledge, related to the previous one, is "history," meaning a set of reports. As Galen writes,

The first and foremost criterion of true history, the empiricists have said, is what the person who makes the judgment has perceived for himself. For, if we find one of those things written down in a book by somebody which we have perceived for ourselves, we will say that the history is true. (Walzer and Frede, trans., 1985, 35)

For our purposes it does not matter whether in fact there were Empiricists who held these views; we only wish to show that views of this kind were in circulation at the time of Ptolemy, and Galen's text is sufficient to demonstrate this (cf., e.g., Walzer and Frede, trans., 1985, 23). Similarly, it is not necessary to claim any interaction between the medical and the astronomical traditions, for the evidence that Ptolemy has appealed to these two sources of knowledge is apparent in the *Almagest*. And we are certainly not claiming that Ptolemy was an Empiricist.

21. Toomer (1984), 211.

22. Toomer (1984), 181.

23. It is often said that Ptolemy depended on the theory derived by Hipparchus from a set of observations, and that his own observations served, at best, merely to confirm the received theory. For moderate views, see, e.g., Neugebauer (1975), 315, 318, *et passim*; Toomer (1973); Jones (1991a, 1991b).

24. Jones (1991a, 110 ff.). For such an arithmetic scheme in Greek texts, see Jones (1983, 1997); for evidence that Hipparchus used an arithmetic scheme for lunar longitude, see *Alm.* v.3; Toomer (1984), 224 and 224n14.

25. The claims in Toomer (1973) for a Hipparchan table of chords and use of trigonometric methods cannot be sustained for various reasons, and the assertion by Theon of Alexandria that Hipparchus had written a work on chords in twelve books is dubious at best. Toomer (1984, 215n75) acknowledged some errors in his previous article, and added that he was no longer certain that Hipparchus used a chord table with base $R = 3438$ while still claiming that he was essentially right in 1973. Jones (1991b), 442n5, accepted Toomer's original argument.

   On Vettius Valens's report of Hipparchan solar tables, see Jones 1991b, 446 f.

26. See Bowen and Goldstein (1996). The same point is made in a somewhat different context in Neugebauer (1969), 50, *ad* 19.

27. Cf. Jones (1991a), 105 ff. Ptolemy also cites three lunar eclipses observed by Hipparchus, but he does not use them to construct his lunar theory—two appear in *Alm.* iii.1 (Toomer 1984, 135) in the course of an account of the length of the year (cf. vii.2; Toomer 1984, 327n50); and one in vi.5 (Toomer 1984, 284) as part of a determination of eclipse limits.

28. Our translation; cf. Toomer (1984), 216.

29. Cf. Jones (1991a), 103.

30. Toomer (1984), 135.

31. Jones (1991a), 108.

32. Aujac (1975), 5.

33. Aujac (1975), 94–98; cf. Bowen and Goldstein (1996).

34. See Toomer (1984), 225 n17. The elongations in these observational reports are clearly inferences from stated solar and lunar positions, but there is no credible way to decide whether these inferences belong to the reports themselves.

35. *Alm.* v.3 and v.5: cf. Britton (1992), 101–4.

36. We accept the argument in Britton (1992), 101n35, somewhat against Toomer, as Britton notes. This interpretation of the Hipparchan reports neither entails nor warrants the assertion (cf., e.g., Toomer 1978, 213) that Hipparchus's reasons for making these observations can be determined from the use Ptolemy makes of them.

37. A simpler instance where Ptolemy derives a basic parameter without appealing to any of his own observations is found in *Alm.* v.14. In this case, Ptolemy has accepted the Babylonians as reliable observers of lunar eclipses (as he showed in the discussion of his first lunar theory), and derives a value for the apparent diameter of the Moon from two Babylonian lunar eclipse reports (cf. Toomer 1984, 253 f).

38. *Alm.* vii.2; Toomer (1984), 328. Cf. Britton (1992), 99.

39. *Alm.* ix.10; Toomer (1984), 461. Cf. Britton (1992), 114–15.

40. Properly speaking, Ptolemy has only one lunar theory, though he presents it in two parts for, as we have stipulated, a theory has to account for positions of a celestial body at all times. This theory is what we have designated as his second lunar theory: it is appropriate for finding the lunar positions at all times, and it reduces to what we have designated as his first lunar theory at syzygies.

41. As a number of recently discovered papyri (notably by Jones) have disclosed, Babylonian tables based on arithmetic schemes were available in Greek. But, while Ptolemy was probably aware of these Babylonian arithmetic schemes, he does not appeal to them at all (cf. Jones 1991a, 122; Jones 1997).

REFERENCES

Aaboe, A, J. P. Britton, J. A. Henderson, O. Neugebauer, and A. J. Sachs 1991. *Saros cycle dates and related Babylonian astronomical texts.* Transactions of the American Philosophical Society 81:6. Philadelphia: American Philosophical Society.

Aujac, G. 1975. *Géminos: Introduction aux phénomènes.* Paris: Les Belles Lettres.

Bowen, A. C. 1999. The exact sciences in Hellenistic times. See Furley (1999), 287–319.

Bowen, A. C., and B. R. Goldstein 1991. Hipparchus' treatment of early Greek astronomy, *Proceedings of the American Philosophical Society* 135:233–54.

——— 1996. Geminus and the concept of mean motion in Greco-Latin astronomy, *Archive for History of Exact Sciences* 50:157–85.

Britton, J. P. 1989. An early function for eclipse magnitudes in Babylonian astronomy, *Centaurus* 32:1–52.

————— 1992. *Models and Precision: The Quality of Ptolemy's Observations and Parameters.* New York: Garland.

Furley, D. J., ed. 1999. *Routledge History of Philosophy: II. From Aristotle to Augustine.* London: Routledge.

Gillispie, C. C. 1970–80. *Dictionary of Scientific Biography.* New York: Scribner.

Goldstein, B. R. 1997. Saving the phenomena: The background to Ptolemy's planetary theory, *Journal for the History of Astronomy* 28:1–12.

Goldstein, B. R., and A. C. Bowen 1991. The introduction of dated observations and precise measurement in Greek astronomy, *Archive for History of Exact Sciences* 43:93–132.

Huby, P., and G. Neal 1989. *The Criterion of Truth: Essays in Honour of George Kerford together with a Text and Translation of Ptolemy's* On the *Kriterion* and *Hegemonikon.* Liverpool: Liverpool University Press.

Jones, A. 1983. The development and transmission of 248-day schemes for lunar motion in ancient astronomy, *Archive for History of Exact Sciences* 29:1–36.

————— 1991a. Hipparchus's computations of solar longitudes, *Journal for the History of Astronomy* 22:101–25.

————— 1991b. The adaptation of Babylonian methods in Greek numerical astronomy, *Isis* 82:441–53.

————— 1997. Studies in the astronomy of the Roman period: I. The standard lunar scheme, *Centaurus* 39:1–36.

Neugebauer, O. 1969. *The Exact Sciences in Antiquity.* 2d ed. New York: Dover.

————— 1975. *A History of Ancient Mathematical Astronomy.* Berlin: Springer.

Petersen, V. M. 1969. The three lunar models of Ptolemy, *Centaurus* 14:142–71.

Sachs, A. 1974. Babylonian observational astronomy, *Philosophical Transactions of the Royal Society of London* A 276:43–50.

Sachs, A., and H. Hunger 1988–96. *Astronomical diaries and related texts from Babylonia.* vols. 1–3. Vienna: Verlag der Österreichischen Akademie der Wissenschaften.

Shapiro, A. 1975. Archimedes's measurement of the Sun's apparent diameter, *Journal for the History of Astronomy* 6:75–83.

Toomer, G. J. 1973. The chord table of Hipparchus and the early history of Greek trigonometry, *Centaurus* 18:6–28.

————— 1978. Hipparchus. See Gillispie (1970–80), 15:207–24.

————— 1984. *Ptolemy's* Almagest. New York: Springer.

Walzer, R., and M. Frede, trans. 1985. *Galen: Three Treatises on the Nature of Science.* Indianapolis: Hackett.

12

Theon of Alexandria and Ptolemy's *Handy Tables*
Anne Tihon

Theon of Alexandria is a well-known figure of ancient Greek astronomy. He flourished in Alexandria around 364 A.D. and wrote several commentaries on Ptolemy's astronomical treatises—the *Almagest* and the *Handy Tables*. Evidently devoted to his daughter, the famous philosopher Hypatia, he put her name in the title of Book III of his *Commentary on the Almagest*. Apart from that, we know nothing of his life except that he himself observed a solar eclipse in Alexandria on the 16th of June 364 A.D.

The notice by Suidas[1] attributes to him many works, but the preserved works of Theon are the following (listed here in chronological order)[2]:

- a Commentary to the *Almagest* (= CA)
- a "Great Commentary" on the *Handy Tables* in five books (= GC)
- a "Small Commentary" on the same *Handy Tables* in one book (= PC)

I have edited the "Small Commentary" and Books I to III of the "Great Commentary." The edition of Book IV is now finished. Unfortunately this Book, which covers the planetary theory, is incomplete, and Book V is missing. The preserved part of Book IV deals with the mean motions of the five planets and to some extent with the tables of planetary anomalies. But the most interesting part of the planetary theory in the *Handy Tables*—the latitude theory—is missing.

In this paper I will try to summarize Theon's approach to Ptolemy's *Handy Tables*. Ptolemy himself did not leave us a clear account of the structure of the *Handy Tables*: the short treatise edited by Heiberg among the *opera minora* of Ptolemy[3] merely explains the use of the tables, without saying anything about how they were calculated. Theon's "Great Commentary" is thus the only ancient work we have that provides fully detailed comments on the structure of the tables.

Having edited almost completely the two commentaries on the *Handy Tables*, I have come to three conclusions, based mainly on the "Great Commentary":

(1) Theon did not introduce any change in the *Handy Tables*;

(2) Theon explains the tables as they are preserved in the extant manuscripts;

(3) Theon had no special information concerning the structure of the tables, that is, nothing other than what we have today—Ptolemy's short treatise and the tables themselves.

(1) The opinion that Theon himself has modified Ptolemy's *Handy Tables* is widespread in many modern studies. I suppose that it has been prompted by the Arabic tradition.[4] But there is nothing in Theon's treatises that supports this idea. Indeed, in some cases one can see clearly that Theon did not understand exactly how a particular table has been calculated. I will show some examples below. Of course Theon might have introduced something into the checking of the numerical values recorded in the tables, but not into their structure.

(2) In my comments of Theon's work, I have tried to compare as far as possible what Theon says with the tables as preserved in the most ancient manuscripts. I have never discovered any discrepancies except in small details. Of course, I could not check each table entirely, and this conclusion will perhaps have to be modified after further study. It is important to note that the *Handy Tables* cannot be considered as a single whole: each table needs a separate and careful analysis.

(3) In the preface to the "Great Commentary" Theon asserts that he is the first who tried to write such a commentary. Such statements are common in the ancient literature and cannot be taken too seriously. We know that there were people who dealt with the *Handy Tables* and especially with the concordance (συμφωνία) between the *Almagest* and the *Handy Tables*, as appears from a fragment of 213 A.D. reedited recently by A. Jones.[5] But the only author other than Ptolemy quoted in the "Great Commentary" is Serapion in a passage that I consider an interpolation.[6] For the rest, Theon's method seems to be pure deduction based on use of the tables themselves, comparison with the *Almagest*, or Ptolemy's instructions.

These conclusions might seem disappointing, and one might well ask: what is Theon's contribution to our understanding of the *Handy Tables*?

Generally speaking one can say that Theon employs, for better or worse, an ancient method of explaining the tables. However, a summary judgment of the "Great Commentary" is not really possible, because some parts are in a finished form, while others appear to be early drafts. Moreover, it is hard to say what is due to a corrupted manuscript transmission and what is due to the author himself.

Theon's aims, as he himself explains in his preface, are to explain:

• how the numbers written in the tables have been calculated, in order to correct them in case of errors in the copies; and

• the differences between the *Almagest* and the *Handy Tables*.

He only partly completed this program, as can been seen from the following examples.

(1) The first example is taken from the parallax tables.

A complete description of the parallax tables in the *Handy Tables* has been given in a recent paper by J. Chabás and myself.[7] I summarize here our conclusions.

The difference between lunar and solar parallaxes in longitude and latitude are given in columns 2 and 3 respectively. These values have to be calculated from the vertical parallaxes given in the *Almagest*. Theon's method here is clearly and neatly presented. Chabás and I have checked the complete parallax tables according to Theon's method. The results of our calculations are very satisfactory: they generally agree with the values given in the manuscript and also with the results obtained by the modern formula.

For column 4 the situation is completely different. This column gives an arc of the horizon which is used for calculating the "prosneusis" in the eclipse calculations.[8]

This has no equivalent in the *Almagest*, and Theon's method, given in Book III of the "Great Commentary,"[9] is extremely long. His method is illustrated by an example that consists of calculating the value of column 4 for Leo 0°, 3 equinoxial hours P.M., in the climate of Rhodos. After a calculation that fills 18 printed pages, with several mistakes here and there, Theon reaches a final result of 99;45°, which corresponds, he says, to the 100° written in the table for the chosen example. Of course Theon's method could not be used for checking the 1158 entries of the parallax tables. For this column we had to use a modern formula[10] and the results are very unsatisfactory. They agree with the manuscript values in only 332 cases. This means that the modern formula does not exactly represent the ancient method, and that the manuscript transmission for this column has been corrupted; in fact, there are numerous textual variants for this column. This is no accident: it corresponds to a problem for which Theon was unable to propose a quick and efficient method and for which the *Almagest* offers no help.

(2) A second example is the table of correction to the parallax.[11] Here it is clear that Theon did not exactly understand the structure of the table.

In the *Handy Tables*, columns 3 and 4 of the "table of correction" to the parallax must be used for correcting the values given in the parallax tables both for longitude and latitude, when the Moon occupies an intermediate position between the apogee and perigee of the epicycle (column 3), or when the epicycle itself is found at an intermediate position between the apogee or perigee of the eccentric (column 4).

The values given in the *Handy Tables* correspond to those of the *Almagest* according to the following formula:

column 3 (HT) = column 7 (Alm) × 0;12

column 4 (HT) = column 9 (Alm) × 0;32

The constant values 0;12 and 0;32 are simplifications introduced by Ptolemy. But Theon says nothing of these constant values. For him they have to be calculated each time, and they represent the ratio:

(maximum parallax − minimum parallax) / minimum parallax

In the case of column 3, this formula will always produce the constant value of 0;12 and it would be a waste of time and effort to recalculate it for each entry in the table as Theon recommends one to do. But in the case of column 4, the formula given by Theon does not give a constant value of 0;32, but a range of values between 0;27 and 0;41. Theon has chosen an example which gives exactly 0;32: so his method seems to coincide exactly with Ptolemy's calculation, although it cannot be applied to the whole table. Ptolemy's choice of the value 0;32 is not clear and Theon's purely logical explanations do not represent the real structure of the table.

(3) Another example can be found in the calculation of syzygies. In the *Almagest*, the moment of the true syzygy is calculated by this formula:

$t = (13/12)\Delta\lambda/v_L$

in which t represents the time between the mean and the true syzygy, $\Delta\lambda$ the difference between lunar and solar longitudes at the moment of the mean syzygy, and v_L the lunar velocity.

This simple formula supposes among other simplifications that the lunar velocity remains unchanged between the mean and the true syzygy. However the lunar velocity can vary from 0;30° to 0;36° approximately and this might produce the wrong estimate of the moment of the true syzygy. In his treatise on the *Handy Tables*,[12] Ptolemy introduces a correction consisting of the following steps:

(a) take $\frac{1}{4} + \frac{1}{30}$ of the distance between longitudes of Sun and Moon;

(b) the result is added to (or subtracted from) the "center of the Moon," i.e., the position of the Moon on the epicycle, or anomaly;

(c) with help of this corrected anomaly obtain a new correction (*prosthaphairesis*) which is applied to the longitude of the Moon. Then repeat the calculation of the syzygy.

This procedure, which I cannot discuss in detail here,[13] has embarrassed modern commentators and in my opinion remains unclear. Theon explains it as a way of taking into account the fact that between the mean and the true syzygy the lunar epicycle does not remain at the apogee of the eccentric as implied in the procedure in the *Almagest*.

This will appear clearly from the following figures. Figure a shows the situation of the lunar eccentric and epicycle at the moment of the mean syzygy:the centre of the epicycle (C) is exactly at the apogee (A) of

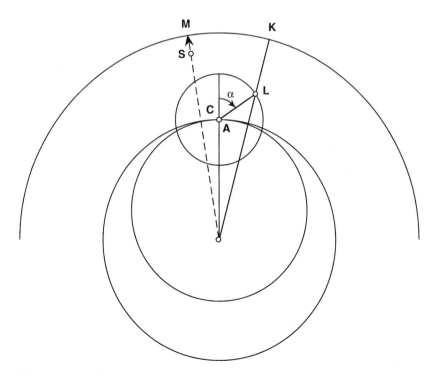

Figure a

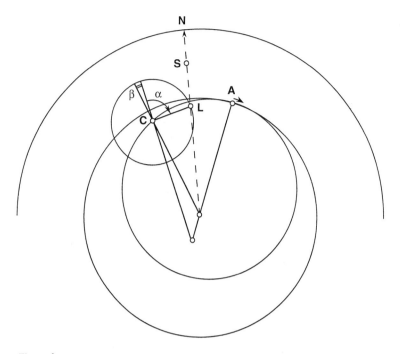

Figure b

the eccentric. But there is some longitudinal interval between the Sun (S) and Moon (L) due to their positions on their epicycles. The interval KM is shown on the ecliptic.

At the moment of the true syzygy, Sun and Moon occupy the same position N on the ecliptic, but the lunar epicycle is no longer at the apogee of the eccentric. So, according to the complete lunar model described by Ptolemy, a correction must be applied to the position of the Moon on its epicycle (α), as shown in figure b by the small angle β. This angle is tabulated in the table of complete lunar anomaly col. 3 (in both *Alm.* and *H.T.*).

According to Theon's explanations, $\frac{1}{4} + \frac{1}{30}(= \frac{17}{60})$ of the distance KM represents the same value as the angle β. Thus, steps (a) and (b) of the procedure discussed here are the equivalent to correcting the position of the Moon on its epicycle by adding (or subtracting) the correction β:

$$\alpha \pm (17/60 \times \Delta\lambda) = \alpha \pm \beta = \alpha'$$

This is roughly correct if one considers only the situation of a syzygy, when the distance between Sun and Moon remains quite small. Al-Battânî describes the same procedure, but replaces Ptolemy's and

Theon's value $\frac{1}{4} + \frac{1}{30}$ by the more precise value $\frac{1}{4} + \frac{1}{24}$.[14] One can certainly accept this part of Theon's comments.

In the "Great Commentary" only steps (a) and (b) of the procedure are explained, while the explanation of the last step (c) is simply dropped. Apparently step (c) seemed obvious to Theon, but for us it appears rather odd. For when the angle β of the second lunar anomaly must be introduced into the calculation, the Moon has moved on the epicycle. But the whole procedure seems to proceed exactly as if the Moon remained at a fixed position on its epicycle, which is obviously wrong. This particular point is without any sound justification.

Thus Theon's explanations remain quite unsatisfactory. This is perhaps not due to Theon himself. The text of this chapter indeed seems to be dramatically corrupted. However, one can see that this correction was already discussed in Theon's time, since the text of the manuscript contains this remark: "(we will show) that this value is indeed $\frac{1}{4} + \frac{1}{30}$ of the distance, and not as some people think. . . ." Unfortunately there is a lacuna just here and unless we discover a new manuscript we will never know what value had been proposed by "some people"!

In the calculation of a syzygy, one has to divide $\frac{13}{12}$ of the distance between Sun and Moon by the hourly motion of the Moon in order to find the time interval from the mean to the true syzygy. The lunar velocity is found with an argument equal to the position of the Moon on its epicycle, that is its anomaly. In the *Almagest*, the lunar velocity is calculated with help of the mean anomaly α. But in the *Handy Tables* we have now a corrected anomaly α′, which is normally used for the calculation of the lunar velocity. However, in Ptolemy's treatise on the *Handy Tables*, this point is not clearly established,[15] nor is the "Great Commentary" clear in the passage concerning the calculation of lunar velocity.[16] But the "Small Commentary" is more explicit: in the case of a syzygy the velocity of the Moon is calculated with the anomaly α′, that is, the anomaly α corrected by step (b).[17] The Byzantine syzygy calculations which follow the Ptolemaic tradition of the *Handy Tables* also use the corrected anomaly α′ for the calculation of lunar velocity.[18] So one can see that in the procedure of the *Handy Tables* the lunar velocity is calculated on basis of the true anomaly, as one will find later in Regiomontanus.[19]

These examples are cases in which Theon's explanations are quite unsatisfactory, and they might lead to an entirely negative judgment on Theon's work. This would, however, be unjustified.

Table 1

Planet	Parameter	Théon GC IV	Hyp. Pl. (Greek)[a]	Hyp. Pl. (Arabic)[b]
Saturn	u	228;24	228;24	228;24
	k	211;2	210;38	210;38
Jupiter	u	156;23	156;24	156;24
	k	292;23	292;43	292;20·
Mars	u	110;54	110;44	110;54
	k	356;7	356;20	356;7
Venus	u	50;24	50;24	50;24
	k	177;17	177;12	177;16
Mercury	u	185;24	185;24	185;24
	k	42;16	52;16	42;16

a. Ptolemy, op. min. 71–145
b. Morelon (1993)

For many other problems Theon's account appears to be perfectly clear: all the ἐποχαὶ of the mean motions as well as their increments are clearly established. This is especially interesting because we can compare them with the data given in Ptolemy's *Planetary Hypotheses*. The ἐποχὴ of the *Hypotheses* is the same: the first year of Philipp's era, the 1st of the Egyptian month Thoth at noon. But there are many discrepancies between the Greek text and the Arabic tradition. A comparison with the "Great Commentary" gives the following results in table 1, where:

u = apogee of the eccentric from Aries 0° in direct sense

k = center of the epicycle from the apogee in direct sense

Theon's values are derived strictly from the *Almagest* except that in some cases there is a difference of 1 or 2 minutes. They agree better with the Arabic version than with the Greek text as edited by Heiberg. I think that the figures given in the Greek text of the *Hypotheses* are badly corrupted: in the case of the "Great Commentary," all the values are established by careful and detailed calculation. That is probably the reason they were preserved fairly well in the manuscripts. In the *Hypotheses* on the contrary there is no explanation and so there is no way to check any corruption.

Let us take the case of the center of the epicycle of Saturn. Theon's value 211;2° was written in Greek CIA B; in the uncial script this can be easily corrupted in CI ΛB (210;32°) ; later in the minuscule handwriting, β (= 2) is often confused with η (= 8). Hence the strange result found in both Arabic and Greek: 210;38° instead of the correct value 211;2°.

Almost all the differences between Theon and the *Hypotheses* might have resulted from similar confusion. A new critical edition of Ptolemy's text would be welcome.

Theon is the only ancient author who explains a curious method used for interpolating the tables.[20] This happens when a table in the *Almagest* progresses by 3° or 6°, while the corresponding table in the *Handy Tables* progresses by 1°.

If one has for example a difference of 15′ corresponding to a space of 6°, how does one find the values corresponding to 6 steps of 1°?

First one divides the 15′ into two groups corresponding each to an argument of 3°:

3°　　　+7′ + 1′

3°　　　+7′

Then each of them is resolved into:

1°　　　+2′ + 1′

1°　　　+2′ + 1′

1°　　　+2′

and

1°　　　+2′ + 1′

1°　　　+2′

1°　　　+2′

In each section of 3°, one must reproduce the general scheme of the table: if the increases are bigger toward the apogee, the excess over the uniform progression must be given to the arguments that are nearer to the apogee. If we draw a graph of a table calculated that way, the result will not be a curve or a straight line but a progression with definite steps, which is the reason why I called such a procedure "stepwise interpolation."[21]

This procedure gives a result in no way better than a linear interpolation, as Theon himself asserts in book IV, and it is only a very quick and ingenious method that avoids dealing with sexagesimal fractions less than one minute. It was certainly not invented by Theon, although its origin is unknown (perhaps Egyptian?).

It is time to conclude. Theon was certainly no genius, but he performed a very useful task that helped generations of students in astronomy to progress to Ptolemy's level, in the Alexandrian schools, in the Byzantine period, and in the Arabic world. In spite of its limitations Theon's

"Great Commentary" certainly contributed to the sound transmission of the *Handy Tables* within the Alexandrian schools, and from them to us through the manuscripts. Unfortunately the "Great Commentary" has reached us in only a corrupted and incomplete tradition that goes back up to the unique *Vat. gr.* 190. When I look at the apparatus of my edition, I feel depressed on reading all the *correxi, conieci, supplevi, delevi, addidi, cancellavi* ... which would be superseded in case a better manuscript is discovered. One can hope at least to discover some fragment of the missing parts in the abundant scholia to Ptolemy's *Almagest*.

Let us turn now to the *Handy Tables* themselves. A critical edition is greatly needed. But this will be a heavy task: the *Handy Tables* are preserved in approximately 45 manuscripts, which must be carefully described and if possible placed in a *stemma codicum*; the complete set of tables is very long, and as I said before, each table requires a separate and careful analysis. The method of editing such tables gives rise to problems. For example, in the paper quoted before, Chabás and I have proposed a kind of "ideal" reconstruction of the tables of parallaxes. But in the case of an edition, it would be very artificial to present such a reconstruction. No ancient or medieval user of the *Handy Tables* ever met such a perfect table! So it would probably be better to present the tables as they appear in the most ancient manuscripts—the four copies in uncial script—but it is not easy to decide which one must be considered as "the best one."

For the present I have an elaborate classification of the tables that helps me to master all the material found in the manuscripts.[22] I have tested it against 30 of them. The classification is:

A = astronomical tables

B = tables for Byzantium

C = chronological tables

G = geographical tables

S = supplementary tables, this last category being destined to grow endlessly!

Thereafter one will have to compare the manuscripts' tables to the fragments preserved in the papyri, a task that will be possible thanks to the intended edition of A. Jones.

Now that the edition of Theon's "Great Commentary" is finished, we can hope to make some progress in the edition and analysis of the *Handy Tables* themselves.

ABBREVIATIONS

JHA = Journal for the History of Astronomy
MIDEO = Institut Dominicain d'Etudes Orientales du Caire. Mélanges
RHT = Revue d'histoire des textes

NOTES

1. Suidas (ed. Adler) II, 702.

2. All references concerning the editions of Theon's works are given in the bibliography at the end of this paper (see "Theon").

3. Ptolemy, *op. min.*, 159–185.

4. Many ancient Arabic astronomers refer to the *Handy Tables* under Theon's name: see, for example, al-Battânî (ed. Nallino) I, 69; 82; 84; 105.

5. Jones (1990).

6. See discussion in GC I, 288ss.

7. Chabás-Tihon (1993).

8. See the figure and explanation in GC III, 322 ss. Also HAMA, 992ss.

9. GC III, 207, 15ss (Greek text); 262 ss (French translation); 322 ss (commentary).

10. Given by Neugebauer in HAMA, 993.

11. GC II, 118–142; HAMA, 994–996.

12. Ptolemy, *op. min.*, 176–177.

13. A full discussion is given in GC II, 154–168. See also HAMA, 1001 and van der Waerden (1958), 74.

14. Al-Battânî (ed. Nallino), I, 93 ($\frac{1}{6} + \frac{1}{8} = \frac{7}{24} = \frac{1}{4} + \frac{1}{24}$).

15. See my remarks in GC II, 164 note 27.

16. GC II, 55, 12ss (Greek text); 88 (translation); 168 (commentary).

17. PC, 265, ll. 6–7 (Greek text); 333 (translation). See also the example, PC, 273 (Greek text); 335 (translation).

18. For exemple in Nicephorus Gregoras, 179 (§61) and also in several unedited Byzantine texts.

19. Goldstein (1992). One wonders if Regiomontanus had been influenced in some way by Theon or al-Battânî. It is interesting to note that Regiomonanus had access to

Theon's "Great Commentary": this treatise is contained in the *Norim'bergensis* Cent V app. 8, a manuscript which was given by Bessarion to Regiomontanus, but there is no annotation of Regiomontanus's hand in the margins of Theon's text (see description of that manuscript in GC I, 16–17). On the other hand, we know that Bessarion himself had studied astronomy in the school of Plethon in Mistra. In Plethon's school, al-Battânî's astronomy was very influential as appears from the astronomical tables compiled by Plethon on basis of al-Battânî (cf. the edition by A. Tihon and R. Mercier, in the Corpus des Astronomes Byzantins IX).

20. See GC I, 254 ss. Also CA III, 894 (note 3).

21. See GC I, 259. A. Rome, who was the first to draw attention to this method, comments on it as follows: "Il y a là un essai intéressant de faire entrer en ligne de compte l'allure générale de la fonction" (Rome [1939], 219).

22. Tihon (1992).

REFERENCES

ANCIENT AND MEDIEVAL AUTHORS

al-Battânî = Al-Battânî sive Albatenii opus astronomicum, ed. C. A. Nallino, 2 vols., Milan, 1903–1907 (reprint Frankfurt Minerva 1969).

Gregoras = Nicéphore Grégoras. Calcul de l'éclipse de Soleil du 16 juillet 1330 par J. Mogenet (†), A. Tihon, R. Royez, A. Berg, Corpus des Astronomes Byzantins I, Amsterdam, Gieben, 1983.

Pléthon = Georges Gémiste Pléthon. Manuel d'astronomie, par Anne Tihon et Raymond Mercier, Corpus des Astronomes Byzantins IX, Louvain-la-Neuve, Academia-Bruylant, 1998.

Ptolemy, Alm. = Claudii Ptolemaei opera quae exstant omnia. Vol. I: Syntaxis Mathematica, ed. J.-L. Heiberg, 2 vols., Leipzig, Teubner, 1898–1903.

Ptolemy, op. min. = Claudii Ptolemaei opera quae exstant omnia. Vol. II: Opera astronomica minora, ed. J.-L. Heiberg, Leipzig, 1907.

Suidas = Suidas Lexicon, ed. A. Adler, I–IV, Leipzig, 1928–1938.

THEON

CA = A. Rome, Commentaires de Pappus et de Théon sur l'Almageste, t. II–III, Studi e Testi 72 et 106, Città del Vaticano, 1936–1943.

GC I = J. Mogenet, Le "Grand Commentaire" de Théon d'Alexandrie aux Tables Faciles de Ptolémée, livre I (commentaire par A. Tihon), Studi e Testi 315, Città del Vaticano, 1985.

GC II–III = A. Tihon, Le "Grand Commentaire" de Théon d'Alexandrie aux Tables Faciles de Ptolémée, livres II–III, Studi e Testi 340, Città del Vaticano, 1991.

GC IV = A. Tihon, Le "Grand Commentaire" de Théon d'Alexandrie aux Tables Faciles de Ptolémée, livre IV, Studi e Testi 390, Città del Vaticano, 1999.

PC = A. Tihon, Le "Petit Commentaire" de Théon d'Alexandrie aux Tables Faciles de Ptolémée, Studi e Testi 282, Città del Vaticano, 1979.

Modern Studies

Chabás, J., and Tihon, A. (1993). Verification of Parallax in Ptolemy's Handy Tables. JHA XXIV, 123–141.

Goldstein, B. (1992). Lunar velocity in the Ptolemaic tradition. The Investigation of Difficult Things, 3–17, Cambrige University Press, 3–17.

Jones, A. (1990). Ptolemy's First Commentator. Transactions of the American Philosophical Society, vol. 80, part 7, Philadelphia.

Morelon, R. (1993). La version arabe du Livre des Hypothèses de Ptolémée, MIDEO 21, Louvain-Paris.

Neugebauer, O.: see HAMA

Rome, A. (1939). Le problème de l'équation du temps chez Ptolémée. Annales de la Société scientifique de Bruxelles, LIX, série 1, 211–224.

Tihon, A. (1992): Les Tables faciles de Ptolémée dans les manuscrits en onciales. RHT XXII, 47–87.

van der Waerden, B.L. (1958): Die handlichen Tafeln des Ptolemaios. Osiris 13, 54–78.

Notes on Contributors

Asger Aaboe is Emeritus Professor of Mathematics, History of Science, and Near Eastern Languages and Civilizations at Yale University. His work deals with the history of the exact sciences, mostly in antiquity, and especially with the arithmetical astronomy of the Late-Babylonian period.

Alan C. Bowen is director of the Institute for Research in Classical Philosophy and Science. His research interests and publications concern ancient harmonic science and astronomy in Greek and Latin sources. He has just finished *Physics and Astronomy in Later Stoic Philosophy: Cleomedes, "On the Heavens"* with Robert B. Todd for the University of California Press, and is currently writing a book on pre-Ptolemaic planetary theory.

Lis Brack-Bernsen teaches history of science at the universities of Frankfurt and Regensburg. Her research is concerned with the history of the exact sciences, particularly of ancient astronomy. She is author of *Zur Entstehung der babylonischen Mondtheorie: Beobachtung und theoretische Berechnung von Mondphasen* (Stuttgart: Franz Steiner Verlag, 1997).

John Britton received his Ph.D. in the history of science from Yale University. His published research includes topics in Babylonian astronomy and mathematics and editions of related cuneiform texts. He lives and works in Wilson, Wyoming.

Bernard R. Goldstein is University Professor Emeritus at the University of Pittsburgh. In addition to collaborating on a series of articles concerning ancient Greek science with Alan C. Bowen, he has been engaged for many years in the study of medieval and early modern astronomy based on sources in Arabic, Hebrew, and Latin. Among his numerous publications is *The Astronomy of Levi ben Gerson (1288–1344)* (New York and Berlin: Springer-Verlag, 1985).

Gerd Graßhoff is professor for history and philosophy of science at Berne University, Switzerland. He held an assistant professorship at Hamburg University, was member of the Institute for Advanced Study,

Princeton, and was senior researcher at the Max Planck Institute for the History of Science in Berlin. He has published on history of astronomy, methodology of scientific discovery, and natural philosophy.

Hermann Hunger is associate professor of Assyriology at the University of Vienna. He has edited and translated *Astrological Reports to Assyrian Kings* (1992), *Mul.Apin, Astronomical Compendium in Cuneiform* (1989) with David Pingree, and *Astronomical Diaries* (1988, 1989, 1996), completing work begun by Abraham Sachs.

Alexander Jones is professor of classics and history and philosophy of science at the University of Toronto. He studies the exact sciences in the ancient and medieval worlds, centering on Greco-Roman antiquity.

Erica Reiner is the John A. Wilson Distinguished Service Professor in the Oriental Institute and in the Departments of Near Eastern Languages and Civilizations and Linguistics at the University of Chicago, and an editor (until recently editor-in-charge) of the *Chicago Assyrian Dictionary*. Her work has centered on the linguistic and literary analysis of Assyrian and Babylonian texts; most recently she has been engaged, in collaboration with David Pingree, in editing Babylonian astral omens, a project of which three fascicles, *Babylonian Planetary Omens* 1, 2, and 3, have appeared to date.

Francesca Rochberg is professor of history at the University of California at Riverside and author of *Apects of Babylonian Celestial Divination: The Lunar Eclipse Tablets of Enūma Anu Enlil, Archiv für Orientforschung,* Beiheft 22 (1988) and *Babylonian Horoscopes, Transactions of the American Philosophical Society* 88, 1 (1988).

N. M. Swerdlow is professor in the Department of Astronomy and Astrophysics at the University of Chicago. His research concerns the history of the exact sciences from antiquity through the seventeenth century. He recently published *The Babylonian Theory of the Planets* and is currently working on a study of Galileo's astronomy.

Anne Tihon is a lecturer in Byzantinism and history of science at The Catholic University of Louvain (Louvain-la-Neuve, Belgium). She is the scientific editor of the *Corpus des Astronomes Byzantins* and author of several editions of *Theon of Alexandria*.

C. B. F. Walker is deputy keeper of Western Asiatic Antiquities at the British Museum. He has edited *Halley's Comet in History* (1985) and *Astronomy before the Telescope* (1996), and has written on Babylonian astronomy.

INDEX